Environment and Society

This book presents a comprehensive, lucid, and accessible approach to environmental sociology. It traces the origin of environmental sociology and examines the realist–constructionist debate in ecology for a holistic exploration of the field.

The volume:

- Presents a step-by-step systematic approach to the study of environmental sociology
- Includes case studies from Asia, Africa, Europe, and the Americas and introduces theoretical perspectives from Asia, Africa, and South America to provide a more comprehensive view of the field
- Has separate chapters on sustainable development and climate change
- Discusses ecological movements in India and highlights environmental issues of the Global South

A key text for undergraduates, postgraduates, and civil services aspirants, this book goes beyond western scholarship to include indigenous approaches to the field. It will be indispensable for students of sociology, climate change, environmental studies, and sustainable development.

Subhash Sharma is an Independent Scholar and Development Practitioner with over two dozen books published in English and Hindi. His book on human rights was awarded the first prize by National Human Rights Commission, New Delhi, in 2011.

Kshipra Sharma is an Assistant Professor at Indraprastha College for Women, University of Delhi, India.

Environment and Society

Climate Change and Sustainable Development

Subhash Sharma and Kshipra Sharma

LONDON AND NEW YORK

Designed cover image: Getty Images

First published 2023
by Routledge
4 Park Square, Milton Park, Abingdon, Oxon OX14 4RN

and by Routledge
605 Third Avenue, New York, NY 10158

Routledge is an imprint of the Taylor & Francis Group, an informa business

© 2023 Subhash Sharma and Kshipra Sharma

The right of Subhash Sharma and Kshipra Sharma to be identified as authors of this work has been asserted in accordance with sections 77 and 78 of the Copyright, Designs and Patents Act 1988.

All rights reserved. No part of this book may be reprinted or reproduced or utilised in any form or by any electronic, mechanical, or other means, now known or hereafter invented, including photocopying and recording, or in any information storage or retrieval system, without permission in writing from the publishers.

Trademark notice: Product or corporate names may be trademarks or registered trademarks, and are used only for identification and explanation without intent to infringe.

British Library Cataloguing-in-Publication Data
A catalogue record for this book is available from the British Library

Library of Congress Cataloging-in-Publication Data
Names: Sharma, Subhash, author. | Sharma, Kshipra, author.
Title: Environment and society: climate change and sustainable development/
Subhash Sharma and Kshipra Sharma.
Description: First Edition. | New York: Routledge is an imprint of the
Taylor & Francis Group, 2023. | Includes bibliographical references and
index. | Summary:
Provided by publisher.
Identifiers: LCCN 2022048564 (print) | LCCN 2022048565 (ebook) |
ISBN 9781032342061 (Hardback) | ISBN 9781032372808 (Paperback) |
ISBN 9781003336211 (eBook)
Subjects: LCSH: Environmental sociology–Case studies. | Sustainable
development. | Climatic changes–Prevention. | Green movement–India.
Classification: LCC GE195 .S474 2023 (print) | LCC GE195 (ebook) |
DDC 304.2–dc23/eng20230119
LC record available at https://lccn.loc.gov/2022048564
LC ebook record available at https://lccn.loc.gov/2022048565

ISBN: 978-1-032-34206-1 (hbk)
ISBN: 978-1-032-37280-8 (pbk)
ISBN: 978-1-003-33621-1 (ebk)

DOI: 10.4324/9781003336211

Typeset in Sabon
by Deanta Global Publishing Services, Chennai, India

Dedicated
to the memory of
Prof. Radhakamal Mukerjee
(1889–1968)
who pioneered
'Social Ecology'
and
'Region Sociology'

Contents

List of figures	viii
List of tables	ix
List of boxes	xii
Preface	xiv
Acknowledgements	xvi

1	Environmental Sociology: Its Origin and Concerns	1
2	Approaches to Environment and Society	39
3	Sustainable Development: Concept, Principles, and Practice	90
4	Development, Displacement, and Rehabilitation	134
5	Ecological Movements in India	183
6	Global Environmental Politics	223
7	Climate Change and Society	262
8	Global Environmental Issues	307
9	Commons for the Communities: A Critique of the 'Tragedy of Commons'	347

Bibliography	367
Index	384

Figures

2.1	Different societies in different stages of growth. Percentage of use of natural resources/energy over time	62
8.1	Values of Forest Ecosystems	337

Tables

1.1	Ecosystem Services Provided by Grasslands to Humans	18
1.2	Eco-Strategies and Eco-Goals for 'Use Less' Energy	20
1.3	Typology of Social Constructivisms	30
1.4	Comparison of Ideological/Constructionist and Realist Perspectives	35
2.1	Evolution of Integrated Life System	44
2.2	A Comparison of Human Exceptionalism Paradigm (HEP) and New Ecological Paradigm (NEP)	46
2.3	A Comparison of Three Alternative Syntheses of Societal–Environmental Dialectic	49
2.4	Treadmills of Production and Destruction: Core Conditions for Understanding the Distribution of Environmental Inequality	54
2.5	Weak Versus Strong Ecological Modernisation	59
2.6	Pattern Variables in Traditional and Modern Societies	60
2.7	Features of Societies in Various Stages of Growth	61
2.8	Society's Restrictions on Women in Rural Areas in India	64
2.9	A Comparison of Various Streams of Ecofeminism	68
2.10	Paul Robbins' Five Theses of Political Ecology	74
3.1	Collapse of Civilisations/Cities due to Climate Change/Environmental Degradation	92
3.2	Different Occupations in Various Ecozones in Tamil Nadu (India)	92
3.3	Contrasting Perceptions of People Residing in Forests and Settlements	93
3.4	Broad Scope and Contents of Brundtland Report (UN)	96
3.5	Main Selected Principles of Rio Declaration (1992)	100
3.6	Criticism of Sustainable Development	102
3.7	Four Perspectives on Interactions Between Social, Economic, and Environmental Processes	105
3.8	Multiple Conditions for Holistic Sustainability Framework	106
3.9	Millennium Development Goals (MDGs) (UN)	107
3.10	Sustainable Development Goals and Selected Targets (2015–2030)	109
3.11	Top Fifteen Most Sustainable Countries in the World (2018)	116
3.12	Twenty Worst Sustainable Countries in the World (2018)	117

x   Tables

3.13 A Comparison Between Payments for Ecosystems (PES) and
    Compensation for Ecosystem Services (CES) Models (Mexico)   120
3.14 Conceptualisation of the Environmental State and Social Change in
    Four Traditions   129
3.15 Alternative Justice Theories and Conceptualisations for Sustainability
    Transition   130
3.16 Plan of Action for Alternative Sustainability Transition   131
4.1 Top 10 High (Above 0.8) Human Development Index (HDI) Ranking
    Countries (Human Development Report [HDR] 2019)   145
4.2 Lowest (Below 0.5) 10 Human Development Index (HDI) Ranking
    Countries (Human Development Report [HDR] 2019)   146
4.3 Medium (0.5–0.799) and High (Above 0.8) Human Development
    Index (HDI) Ranks of Selected Countries (Human Development Report
    [HDR] 2019)   146
4.4 Important Differences Between Voluntary and Involuntary Migrants   151
4.5 Distinctions Between Environmentally Induced Migrants as per Force
    or Choice   152
4.6 Principal Causes of Environmentally Induced Migration   152
4.7 Types of Environmentally Induced Migration and Mobility Response   153
4.8 Labelling of Migrants and Displaced Persons   153
4.9 Labelling of Settlers and Resettlers   156
4.10 Causes of Dam Failures in the World (1900–1975)   163
4.11 Comparison of Old and New Laws of Land Acquisition in India   170
4.12 Displaced Population in Different Development Projects in India   175
4.13 Best Global Practices for Sustainable Livelihoods of the Displaced People 177
4.14 A Profile of Area Submerged and Displaced Population due to Large
    Dams in Jharkhand (India)   178
4.15 Different Perceptions on Dam-Building, Displacement, and
    Rehabilitation in Suvarnarekha Multipurpose Project (SMP)
    (Jharkhand), India   179
5.1 Categories of Environmental Movements by Issues and Examples in India   185
5.2 A Comparison of Worldviews of the Chipko Movement and the Forest
    Department, India   195
5.3 Narmada Water Distribution to Four States in India   200
5.4 A Comparison of Chipko, Silent Valley, and Save Narmada Movements
    (India)   218
6.1 Leading Green International NGOs   226
6.2 A Comparison of Three Ecological Schools about Ecological Crisis
    (1960s–1980s)   228
6.3 Top 20 Countries Used as Dumping Grounds for the World's Trash   236
6.4 Major Environmental Events Organised by the UN   240
6.5 A Comparison of Environmentalism in the North and the South   245
6.6 Ideological Orientations and Generic Modes in Indian Environmentalism   253

6.7	Environmental Thought and Action at Different Levels	255
6.8	A Comparison of Deep Ecology and Shallow Ecology	256
6.9	A Typology of Participation in Environmental Action	258
6.10	A Comparison of Theories of Politics of Global Environmentalism	261
7.1	Ten Hottest Years Globally (in a Sequence)	271
7.2	Major Global Events of Climate Change	272
7.3	Ten Costliest Disasters in the World in 2020	273
7.4	Deaths and Damages Caused by 2019 Floods in India	277
7.5	Statewise Deaths in India due to Extreme Weather Events in 2020	279
7.6	Percentage of Urbanisation in Different Countries of the World (1950–2019)	281
7.7	Per Capita GHG Emissions (Including Land Use Change and Forestry) in the World (2014)	283
7.8	Annual GHG Emissions (Including Land Use Change and Forestry) in the World (2014)	283
7.9	Sources of Global CO_2 Emission in 2016	284
7.10	Highest and Lowest Energy Consuming Countries in the World (2018)	285
7.11	Percentage Shares of Types of Fuel in Total Energy Consumption in India and the World (2018)	286
7.12	Average Per Capita Electrical Energy Consumption in Selected Countries	287
7.13	Multiple Shocks from Climate Change and Recoverability	288
7.14	People's Strategies to Cope with Climate Variations	300
8.1	Top 11 Most Polluted (PM 2.5 µg/m³) Cities in the World (2016)	308
8.2	Top Ten Most Polluted Cities in the World (2019)	309
8.3	Air Pollution Level in Various Parts of Delhi (1 November 2017)	310
8.4	Major Causes of Air Pollution in Delhi	311
8.5	Loss in Water Bodies Due to Rapid Urbanisation in Major Indian Cities (2000–2016)	315
8.6	Air Pollutants as People Breathe in Different Cities in India (2 October 2018)	320
8.7	Health Hazards Caused by Various Pollutants (India)	321
8.8	Permissible Noise Levels in Different Areas/Zones (India)	323
8.9	Declining pH level in Rainwater at Different Places in India (1981–2012)	330
8.10	Differentiation in Forest Cover in India (2014)	334
8.11	Diversion of Forestlands in India 1990–2012	335
8.12	Loss of Primary Forest and Tree Cover in India (2001–2018)	336
8.13	Tree Cutting in Different States During 2016–2019	337
9.1	Extent of Households' Dependence on CPRs in Dry Regions of India	349
9.2	A Comparison of Institutional Mechanisms for Managing CPRs in Different Areas	353

Boxes

1.1	God Tane and Three Knowledge Baskets in Maori Tribe's Tradition (New Zealand)	11
1.2	Falling of Groundwater Table in Ghaziabad–Noida (UP)	15
1.3	Use Less, Not More Energy	20
1.4	Justice D. Shashadri Naidu's Requiem for a Falling Tree	22
1.5	Protest against Environmental Injustice Done to the Ogonis, Nigeria	36
2.1	Widening of Jawaharlal Nehru Marg (Bailey Road), Patna, Bihar (India), by Felling Trees	40
2.2	Social Exchange of Land for Subhash Path in Sahebganj District, Jharkhand (India)	43
2.3	Cases of Environmental Racism and Classism (Inequality)	52
2.4	Limits of Hazards Research Approach	75
2.5	Failure of Dairy Development Project in Odisha	77
2.6	Maya Culture Identity and Ecology: 'Decolonising Development'	81
2.7	Victory of People and Defeat of the Mining Giant Vedanta: Environment, Health, and Livelihood	86
3.1	Broad Areas of Agenda 21 (UN)	100
3.2	Sustainable Mining or Cultural Genocide in Niyamgiri, Odisha (India)	114
3.3	How Dominant Pathway Damages Environment (India)	120
3.4	Ecological Restoration at Cultural Heritage Complex, Nizamuddin, New Delhi (India)	124
3.5	Madagascar's Small Fisher Folk Live with the Sea (Madagascar's)	125
3.6	Green, Healthy, and Happy, Pennsylvania (USA)	126
4.1	Consequences of a Single Social System of Capitalism	143
4.2	Delhi–Doon Express Way's Extension Leading to Cutting of 2,500 Old Trees, Uttarakhand (India)	150
4.3	Meaning and Significance of Land (Vanuatu)	157
4.4	Multi-dimensional Effects of Farakka Barrage on Bangladesh	158
6.1	Consumerism Generates More Wastes	234
6.2	Save Khejri Tree Movement by the Bisnois in Rajasthan (India)	247
7.1	The New Normal due to Climate Change	264
7.2	Climate-induced Migration in South Asia	265

7.3	Oceans and the Impact of Climate Change	266
7.4	Climate Change Over the 2010s in Six Ways	267
7.5	Everyone Loves a Good Drought!	289
7.6	Questions Regarding Innovation in Agro-ecology	297
7.7	Levels of Agro-food System Changes through Agro-ecological Approach	298
7.8	Green India Mission: Policy and Reality	303
8.1	Status of Water Situation in India	313
8.2	Effects of Oil Spill/Pollution on Marine Environment	317
8.3	How Cuninico Women Fought for Community Rights (Peru)	318
8.4	Changing Behaviour of Birds Due to Noise (World)	324
8.5	The Bittersweet Story of How We Stopped Acid Rain in North	329
8.6	Rajaji Tiger Reserve Ranger Wins International Award in Uttarakhand, India	339
9.1	Social Conflicts over Commons: A Case of Narmada River (MP and Gujarat), India	350
9.2	Holistic Transformation of Barren Land in Mawlyngkot, Meghalaya (India)	356
9.3	Restoration of 178 Water Bodies in Karaikal District, Puducherry (India)	357
9.4	Proper Use of Commons in Shahpur Nanemau Village in Sultanpur District, UP (India)	358

Preface

Environment and Society are interlinked in various ways. Both natural and 'built' environments have multi-dimensional linkages and interdependence with human society historically. Prof. Radhakamal Mukerjee (1889–1968), of Lucknow University, was the pioneer of 'Region Sociology' (1926) and 'Social Ecology' (1945) much before the emergence of environmental sociology in the 1970s. He talked of human 'regions' as the unit of sociological analysis and identified three elements in a region or ecological zone: 'inanimate objects' (the physical environment), 'biotic element' –plants and animals, having life and being capable of migration – and humans, having life and 'culture bearing'. He also talked of an integral unity of ecological, economic, and sociological elements. Hence, this work is a humble tribute to him.

Multiple knowledge systems or 'knowledge baskets' (A. T. P. Mead) and perspectives are linked with nature in complex ways. Traditional knowledge systems of the peasants and tribals are quite visible in their agricultural practices, forms, and variety of technologies and apparatuses used, water conservation and harvesting structures, irrigation methods (e.g. *Ahars and pynes* in Bihar), construction of earthen dams, cane/bamboo bridges in hills and waterlogged areas, growing and use of herbs, social idioms and sayings about weather, soil, bullocks, blowing of wind, clouds, rains, and drought ('Ghagh and Bhaddari in North India'), healing practices, conserving, regulating, and sharing of local common natural resources, relationship of various plants and animals/birds with livelihood, cultural festivals, and religious rites as well as their indigenous modes of imparting education (oral and written) and skills. Thus, there is 'environment in society' and 'society in environment' – 'Prakriti' (nature) and 'Sanskriti' (culture) are not binary, but rather interconnected and inter-dependent, and hence environmental sociology sees them as supplementary – no either–or, no if–but, and no two poles/extremes. Ecosystems serve in different ways: 'resource depot', 'waste repository', and 'living space' for humans and other creatures (Dunlap 1997).

Over the decades, the frequency (number) and intensity of extreme weather events and conditions has increased tremendously – e.g. Long spell of floods (more rain in less duration), long spell of drought (less rain in more duration), more rain in more duration, hazardous earthquakes, tsunamis, cyclones, storms, etc. are resulting in the huge loss of lives of humans and animals as well as that of natural resources and other assets. Earlier many environmental problems like global warming, acid rain, ozone depletion, extinction of animal and plant species, decline of the underground water table, extinction of seeds, poisonous elements in fertilisers, insecticides/pesticides, and weedicides, desertification and erosion of soil, water stress, high level of various types of pollution, massive deforestation (including a huge decline of rare evergreen rainforests), and climate change were considered only as remote theoretical problems, especially in developed countries. However, in the twenty-first century these problems have been highly accentuated and are quite visible in our everyday life in almost all countries, regions, and continents. Inter-Governmental Panel on Climate Change (IPCC), in its several reports, has given authentic evidence about most of such environmental problems, and hence one and all are experiencing these problems in some way or another. Now due to their gigantic forms and complexities, environmental problems are interdisciplinary subject matter, especially economics, sociology, social ecology, eco-politics, environmental studies, and so on.

Therefore, the significance of this book *Environment and Society: Climate Change and Sustainable Development* increases itself as both theoretical and empirical dimensions of such problems have been dealt with here in detail. Our efforts are to cover it in nine chapters holistically, namely 'Environmental Sociology: Its Origin and Concerns', 'Approaches to Environment and Society', 'Sustainable Development: Concept, Principles, and Practice', 'Development, Displacement, and Rehabilitation', 'Ecological Movements in India', 'Global Environmental Politics', 'Climate Change and Society', 'Global Environmental Issues', and 'Commons for the Communities: A Critique of the "Tragedy of Commons"'. This will not only enhance the 'art of listening' (to the system and power that be) but also rekindle a 'spirit of inquiry', curiosity and courage of questioning what has happened and what is happing in the ecology as a whole nowadays (sometimes explicit, other times implicit). This will ultimately be creative, reflexive, and dialogic in the real sense of the terms. We hope that intelligent readers, teachers, students (both Honours UG and PG), NGOs, movement activists, and government officials will like it and benefit in some way or the other by sociologically re-imagining much more than what this book manifestly provides. We welcome any concrete suggestions for further improvement of this book.

New Delhi,
20 January, 2023

Dr. Subhash Sharma
Dr. Kshipra Sharma

Acknowledgements

For writing a serious book like this, one gets inspiration from one's renowned teachers, intellectuals, public icons, and senior colleagues as well as cooperation from various quarters of professional and personal life. The first author got inspiration from his teachers like Dr. David Hulme, and Dr. Philip Woodhouse (from the Institute for Development Policy and Management, The University of Manchester, UK), Prof. Yogendra Singh, Prof. K. L. Sharma, and Prof. R. K. Jain (eminent sociologists from JNU, New Delhi) as well as from public intellectuals like Prof. P. C. Joshi and Prof. Manoranjan Mohanty (of Delhi University). He expresses his gratitude to them from the core of his heart. In professional life the first author was always inspired by Shri V. S. Dube (former Chief Secretary of Bihar and Jharkhand), Shri K. B. Saxena (former Secretary to Govt. of India, and currently a professor at Council of Social Development, New Delhi), Shri S. P. Keshav (IAS), Shri Sunil Kumar Singh (IAS), and Shri Anant Kumar Singh (IAS). He is thankful to all of them. Prof. Anand Kumar, Dr. Bijoy K. Choudhari, Prof. R. N. Sharma, Prof. B. K. Singh, Prof. Avijit Pathak, Anant Kumar Singh, Jawahar Pandey, Mahipal Singh, R.R. Giri and many others, too, indirectly helped and their cooperation is duly recognised. However, often words are unable to express the heartfelt gratitude. Second author, too, is grateful to her teachers and colleagues. During the preparation of this serious work, K. K. Sahu was very helpful in collecting various sources and references. He deserves genuine thanks. For secretarial assistance, Santosh Ram performed his tasks very efficiently and timely. We thank him too. We also thank our family members, relatives, and native people for their best wishes.

Dr. Subhash Sharma
Dr. Kshipra Sharma

Chapter 1

Environmental Sociology
Its Origin and Concerns

Here we will briefly discuss the definition and origin of sociology in general and environmental sociology, in particular, and the latter's concerns like the realist–constructionist debate so that a clear perspective of environmental sociology may emerge.

Introduction

Before attempting to specifically address the origin and concerns of the specialised branch of environmental sociology, let us briefly discuss what sociology is. Sociology, as a discipline of social sciences, emerged in the nineteenth century due to the emerging issues from the industrial revolution in Western Europe and North America. Sociology is generally defined as the 'science of society', or 'a study of human communities', implying that it studies social behaviour of human beings in specific groups (primary and secondary), communities, neighbourhoods, sub-national entities, nations, regions, or the world, (at micro-, meso-, and macro-levels) in totality, in different contexts (time and place). Though it sounds narrow, sociology may be defined in one word: 'context'. French sociologist Auguste Comte is credited for the coining of the term 'sociology'. In his work, 'The Course in Positive Philosophy' (1822), he mentions that society as a whole, and each particular science, develops through three conceived stages: (a) The theological stage, (b) the metaphysical stage, and (c) the positive (scientific) stage. His positivism is a system that is confined to the data of experience and excludes a priori or metaphysical speculations. Thus, he emphasised the empirical scientific evidence for studying social phenomena. To him, the sciences emerged in strict order beginning with mathematics and astronomy, followed by physics, chemistry, and biology, and culminating in the new science of sociology. It is like infancy, adulthood, and old age (maturity). After Comte, Emile Durkheim focussed on society as the phenomenon *Sui Generis*, i.e. social reality cannot be reduced to – it is 'out there'. Further, he added that social facts are 'things' having an objective existence outside of humans, like natural phenomena. Thus location, social situation, defines and determines human actions/behaviours according to the positivists like Comte and Durkheim. Further, in his work, 'The Rules of Sociological Method', Durkheim states five stages of scientific investigation (Rex 1976):

(a) Definition of the subject matter in terms of some observable characteristic
(b) Description of normal types after a study of many cases
(c) Classification into species, genera, etc.

DOI: 10.4324/9781003336211-1

(d) Comparative and causal investigation of the reasons for variation
(e) Attempt to discover any general law emerging in the course of various stages

Thus he emphasised that the empirical sociological dimension (causal explanation) contrasted with the psychological one, but his 'thingification' of social facts, and 'average type' (as against Max Weber's 'ideal type') are not usually accepted by most of the scholars. Karl Popper has criticised positivists like Durkheim because a scientific hypothesis/theory needs to be falsifiable (unlike theological or metaphysical statements). Durkheim, in his study of suicide, and Max Weber, in his analysis of relations between protestant ethic ('time is money') and capitalism, had established causal connections and functional correlations between different social phenomena. But Karl Popper distinguished between (natural) scientific and historical causal explanations – the former focusses on the relation between a particular fact and a universal law, while the latter emphasises the relation between a particular fact and 'specific initial conditions'. Historical and comparative methods are used in sociology to a large extent, and there is a new trend of unity of social sciences interdisciplinary – for a better understanding of various problems of society. But his societal determinism and methodism were also challenged by role theorists like Charles Cooley and George Herbert Mead – a role in society has a certain identity that is not given but is 'bestowed in acts of social recognition'. Cooley talks of the self as a reflection in a looking glass. Further, G. H. Mead remarks, during socialisation a child comes to know both what he is and what society is. The child first learns roles from 'significant others' (the intimate persons) and later, at the higher level of abstraction, discovers the 'generalised other' (society tells him what to do and what not to do – norms and values). Thus 'self' and 'society' are, in fact, the two sides of the same coin. In this regard, Peter L. Berger (1963) talks of 'men in society' ('out there') and 'society in man' (the inner-most part of being). To him, every society can be understood in terms of (a) its social structure, (b) its socio-psychological mechanisms, and (c) its world view (common universe of its members). Karl Marx, too, emphasised the 'scientific' (against the 'utopian') analysis of nineteenth-century society. V. Pareto also talked about 'logico-experimental' methods, while Max Weber applied the 'Verstehen' (Understanding) method for history and social studies. A famous sociologist C. Wright Mills, in his classic work, 'The Sociological Imagination' (1959) mentions three pillars of sociology (or 'sociological imagination'): Biography, history, and society:

> every individual lives, from one generation to the next, in some society; that he lives out biography, and lives it out within some historical sequence ...he contributes, however minutely, to the shaping of this society and to the course of its history, even as he is made by society and by its historical push and shove.
>
> (1959, 2)

His sociological imagination distinguishes between '*the personal troubles of milieu*' and '*the public issues of social structure*' – thus a public issue goes beyond the immediate relations of an individual with others and involves a crisis in the institutional arrangements, and often 'contradictions' or 'antagonisms'. Various collective human behaviours are socially and culturally defined through different processes of socialisation (internalisation of the social world in the child). Primary socialisation takes place in one's family and peer groups (of similar age and circumstances) through a 'face-to-face' (direct) relationship with a high degree, range, and depth (intensity) of the bond. On the other hand, secondary socialisation takes place outside such a close sphere in schools, clubs, work associations, formal institutions, and so on.

Various instances of the wolves, rearing human children in their dens, have been recorded historically in different parts of the world and their commonalities have been these: They walked on four limbs (two legs and two hands) just like wolves; they preferably ate raw meat of various animals, and they howled like wolves but could not speak any word from any human language. Thus their walking, eating, and communicating styles and habits were animal-like, not human-like. Thus, it is obvious that one does not become a social human being simply by birth, rather the social group (collective) like family, community, school, and the larger society make them so. Hence the famous Greek philosopher Aristotle has rightly observed: 'Man is a social being'. A man or woman is a social being due to the following factors:

(a) He/she speaks one or more languages to communicate with others.
(b) He/she is guided by certain social norms (informal rules and guidelines) and values; for instance, those related to commensality (eating), sexual relation and marriage (exogamy and endogamy), religious rites, dress code, etc.
(c) There is an inherent desire and need for a human social company, responsive reciprocity (exchange of various sorts), compassion for others and interdependence in a specific context (time and place) – during the COVID-19 pandemic it was re-confirmed.

Such a society consists of various sectors, spheres, and arenas of life. For instance, economy, polity, culture, and social structure are major structural sectors in any society and those interact, mutually support, and assist each other in different ways in different contexts. Similarly, there are three major social spheres of society: Market, civil society, and State (political society). That society prospers and flourishes the most in which the sphere of civil society is wider and freer than either market or State. In reality, there is hardly a fully free market or fully humane State even in a democracy. Regarding the relationship between individual and society, two points are noteworthy: First, the degree and extent of mutual bond vary from society to society; for instance, in the USA there exists much individual assertiveness (from free sex to crime) leading to '*individualism*', while there is far less genuine concern for family, collectivity, community, and society – even marriage is 'a matter of convenience' more than a permanent relationship; hence, marriages are often broken, resulting in frequent divorces. Therefore, there are more 'single mothers'. On the other hand, in Europe there is a co-existence of part individual assertiveness and part genuine concern for family, collectivity, community, and society as a whole, and hence there is the prevalence of '*individuality*' (not individualism) – that is, love marriages and nuclear families are more stable than those in the USA. Still in others, especially in developing countries, like India and China, there are more genuine collective concerns than individual concerns (the prevalence of more 'arranged marriages' than love marriages, and more 'extended families' than nuclear families, etc.). Needless to say that in every society there are both social cooperation and social conflicts, though varying in degree, range, and intensity, at individual, group, faction, region, sex, language, caste, class, race, religion, nation, and ethnicity levels. Often conflicts arise due to the real or perceived, de facto or de jure, discrimination and/or deprivation in one or more indicators of social development. However, every sort of difference is not discrimination, and some differences may finally kindle enthusiasm, motivation, and healthy competition for various kinds of achievements (more quality education, better health, more income, better livelihood, etc.). Thus sociology is holistic in perspective and vision, and it comprises economic, social, cultural, and political sub-systems as an integrated whole. The following aspects distinguish a sociologist from other disciplines (Berger 1963): (a) Democratic focus of interest about everyday life; (b) listening to others, without expressing his/her views; (c) responsibility for evaluating his/her findings, without his/her likes and dislikes, hopes, or fears; and (d) a symbiotic relationship with history.

Origin of Environmental Sociology

It emerged and developed in the latter half of the twentieth century, especially in the 1970s when environmental problems became more visible, intense, and substantially dangerous. The term 'environmental sociology' is to be preferred to social ecology because here environment is seen from a sociological perspective (impact of environment on society and vice versa). Catton and Dunlap (1978) defined environmental sociology as 'the study of interaction between the environment and sociology'. They argued to overcome classical sociology's reluctance to accept the significance of the physical environment, especially Emile Durkheim's view of explaining social phenomena only in terms of other 'social facts'. Further, due to the new trends of rapid urbanisation and industrialisation (through massive use of natural resources and technologies), sociologists (as well as economists) wrongly presumed that, at least in industrial societies, human life was becoming independent of the physical environment. That is to say, the exceptional features of humans (like language, culture, science, and technology) presumably *exempted* them from the constraints of the physical environment. But they also reminded us of the new problems like the oil crisis of 1973–1974 (and even later, too, in 1979) that necessitated the sociologists as well as other social scientists (especially economists) to shift towards a new ecological paradigm that accepts the ecosystem dependence of all human societies. The American Sociological Association, in 1976 itself, established a new section of environmental sociology (whose membership increased from 290 in 1976 to 321 in 1979) within it. As Buttel (1987, 466) perceptively remarks about it: 'Environmental sociologists sought nothing less than the reorientation of sociology toward a more holistic perspective that would conceptualise social processes within the context of biosphere'.

In fact, before the 1970s, rural sociologists did study natural resources like land, water, forest, etc. in village habitats and settlements, on the one hand, and urban sociologists also studied 'built environments' (parks), on the other. But environmental sociology in the 1970s responded primarily to various environmental problems, focussing on environmental movements, people's views on environmental issues, environmental policy-making, and environmental quality as a social problem (e.g. pollution of air, water, soil) by applying 'mainstream sociological perspectives'– a 'normal science' approach, and hence it was called 'sociology of environmental issues' (Dunlap and Catton 1979). Beginning with the 'Limits to Growth' (Meadows et al. 1972) and the oil (energy) crisis of 1973–1974 (and later, too, in 1979), focus was more on the scarcity of resources than on the social forces affecting the use of energy. Thus the fact of ecological limits was recognised, and hence environmental sociology emerged as a distinct specialised field of study, pointing out that environmental conditions do affect human societies, and so the environment–society interaction went beyond the classical boundary of sociology fixed by Emile Durkheim (that causes of 'social facts' must be explained by other social phenomena). Thus there was a transition from 'sociology of environmental issues' to 'environmental sociology'. Various sociologists, studying housing and built environment, energy, and natural resources, social impact assessment (SIA), natural hazards, and environmentalism, by and large, joined this specialised field of study.

However, the 'ecological limits' could not convince and sustain the American psyche (of individualism and free enterprise) and ethos of '*more, faster*, and *bigger*' (reflecting abundance and prosperity) in the 1980s. US President Ronald Reagan rejected the idea of ecological limits and promised 'to make America great again' – it contrasted with President Jimmy Carter's energy policy with limits and his sponsorship of the 'Global 2000 Report'. Hence, consequently various government restraints on markets were removed in the USA. Julian Simon (1981) talked of population growth as desirable as humans are 'the ultimate resource'! This logic suited Reagan's view of growth without ecological limits. Later Julian Simon and Herman Kahn (1984) wrote the book 'The Resourceful Earth' that was used by the US administration to reject Jimmy Carter – sponsored the Global 2000 Report (Boggs 1985). This affected the growth of the American Sociological Association Section on Environmental Sociology – its membership declined to 274 in 1983. While these books by Julian Simon became popular text books, the courses of environmental sociology and the associated appointments declined in the 1980s. However, on the other hand, various exposures to toxic and other hazardous wastes like the controversy over Love Canal, and major accidents like Three Mile Island (1979), Bhopal (India) gas tragedy (in 1984), and the nuclear accident at Chernobyl (1986) led to massive human protests. Thus such hazards, affecting the deprived lower classes and minority communities, stimulated sociological research on local, grassroots environmental action (Bullard 1983). However, often, the local hazards were studied by the environmental sociologists at the micro-level by linking the existing social stratification and political power, and often the environmental conditions were viewed as 'socially constructed' without raising the issue of the physical limits to growth (as done in the 1970s): As rightly remarks R. E. Dunlap (1997, 25):

> Yet, as in the 1970s, with the emphasis on the social impacts of scarcity, work in the 1980s was initially more concerned with understanding the impacts of environmental conditions on humans (especially as mediated by perceptions, collective definitions, and community networks) than with the impacts of humans on the environment.

During the late 1980s and early 1990s, there was a revitalisation of environmental sociology because various environmental problems like contamination of the Atlantic coast with hospital wastes, acid rain, ozone depletion, rainforest destruction, and global warming were highlighted in America. *Time* magazine, due to the hotter summer of 1988, named 'Endangered Earth' as 'Planet of the year' in lieu of 'man of the year' for 1988! Now the quality of the environment (as the oil crisis in the 1970s) was perceived as deteriorating. The 20th Earth Day (22 April 1990), and Earth Summit at Rio de Janeiro (1992), along with the Exxon Valdaz oil spill, tropical rainforest destruction, as well as environmental devastation in Eastern Europe led to the massive interest of people in environmental issues in the 1990s, in the USA in particular and in the whole world in general (Dunlap 1997). In fact, the nature of various contemporary environmental problems differs from that of earlier ones (in frequency, scale, and seriousness) in the following ways (R. E. Dunlap 1997):

(a) The scale of environmental problems increased from *localised* level (pollution of urban air, rivers, etc.) to the *regional* level (e.g. acid rain), to even the *global level* (e.g. ozone depletion, hence affecting more people.
(b) Localised problems like contaminated water supplies and inadequate solid waste repositories occur with *more frequency* for them to be perceived as generalised problems.
(c) Environmental problems' origins are not properly understood, and it is difficult to detect and predict their consequences, and hence these seem 'riskier' than earlier ones.
(d) Impacts of several environmental problems have serious consequences for the health and welfare of humans (including their future generations), and other species too; unfortunately, some impacts are irreversible.

Thus, in the words of R. E. Dunlap (1997, 27),

> whereas in 1960s and early 1970s environmental degradation often seemed an aesthetic issue (or, at most, an irritant affecting outdoor activities), it is increasingly seen as a direct threat to human health and well-being, from the local level (for example, toxic wastes) to the global level (for example, ozone depletion).

Thus human–environment interactions not only take place at the symbolic (or cognitive) level but also have direct impacts on human behaviour and well-being – the core of environmental sociology. Now there is a broad consensus that anthropogenic activities have resulted in ozone depletion, loss of biodiversity, and global climate change. The ecosystems in everyday life serve in three ways and are increasingly in conflict with each other (Dunlap 1997):

(a) 'Resource depot'
(b) 'Waste repository' for human societies
(c) 'Living space' for humans, and other creatures

In the early 1990s environmental sociology was revitalised in the USA, and its membership reached 400 in 1993 along with the publication of new text books, and more students opted for this course in the USA. Environmentalism, green politics, environmental policy-making, political economy of environmental problems, environmental hazards, NIMBY (not in my backyard) syndrome, 'environmental racism' (location of hazards in predominantly minority areas), and 'environmental justice' movement became the emerging areas of study in environmental sociology in the 1990s.

Features of American Environmental Sociology

Environmental sociology emerged in the USA and has the following salient features (Dunlap 1997):

(1) **Empirical Orientation** – Fields from natural resources to environmentalism, socio-cultural factors influencing both energy consumption and conservation, disproportionate location of landfills in lower-income/minority community areas, leading to environmental racism, and 'environmental justice' movements, technological disasters (leakage from toxic waste sites) differing from natural disasters (floods, earthquakes, hurricanes) – a small sampling of empirical research involves both 'sociology of environmental issues' (e.g. studies of environmental attitudes), and 'environmental sociology' (examining the relationship between social variables like race, status, class, gender) and environmental variables (like pollution) – now environmental sociology consists of sociological works conducted on environmental topics (Buttel 1987).

(2) **Cautious Constructionism** – It avoids a strong social-constructionist perspective –that sees the environment as purely a social construction – 'simply a product of language, discourse and powerplays' (Dickens 1996, 71). However, weaker forms of constructionism, analysing roles played by actors like activists, scientists, and policy-makers in defining environmental conditions as problems, without denying the objective existence of such problems, have been widely used in the USA. On the contrary, most of the sociological works on the global environment in Europe are confined to theoretical analyses and investigations of the roles of claims-makers (scientists and environmentalists).

(3) **Insular and Atheoretical** – Environmental sociology in the USA has remained isolated ('Balkanisation') from the larger discipline of sociology and is highly atheoretical (Buttel 1987). On the contrary, in European environmental sociology, there is theorisation by Giddens, Luhmann, Beck, and Touraine. Dunlap (1997) admits that since mainstream sociology ignored the significance of the physical environment, American environmental sociology did not give due importance to theories regarding environmental issues. He pleads for the joining of hands by the strong empirical orientation of American environmental sociology, and the strong theoretical orientation of European (and elsewhere) environmental sociology for a better future of the field.

Dunlap and Catton (1978) also formulated a 'paradigm shift' from Human Exceptionalism Paradigm (HEP) to New Ecological Paradigm (NEP) and thereby argued for more ecologically sensitive or 'greener' versions of Marxist, Weberian, or other perspectives, not to replace these perspectives. Their intention was to incorporate environmental variables or 'non-social facts' into sociological analyses. Since the 1970s, ecological matters have increasingly become a part of sociological studies in different ways.

Indian Context

In the Indian context, Patrick Geddes (1854–1932), who was a Scottish internationalist, was a Professor of Sociology and Civics at Bombay University and spoke against the debris left by the 'carboniferous capitalism' while he condemned polluting technologies of the nineteenth century as 'polytechnic' based on non-renewable and 'dirty' materials (coal and iron). He argued for clean and renewable energy sources. He took a historical–ecological approach in his book 'Cities in Evolution' (1915). He propounded two concepts in town planning: (i) 'diagnostic survey' – intensive walking tour to acquaint with the town; and (ii) 'conservative survey' – the practice of gentle improvements with minimal disruptions to people and their habitat. He also argued for 'the change from a mechanocentric view and treatment of nature and her processes to a more and more fully bio-centric one' (cited in Guha 2006, 39). He recommended for following provisions in town planning:

(a) Open space for recreation, tree planting, and protection, especially around sacred places
(b) Provision of a sustainable and safe water supply
(c) Preservation and maintenance of tanks and reservoirs
(d) Protection against flooding after heavy rains
(e) Source of an assured supply of water
(f) Narrow public thoroughfares and leafy courtyards as in Indian tradition, not the wide and dusty streets in modern towns allowing automobiles (emitting carbons)

He also emphasised the participation of the town planner. His follower Radhakamal Mukerjee (1889–1968), Professor of Sociology at Lucknow University, as early as the 1930s and 1940s talked of solidarity between the village and city, in order to 'retain the vitality of life and culture of the village' (cited in Guha 2006, 43). He wrote the essay 'An Ecological Approach of Sociology' (American Journal of Sociology, 1930):

> There is a balance between the natural and the vegetable, and the animal environment, including the human, in which nature delights. It is maintained by chains of actions and interactions, which link man with the rest of his living realm, reaching up and down and all around as his invisible biological and social destiny.
>
> (cited in Guha 2006, 43)

He further adds, 'Human, animal and plant communities are subject to similar rules, though shifting ones, which maintain a balance and rhythm of growth for all. Each community cannot appropriate more than its due place in the general ordering of life' (cited in Guha 2006, 44). However, he admits that the human species is distinguished from animal and plant species in three respects (Guha 2006, 44):

(i) As a tool maker, man alone is capable of a dynamic, interventionist relationship with nature; as the 'super dominant of life's existing communities', humans have an awesome potential for ecological devastation and/or regeneration.
(ii) Humans have a history and traditions in a way that plants and animals do not have.
(iii) The human's response to nature is characteristically collective rather than individual and thus amenable to 'synecology' (the study of communities in their natural environment) rather than an individual-oriented 'auto-ecology'.

Hence Radhakamal Mukerjee also talked of 'social ecology', in lieu of ecological ontology, that will bring together complementary skills of the sociologist and the ecologist. In his words, 'renewal and environment rather than exhaustion and depletion of the region should be man's synecological goal, if he wants to nourish his land-water culture in which civilisation is rooted on a continuous basis for his species' (Mukerjee 1942, 129) He, therefore, wanted a paradigm shift from 'Social regression' to 'Social evolution'. He desired a non-dualistic relationship between ideal and real, and also between natural and cultural. His idea of 'regional balance' proves his inclination to highlight the ecological dimensions of social life as well as the ethics of nature. In his book 'The Dynamics of Morals' (1952) he clarifies: 'social ecology is the biological aspect of ethics or evolutionary ethics'; 'social and environmental interrelationships are necessary for a sustainable world'; our 'ecological relationships and cultural patterns (are) parts of one dynamic region; field, social and moral space 'enable us to engage the problems of regional balance or unbalance (223–225). He further added, 'perhaps some kind of a technological revolution, decentralisation, and regionalisation of industry may be necessary before … a moral change maybe brought about' (1952, 504). He was against the compartmentalisation of social sciences, and on the one hand, he linked economics, sociology, and history, while, on the other hand, he also linked/synthesised ecology and sociology – thus a new field of 'social ecology' was pioneered by him. Various geographical and biological factors worked together to produce an 'ecological zone'. On the other hand, ecology is also conditioned by social, economic, and political factors. According to him, social ecology studies the place, occupation, and time relations of persons and groups in their processes of competition, corporation, conflict, accommodation, and succession.

In fact, in 1926 itself, Radhakamal Mukerjee had written a book entitled 'Region Sociology' in which he defined human ecology as a 'systematic science of plant, animal and human communities, which are systems of correlated working parts in the organisation of the region' (1926, vi). In 1938, his work 'The Regional Balance of Man' and, in 1940, his work 'Man and his Habitation: A Study in Social Ecology' were published. Later his work 'Social Ecology' was published in 1945. Radhakamal Mukerjee, first, talked of human 'regions' as the unit of sociological analysis and identified three elements in a region or ecological zone – (a) *inanimate objects*, without life, the *physical* environments; (b) *biotic elements* – plants and animals, having life and capable of migration; and (c) humans, having life and 'culture bearing'. But his 'regional balance' did not talk of exogenous factors like technology or values though 'both high technology and semitic religious values introduced into a region can fundamentally alter its ecological balance' (Oommen 2015, 18). Mukerjee also talked of an integral unity of ecological, economic, and sociological elements.

Second Mukerjee further observes, 'In rural society, ecologic, social, and moral space tends to coincide' (1945, 214). But, according to Oommen (2015), though in the tribal society, ecological, social, and moral space may coincide, but not in a peasant society (part of wider 'Hindu civilisation'); further while tribal societies are ecologically independent, peasant societies are ecologically dependent as these are part-societies. Hence Mukerjee's statement 'Ecological mobility co-exists with social immobility in both tribal and agricultural societies' (1945, 211) is not admissible, particularly in India (Oommen 2015).

Third, to Mukerjee, ecological relations build up 'fraternal modes of behaviour', but cultural relations build up the super individual plane. But Oommen (2015) disagrees with him, observing that, with technological development, the necessity for cooperation, both in inter-personal and inter-group contexts, 'can reduce or altogether disappear'.

Fourth, to Mukerjee, human habitation is conditioned by ecological factors; but, according to Oommen (2015), it may also be conditioned by sociological factors like apartheid rooted in racism or caste hierarchy; the Greek 'slave', the Indian 'untouchable', the South African 'Black', and the American 'Negro' were physically segregated. Thus human settlement patterns are guided by both ecological and sociological factors.

Finally, to Mukerjee, 'culture and language no doubt are powerful binders but these are less primary than ecological realities and processes which represent the physical base of man's social heritage' (1945, 330). But, to Oommen (2015), culture and language facilitated the formation of Western Europe, following the Treaty of Westphalia in 1648, followed by the rest of the world after World War II – thus ecology did not play a significant role in the formation of nation-States independent of culture. Further Mukerjee is of the view that the human ecological theory, the sociological theory, and the theory of values and symbols have autonomy, but a general theory of society may emerge in their integration. Oommen (2015, 24) tries to tighten the loose ends by arguing that 'western technology and a distortion of Judaic monotheism in combination have been largely instrumental for the estrangement between humanity and nature resulting in ecological devastation'. In fact, in the present context, Oommen rightly talks of three major sources of human security: 'genocide, culturocide, and ecocide' – and the ecocide is done through the massive use of high technology. Hence he also talks of the 'technological continuum' of traditional, intermediate, and high technologies – a specific type of technology is to be used on the basis of the needs of specific sectors in specific societies, not on the basis of their resources or purchasing power. This balanced argument has a significant bearing on the field of environmental sociology.

Major issues in environmental sociology are environmental implications of socio-cultural, economic, and political institutions, linkage of local, national, regional, and global contexts, sustainability, roots of environmentalism, trans-boundary nature of environmental issues like climate change, 'ecological debt' in the South, urban housing, sanitation, health hazards, dumping of waste in the South, ecologically unequal exchange, and pollution. In everyday life, most of the environmental problems are those 'of' human society, 'for' the wider ecosystem and society, and are often created 'by' human society. It is quite useful to see the linkage of knowledge systems with nature.

Multiple Knowledge Systems' Linkage with Nature

Undoubtedly the interconnections and interdependence of human beings and nature are highly complex and shaped through appropriation, consumption, transformation, and exchange in various ways. This could be understood by integrating different methods, as well as knowing the 'multiple knowledge systems and perspectives' as Aroha Te Pareake Mead introduced three 'knowledge baskets' in the context of Maori tribe's tradition of God Tane's ascent through 12 heavens to bring back to earth 3 knowledge baskets (Marsden 1992) (Box 1.1).

Box 1.1: God Tane and Three Knowledge Baskets in Maori Tribe's Tradition (New Zealand)

Tane was the God of forests and all that resides therein. To acquire the baskets of knowledge, Tane had to ascend to the 12th heaven and there be ushered into the presence of the Supreme God, Io-Matta-Kore, to request knowledge that was accepted – knowledge came before humanity. Three baskets of knowledge are te kete tuauri (sacred knowledge), te kate tuautea (ancestral knowledge), and te kete aronui (knowledge before us).

(1) Te Kete Tuauri (sacred knowledge) – this basket contains knowledge of things unknown – rituals, incantations, and prayers. Reverend Maori Marsden describes tuauri as the real world of the complex series of rhythmical patterns of energy, which operate beyond this world of sense perception.
(2) Te Kete Tuatea (ancestral knowledge) – this basket has knowledge beyond space and time, beyond our contemporary experiences – as we have connections with one another, and with the past, knowledge of spiritual realities.
(3) Te Aronui (knowledge before us) – this basket of knowledge of aroha (love), peace, and arts, and crafts that benefit the earth and all living things. It is acquired through careful observation of the environment. Sometimes it is regarded as the basket of literature, philosophy, and humanities.

Wisdom needs that these three baskets of knowledge should be used together, never one in isolation.

Source: M. Marsden (1992) A. T. P. Mead (2016), and R. Taonui (2012), cited in H. Suich et al. (2016)

Thus environmental sociology requires us to respect the multiple knowledge systems giving critical weightage to the local perceptions and cultural processes for a critical holistic understanding of humans, culture, and ecology/environment. Traditional knowledge systems of tribals/indigenous people are visible in agricultural practices, water harvesting structures, irrigation methods (Ahars and Pynes in Bihar), construction of earthen dams, construction of cane/bamboo bridges in hills or waterlogged areas, use of herbs as medicines, meteorological assessments (Andaman tribals during the tsunami in 2004 learnt about the oncoming disaster from the changed behaviour of the fish in the sea), healing practices, equitable sharing of the village and forest commons, protecting biological diversity, etc. Various plants/trees have ethical-religious significance for tribals as well as rural peasants. Therefore, environmental sociology takes into account such treasures of biodiversity as the domains of life and livelihood that are inter-connected and interdependent in everyday social reality.

In this context, however, I remember the wittiest words of Charles Kettering: 'A problem well-stated is a problem half-solved'. Hence a correct statement about environmental problems is this: Though the ecosystem consists of both sentient (life organisms and plants) and non-sentient (mountains, soil, water) which interact with each other and are mutually related and interdependent, most of the environmental problems like climate change, pollution (air, water, noise, soil and radiation), deforestation, floods are created by human beings due to two major factors: First, conspicuous consumption (rather per head over-consumption) of energy/facilities on a huge scale mostly by the upper and middle classes due to travel by air, and use of refrigerators, washing machines, vacuum cleaners, hair driers and hand driers, TVs, computers, mobiles, and other electronic gadgets, mostly in developed countries. Interestingly the former Vice President of the USA, Al Gore, showed much concern for climate change (both verbally and by writing the book 'An Inconvenient Truth'), but he travelled a lot by air, causing, and contributing more to the carbon emission – a gap between his word and deed. Yet he was awarded, along with Inter-Governmental Panel on Climate Change (IPCC), the Nobel Prize for Peace! Second, the higher population growth rate in most of the developing countries, especially among the lower-middle and lower classes of people, is also contributing massively to ecological problems. Both these factors affect the environment due to massive carbon emission, first category by the consumerist life styles of the rich people, though in less number yet in a huge quantity of energy consumption per head, while the second category with less quantity of energy consumption each but a large, and relatively poor, population have a cumulative effect.

Natural Assets, Human Population, and Development

In fact, the relationship between natural resources/assets, growth of human population, and economic development had been highlighted by different social scientists in the nineteenth and twentieth centuries too. As Graham Woodgate (2010) has rightly remarked: 'At the same time as human beings are organically embodied, and ecologically embedded, we are also culturally embodied and socially embedded' (2010, 2). Nowadays there is more human interference in the natural environment and climate to a very large extent locally, nationally, regionally, and globally. However, it is also a fact that, for climate change, some geological processes are also responsible but far less than anthropogenic factors. This has advanced the very 'epoch' of human existence. For instance, the International Union of Geological Sciences (IUGS) has divided the earth's total 4.5 billion-year-old history into Aeons, Eras, Periods, Epochs, and Ages. At present, we live in the Phanerozoic aeon, Cenozoic era, Quaternary period, Holocene epoch, and Meghalayan age. But many scientists opine for its revision as earth has already reached a new epoch called 'Anthropocene'– the age of humans who have 'the most dominant influence on environment and climate', by 'decimating other species', and hence the scientists have termed such an extinction on the earth as 'biological annihilation' and 'sixth mass extinction' (in 'fifth extinction' dinosaurs were wiped out). Presently, first, many species are becoming extinct tens to hundreds of times faster than they would without human influence (Bhushan 2020). Second, he adds further, as per a report by the Inter-governmental Science Policy Platform on Biodiversity and Ecosystem Services in 2019, an average of 25% of all animal and plant species assessed face extinction, many within decades – thus about one million species face extinction. Third, he adds again, as per the 13th meeting of the Convention on Conservation of Migratory Species (CMS COP13), in Gandhi Nagar (Gujarat), the number of most migratory species is declining. Though tiger (India's national animal) population in India increased to 2,967 (2018) (more than double since 2006) and the status of peacock (India's national bird) is quite good, more than a hundred bird species face extinction as per State of India's Birds Report 2020. Fourth, he continues, as per calculation by scientist Vaclav Smil, at present, humans and domesticated animals weigh 99% of all vertebrates (mammals, birds, and reptiles) and the wild vertebrates are only 1% of all vertebrates, while 10,000 years ago (after the settled agriculture) wild vertebrates weighed 99%, and humans and domesticated animals were just 1%. Obviously for the protection of biodiversity this situation needs to be drastically changed, especially by all the conscious humans, scientists, agencies of government, and NGOs.

Moreover, every 24 hours 2,23,000 new persons are added to the world population, which is 7.5 billion (750 crores) at present and out of 200 countries China (22%) and India (17.5%) jointly have about 40% of the total world population, but their cultivable land area is far less than that percentage – while India's more than half of the land area is cultivable, China's only 10% land area is cultivable. In 1600 CE (Common Era), during Akbar's rule, India's population was only 114 million (11.4 crores) and the cultivated area was far less than the area of forests, grass, and marsh lands. But the population momentum began after India's Independence and various government schemes under five-year plans (like irrigation, High-yielding varieties [HYVs] of seeds, fertilisers, pesticides with subsidy) promoted agriculture and industrialisation. Various medical facilities in the public sector drastically reduced the death rates and enhanced the life expectancy at birth of the people. During 1950–2020, in India, the total fertility rate (TFR) has fallen from 5.9 to 2 ('replacement level' is 2.1), on average, even less than the global average TFR of 2.2. Similarly life expectancy at birth has increased from 49.7 years in 1970–1975 to 69.7 years in 2015–2019 in India against the global average of 72.6 years. In fact, the 'doubling time' of the population reduced drastically in the recent past. On the other hand, a huge area of tropical rainforest is cut down for various human activities like agriculture, roads, dams, mines, hospitals, bridges, schools, and town development – resulting in a loss of 525 square km every 24 hours in the world, but only one tree is replanted for every ten trees cut! Further a huge area will become desert by 2100 due to overgrazing, unsustainable agricultural practices, global warming, chemical pollutants, etc. all over the world. The situation will further worsen in African, Asian, and Latin American countries with scarce resources. Undoubtedly, the quantity and quality of groundwater is being badly affected and the water table is hugely depleted in most of the world, especially in developing countries. In Israel, 80% of the rain water is conserved in a sustained way, whereas only 17%–20% of rain water is conserved in India – the rest 80% or more of rain water flows to the rivers and seas in a wasteful manner. There are 15 agro-climatic zones (based on rainfall, soil, temperature, and water resource) in India: Western Himalayan region, Eastern Himalayan region, Lower Gangetic plains region, Middle Gangetic Plains region, Upper Gangetic Plains region, Trans-Gangetic region, Eastern Plateau and Hills region, Central Plateau and Hills region, Western Plateau and Hills region, Southern Plateau and Hills region, East Coast Plains and Hills region, West Coast Plains and Ghats region, Gujarat Plains and Hills region, Western Dry region, and the Islands region (ICAR, Government of India 2016). Though the much acclaimed 'green revolution' in India, especially in Punjab, Haryana, and Western Uttar Pradesh (UP) in the late 1960s, did bring food self-sufficiency and food security substantially (at the national level, of food availability at least) against various obstacles, if we see it in totality, we find following negative environmental, social, and economic consequences of the green revolution in parts of India (especially in Punjab, Haryana, and Western UP):

(a) Groundwater table has gone down substantially, even up to 1,000 feet deep due to more frequent use of water for irrigating paddy and wheat, and free electricity to all private tube wells in Punjab has compounded the problem.
(b) There is much salinisation of soil due to more water use, and desertification due to more use of land for intensive crops continuously, without keeping fields as fallows and also due to not maintaining proper crop cycle.
(c) Too much use of chemical pesticides/insecticides/weedicides, and chemical fertilisers has resulted in the spread of such chemicals in food grains/vegetables/fruits and from there to human blood, resulting in the spreading of serious diseases like cancer – hence the phenomenon of 'cancer express' in Punjab (a film of that name is also worth seeing).
(d) There has been a huge migration of Punjabi youth to Western Europe and North America for better jobs; on the other hand, the affluence and prosperity have also led to the radicalisation of some youth, even resulting in asking for the moon – the demand for Khalistan!
(e) There is substantial mechanisation of agriculture and various activities are done mechanically, but many machines also create new problems or intensify the old ones; for instance, the harvester cuts only the upper portion of paddy plants (grains) and leaves 'puals'/'paralis (the stems) standing in the fields. These are later burnt by the farmers, being in hurry to prepare those fields for the next crops, mainly wheat. This results in huge air pollution affecting many places/regions, including the national capital of Delhi/National Capital Region (NCR) in a very dangerous proportion – air pollution in Delhi reaches 'severe plus' or 'emergency level' (air quality index crossing 500), especially during October–December.
(f) HYVs of seeds, especially the terminator seeds, are very costly yet these seeds' produce cannot be preserved to be used as seeds for the next crops – hence seed security of the farmers is lost.
(g) Many harmful chemical pesticides like dichlorodiphenyltrichloroethane (DDT) have been banned, and even then these are used there, on the one hand, and on the other hand, pests/insects have developed resistance, too, hence not useful either way.
(h) The affluence and prosperity due to the green revolution also resulted in drug addiction among the youth, on a large scale, hence 'Udta Punjab' (flying Punjab due to drug use)! This has a grand nexus of politicians, police, criminals (smugglers), etc.
(i) The green revolution has also encouraged and promoted monoculture, making biodiversity a thing of the past; various indigenous varieties have become extinct – this is a dangerous consequence for both ecology and society.

Unfortunately, on the other hand, the water table in urban areas like Delhi–NCR has also been depleting rapidly in the early twenty-first century (Box 1.2).

Box 1.2: Falling of Groundwater Table in Ghaziabad–Noida (UP)

In Ghaziabad (UP) – that falls under the National Capital Region – first, the groundwater table was depleted about 12 metres during 2016–2020; thus, the rate of depletion is quite high at 3 metres per year. Second, four out of five blocks in Ghaziabad (Ghaziabad city, Bhojpur, Loni, and Razapur) are in 'notified' category – i.e. groundwater has been extracted at a much higher rate than recharged (the fifth block Muradnagar is semi-critical); the extraction rate in Ghaziabad city is 263%, followed by Bhojpur Block (141%), Loni (133%), Razapur (110%), and Muradnagar (88%). Third, till 2019 annual groundwater depletion in Ghaziabad was 1.5 metres, but it doubled (3 metres) in 2020! Fourth, in Noornagar in Sihani area, there is the sharpest fall in groundwater level – from 29.4 metres in 2016 to 48.3 metres in 2020 (a drop of 19 metres). Fifth, out of 39 places in Ghaziabad, where the water level was recorded digitally, 10 spots were 'dry' (recorder could not register the data due to the highly low level); at 12 spots groundwater level was depleted by more than 5 metres.

First, the groundwater situation in the adjacent city of Noida (UP) is worse than that in Ghaziabad; it fell by 17 metres in Noida during 2016–2020 – from 9.9 to 26.7 metres; in Greater Noida, it fell by about 6 metres during the same period. Second, at dozens of spots, the water level in Noida is 'dry' (below the testing limit); for instance, in Noida, Mamura has the lowest water level at 38.8 metres. Third, in Greater Noida, out of 35 spots tested, at least 5 spots have gone 'dry' since 2016; in Greater Noida, on average, the water table has gone down from 6.6 metres to 12.8 metres during 2016–2020.

Reasons: Following are the main reasons for the falling groundwater table:

(a) Rapid urbanisation with more people leading to a demand for more water
(b) Dwindling/encroaching of water bodies (that used to recharge groundwater) (73% of water bodies in Ghaziabad city were encroached)
(c) Illegal extraction for commercial, construction, and industrial purposes
(d) Non-implementation of rainwater harvesting system
(e) Trend of less rain (due to climate change)

Measures: However, the Uttar Pradesh Groundwater (Management and Regulation) Act, 2019, gives hope against hope by putting a blanket ban on the extraction of groundwater for commercial and industrial purposes, as well as mandating a fine of up to Rs. 20 lakh and imprisonment up to 7 years to the violators. This needs to be strictly followed, and there should be a compulsory provision of rainwater harvesting in all public and private buildings with an area of 500 sq metres or more. This requires a strong political and administrative will with full transparency in governance.

Source: Abhijay Jha (2020 a), (2020 b).

NITI Ayog assessed in 2019 that by 2020, 60 crore (600 million) people will face water stress in 21 Indian cities (including Delhi, Bengaluru, Chennai, Hyderabad,

Ghaziabad), and hence the State governments should conserve rain water, and, by 2030, 40% of India's population will have no access to drinking water.

From the above, one may understand very well the inter-connectedness of the ecology (land, water, air, organisms, plants, etc.) and the livelihood of human communities as well as social, cultural, political, and economic sub-systems in everyday life. Thus, environmental sociology has emerged as an inter-disciplinary field of study, as 'the study of community in the largest possible sense' (Bell 2016). Thus environmental sociology studies both the natural communities (animals, plants, birds, water, air, land, etc.) and the human communities, and their relationship. In fact, many social scientists, drawing from various disciplines like sociology, economics, anthropology, political science, geography, history, etc., are engaged in environmental sociology or social ecology or human ecology with a focus on ecological inter-connectedness. Ecology is obviously a much broader term than environment as it envisions the holistic relationships (cooperation and conflict) between humans, animals, natural plants, rocks, water, land, air, forests, and minerals as a dynamic process. Since various natural resources and assets are used, owned, shared/distributed, and disposed of unequally due to social differentiation and inequality, hence, there arise socio-economic, political, and environmental conflicts and consequently protest movements get organised. George Orwell's famous observation holds good even today more or less in almost all societies: All are equal but some are more equal! 'More equal' obviously implies social inequality, favouritism, ecological imbalance, etc.

The relationship between environment, poverty, technology, and development was perceptively pointed out by the then-Prime Minister of India Shrimati Indira Gandhi (1982) at the UN Conference on Human Environment (Stockholm, Sweden) held on 14 June 1972 in these words:

> The rich look askance at our continuing poverty; on the other hand, they warn against our methods. We do not want to impoverish the environment any further and yet we cannot, for a moment, forget the grim poverty of a large number of people. Is not the poverty and need the greatest polluters? The environment cannot be improved in the conditions of poverty. Nor can poverty be eradicated without the use of science and technology.

Here two caveats seem quite necessary: First, poverty (mostly in developing countries) is as much a polluter as consumerism (per capita very high consumption mostly in developed countries) is, but actually 'need' hardly pollutes as Mahatma Gandhi rightly observed long back in the early twentieth century: 'Earth has everything to fulfill every body's need, not anybody's greed'. That is, need does not pollute the environment due to the prudent and limited use of natural resources, while the ambitious 'greed' of the upper class usually pollutes the environment deliberately due to the massive exploitation of natural resources as well as huge consumption of energy. As John Mc Neill (2000) has rightly calculated, first, the energy used by all the humans of the world in the 100-year period of the twentieth century was more than that used by all our human ancestors in the last 10,000 years. Interestingly the first agricultural revolution started in 10,000 BCE (Before Common Era), that is, 12,000 years ago. Second, all types of science and technology do not eradicate poverty in a real sense, as many technologies (like automatic ones) are labour-replacing, or using fossil fuels (coal, petrol, gas) in thermal power generation and vehicles for transport, resulting in the depletion of natural resources and massive pollution due to the huge amount of carbon emission. Similarly most of the chemical fertilisers, pesticides/insecticides, etc. have poisoned our food grains, pulses, oilseeds, fruits, and vegetables, resulting in various health hazards (including cancer). As Rachel Carson (1962) pointed out, long back in her pioneering work 'Silent Spring', that due to the over-use of pesticides, the birds disappeared, hence no chirping even during the spring. Therefore, only 'public interest' science and technology (say solar power, wind power, organic farming) is useful, in a real sense, to both humans and the environment. Technology cannot solve ecological problems. Outdated and obsolete technologies, made available by the western developed countries to the developing countries, have added more to the carbon emission, and hence environmental pollution. The twin processes of urbanisation and industrialisation have also contributed to environmental problems in most of the developing countries because of factors like higher population density, limited urban space, and the tendency to have more (quantity), bigger (size), faster (speed), and looking smarter (fashion/design/appearance/brand) – e.g. my shirt is whiter than yours, my TV's brand is rated/valued higher than yours, my house is bigger and more shining than yours, my bike is faster than yours, I have more shoes than you, especially in the developed world.

For a critical and holistic understanding of the environment–society relationship, it is necessary to have ecological, socio-cultural, and political understandings, too, in addition to the scientific and economic understanding (which have dominated the national and global scenes). In Hindukush Himalayas, there are eight countries: India, Afghanistan, Pakistan, Bangladesh, Myanmar, Bhutan, Nepal, and China. The largest land use system here in this region is the grassland or rangeland (60%) on which the indigenous vegetation consists of grasses, grass-like plants, ferns, and shrubs that provide ecological, economic, and cultural–spiritual services at local, regional, and global scales – forage for livestock grazing, wildlife habitat sustaining flora and fauna essential to support human well-being; store and supply water resources by serving the sources of major rivers in Asia (including Ganga, Tarim, Yangtze, Yellow, Lancang-Meckong, Nu-Salween, Dulong-Irrawaddy, Yarlung Zangbo-Brahmaputra, and Indus rivers); they maintain the stable and productive soils; they provide mineral resources and products; they sequestrate and store the carbon; and they also create natural beauty. Among these, biodiversity habitat maintenance, carbon storage, and water regulation are termed as 'primary ecosystem services' from the grasslands to human beings (C. Dutilly-Diane et al. 2007, cited in Shikui Dong 2017). This may be seen in Table 1.1.

However, Rajni Kothari (1980, 427) had given an innovative and holistic vision of a sustainable environment:

> The urgent need for re-discovering the other traditions that take an integrated and holistic view of life as a whole, in which science and technology, and development and environment all merge in a symbiotic relationship. This entails a search for an alternative concept of both development and technology as well as of lifestyles, so as to ensure diversity in consonance with local resource endowments (human, material & technical), foster self-reliance and autonomy, and promote equity and participation, not only in economic and political processes, but also in giving meaning and content to human dignity at various levels.

Table 1.1 Ecosystem Services Provided by Grasslands to Humans

Scales	Ecosystem Services	Benefits	Beneficiaries
Local	(a) Improved hydrological function (b) Improved soil health (c) Higher plant biomass	Higher pastoral activity	Local pastoralists and agro-pastoralists
Regional	(a) Underground water recharge	Increased water availability	Water users, hydro-power industries
	(b) Flood reduction	Less damage to infrastructure, agricultural lands, and human lives	State (public infrastructure) downstream populations
	(c) Dust-storms reduction	Improved health, lower maintenance costs for infrastructure and industry, reduced damage to farming systems	Urban populations, governments
Global	(a) Carbon sequestration (b) Plant and animal biodiversity	Mitigation of global climate change Healthier resources for future generations	Global population Conservation groups tourism industry

Source: Modified from C. Dutilly-Diane et al. 2007, cited in S. Dong, (2017).

Similarly Herman Daly (2008) rightly talks of 'Steady State Economy' or 'Zero Growth Economy':

> The growth economy is failing ... the quantitative expansive of the economic subsystem increases environmental and social costs faster than production benefits, making us poorer not richer, at least in high consumption countries .... And even new technology sometimes makes it worse.
>
> (cited in Chaturvedi et al. 2017, 6)

It is notable that extracting from the earth the so-called 'black diamond' or 'black gold' (coal) is highly risky and hazardous to human lives, on the one hand, and is very polluting to the environment due to carbon emission on a large scale which degrades and spoils land and its effluents (toxic ash and sludge) spoil surface water (ponds, lakes, rivers, etc.), thus affecting agricultural crops as well as underground water. Many persons die during legal or illegal mining works in many developing countries like India. Actually the idea of 'clean coal' has not been fructified in reality so far. But, given the policy constraints, relatively cheaper, and the huge demand for energy by a substantial population, coal remains the biggest source of energy in India (two-thirds of total energy production) and the second biggest source of energy at the global level.

Nuclear energy is relatively cheaper than coal and crude oil, yet it, too, is highly risky for humans, animals, and the environment. The first worst nuclear disaster in Chernobyl (in Ukraine, in the former USSR) on 26 April 1986 killed at least 4,000 persons (as per International Atomic Energy Agency (IAEA)) over a long time – though Greenpeace International assessed 90,000 deaths! During the 20-year post-accident period, 6,000 cases of thyroid cancer were registered among those who were younger than 18 years at the time of the explosion. In 2005, in Ukraine, 19,000 families were receiving government assistance due to the loss of a bread-winner whose death was deemed to be related to the Chernobyl disaster– it was because of flawed nuclear design and serious mistakes by the operators. Large areas of Belarus, Ukraine, and Russia were contaminated due to such a disaster. According to M. Gorbachev, the former President of the USSR, the Chernobyl accident was a more significant factor in the fall of the USSR than 'Perestroika' (restructuring) and 'glasnost' (openness). Consequently, there was more reactor safety and increased collaboration between Europe and the USA with substantive investment in improving the reactor. Automatic shutdown mechanisms now operate faster than earlier, and now automated inspection equipment has been installed. In 1979, Three Mile Island (USA) also had environmental and health hazards – but far less than Chernobyl had – and, thereafter, the reactor was destroyed, all radioactivities were contained, and the reactor was redesigned – but there was no death then. In 2011, on 11 March, there was Fukishima Daiichi nuclear disaster (Japan) due to Great East Japan Earthquake and tsunami – three of the six reactors at the plant had severe core damage and released hydrogen and radioactive materials, and impeded on-site emergency response efforts.

There are various risks and hazards of nuclear energy plants:

(a) Radiation causing deaths and injuries to humans, animals and birds, and affecting future generations too
(b) Contamination of land, water, etc. from nuclear wastes which may be harmful even for 1,00,000 years
(c) Playing havoc in the hands of terrorists/extremists
(d) Potential for nuclear wars between enemy countries/blocs/regions
(e) May have marine effects due to earthquakes, tsunamis, cyclones, tornados, hurricanes, etc.
(f) Unforeseen disasters from the malfunctioning of nuclear plants

That is why various NGOs and people's movements have been protesting against the different nuclear plants in India – Jaitapur (Maharashtra), Kudankulam (Tamil Nadu), etc. West Bengal Government refused permission to establish a nuclear plant at Haripur. India's nuclear reactors (22 in number) suffer from low capacity factors due to lack of nuclear fuel. India is trying its best for thorium-based fuels. India is producing only 35 TWh from nuclear power and therefrom supplied 3.22% of India's total electricity in 2017, and 3.4% in 2018. Greenpeace International has been fiercely opposing nuclear plants/testing all over the world, including US nuclear testing in Alaska. For all people 'less energy use' should be the practice (Box 1.3).

The term 'environment' is derived from the French word 'Environia' – to surround. Thus it means the surroundings (material and forces) in which organisms (including humans) live at a given point in time and place. Micro environment refers to the immediate local surroundings of the organisms and, on the other hand, macro environment refers to all the physical and biotic conditions that surround the organisms externally.

Box 1.3: Use Less, Not More Energy

In energy consumption, it is advisable to one and all that 'use less', not more, should be the soul motto of every individual, family, community, province, nation, and the whole world because using less energy tantamounts to saving energy, and saving energy tantamounts to producing energy. It has a multiplier effect; for instance, walking on foot or going on cycle thus prevents pollution due to carbon emission. Hence, it also saves money (to be spent on a taxi/tempo/bus) and finally improves overall health (preventing cardiovascular problems, diabetes, blood pressure, and respiratory diseases). Thus using less energy finally significantly contributes to the ecology (saving energy and zero pollution), social well-being (a healthy person is a more responsible and responsive human being), and economy (zero cost). 'Kabad se Jugad' (making useful things out of the waste materials) is another appropriate instance of 'use less', energy. However, using alternative (renewable) energy is the best option.

Source: Authors

Further, in energy consumption appropriate strategies and goals are to be adapted in view of the local social situation, ecological balance, and economic affordability. In fact, every plant has the potential of being a useful herbal medicine, as every letter has the potential of being a pious 'mantra' (canto), as a Vedic hymn rightly says. Eco-strategies and goals can be vividly seen in Table 1.2.

Thus these seven eco-strategies and six eco-goals need to be adapted to the local situations of India and other countries.

Realist–Constructionist Debate in Ecology

Though in common parlance, the two terms 'ecology' and 'environment' are often used interchangeably, there is much difference between the two in terms of both 'range' and 'depth'. The term 'ecology' was used for the first time by the German biologist Ernst Haeckel in 1866, connoting 'the science of the relations of living organisms to the external world, their habitat, customs, energies, parasites, etc.' Thus it is duly recognised

Table 1.2 Eco-Strategies and Eco-Goals for 'Use Less' Energy

Eco-Strategies	Eco-Goals
1. Reduce	1. Prudent use for minimum needs, not greed
2. Recover	2. Equity and inclusion-justice for all (intra-generational)
3. Reuse	
4. Recycle	3. Efficiency – useful output for every unit of input
5. Replace	4. Stability and sustainability for the future (inter-generational equity)
6. Remediate	
7. Restore	5. Cost-effectiveness – economical
	6. Outcome – orientation

Source: Authors.

that human beings are as much part of the entire ecosystem, the planet earth, as are non-human animals, birds, plants, and fish as well as land, water, sunlight, air, and rocks (thus both living organisms and non-living things). Ecology is derived from two Greek words 'oikos' (house) and 'logos' (understanding, logical argument, science) – thus referring to a study to understand the entire planet. However, there are two different traditions in ecology (D. Worster, cited in Sharma 2016): (a) The Arcadian tradition, and (b) the Imperialist tradition.

First, an Imperialist ecologist tries to discover better ways of 'managing' nature for human benefit (anthropocentric), while an Arcadian ecologist takes the 'deep ecology' (Arne Naess' term) approach of giving to non-human life an independent ethical status (biodiversity) and no hierarchy, no species racism.

Second, while an Imperialist ecologist seeks to exploit nature, an Arcadian ecologist seeks to live in harmony with nature.

Third, an Imperialist ecologist desires unrestrained and unlimited economic growth in a free market situation, while an Arcadian ecologist desires restrained and limited economic growth with a regulated market.

Finally, while an Imperialist ecologist speaks of the infinite nature and natural resources for more and more consumption (greed), the Arcadian, on the other hand, argues that nature and natural resources, rather than assets, are finite and, therefore, should be used prudently to fulfil the need of all the people and the animals/birds, as well as their future generations.

If we look at different streams of ecological thought at the global level, we may safely observe that 'deep ecologist', 'ecocentrist' (bio-ethics or Gaianism, and self-reliant community), 'ecofeminist', 'emancipationist', 'anti-establishment environmentalist', 'sub-statist' (decentralisation), 'beyond farmer first' proponent, and 'radical ecologist' (red-green 'eco-Marxist' – and green-green) may be categorised as Arcadian ecologists. On the other hand, 'shallow ecologist', 'techno-centrist', 'reformist', 'scientific environmentalist', 'ego centrist', 'homo-centrist', 'transfer of technology' proponent (modernisation), 'conservationist', 'statist', and 'suprastatist' may be categorised as Imperialist ecologists. That is to say, the former category emphasises the roots of the environmental problems and hence argues for rapid fundamental changes in the basic structure and process of the socio-economic and political system, while the latter desires only slow and incremental (functional) changes therein without destabilising the system as such. A perfect example of Arcadian ecology may be seen in Box 1.4 wherein Justice D. Sheshadri Naidu, of the Goa bench of Bombay High Court, writes a requiem for a falling tree. Its background lies in a dispute between two neighbours in Goa's Santa Cruz, Tiswadi, about the pruning of branches of a tree, leaning in the neighbour's compound. The petitioners filed an appeal against the order of the Deputy Collector-cum-S. D. O. before the court of Conservator of Forests (appellate authority), under the Preservation of Trees Act (1984). On 10 September 2019 in the appellate court (Conservator of Forests) both parties agreed that the owner would cut the branches of the said tree and both parties would equally share the cost. Hence in the writ petition the said Judge D. Sheshadri Naidu dismissed the case finding it a trivial case, but perceptively observing,

Here, a couple of trees have asserted themselves and grown, as they should have, freely in the direction they liked. They have leaned, too, on the neighbours'

22 Environmental Sociology

compound. That has spelt trouble for them and litigation for their owner. For they faced axe, and their owner court proceedings.

(Mumbai Mirror, 2020)

Box 1.4: Justice D. Shashadri Naidu's Requiem for a Falling Tree

A seed or a sapling believes it owns the earth, so it anchors itself with its roots deep into the ground. It feels it owns the sky, so it tries to grow higher and higher, as if to touch the sky. It also feels even the space between these two belong to it. So it spreads, sways, and hangs from above, as it grows. But it does not know man – almost an alien to planet earth – has invaded it, colonised it. As every coloniser does, he pounds, plunders and pillages it. So man makes laws and the laws are human-centric. He commands the aborigines, the trees, to behave themselves. Poor trees, they do not know how? So the axe falls, for the law is amoral almost; for the law brooks no disobedience-always. Nature expects the man to move away from a spreading tree, but the law wants the tree to move away from the man.

And I am bound by law; though not a tree, I am not free. Therefore, I decide this case, decide it in the man's favour, and against the tree. So it is a requiem for a falling tree and a falling human.

Source: Mumbai Mirror (2020)

The above-mentioned requiem of Justice D. S. Naidu fundamentally reflects the Arcadian (or deep) ecological perspective, criticising the Imperialist perspective inherent in the efforts of the two neighbours of Santa Cruz, Tiswadi (Goa), resulting in a dispute to cut the branches of the tree, not allowing the tree to spread and grow more and more as a part of nature and having a natural right to grow indeed. In fact, as early as 1864, George Perkins Marsh published a book on broad ecology (though the term 'ecology' was coined later in 1866) entitled, 'Man and Nature', proving from his empirical research that humans' activities are adversely affecting the natural environment in various ways. He rightly remarked, 'Man is everywhere a disturbing agent. Wherever he plants his foot, the harmonies of nature are turned to discords' (1864, 36, cited in Kandpal, 2018, 9). He clearly pointed out the destruction of forests and the effects of technology, on the one hand, and, on the other hand, once nature's self-sustaining capability is destroyed, it may not heal on its own, rather massive human efforts and care are required in due course. This book played a pioneering role in evolving and articulating the ecological consciousness and later launching a movement in letter and spirit. Humans did such environmental damages either ignorantly or deliberately. But, in both cases, the responsibility lies with the different classes of people – the lower classes for their ignorance and victimhood of the so-called development and the upper class for their deliberate over-use and exploitation of natural assets.

Needless to say that, on this earth, human civilisation has passed through two major revolutions – agricultural revolution and industrial revolution. The agricultural revolution began about 12,000 years ago – in 10,000 BCE (Before Common Era) after passing through two stages of human evolution – hunting and gathering stage, followed by the animal husbandry stage. At the beginning of the agricultural revolution, there was a common practice of subsistence farming (use of family labour, and production for family consumption), and later mechanised farming developed as a result of advanced technology, use of hybrid or High Yielding Varieties (HYVs) of seeds, the assured irrigation, and the chemical fertilisers, pesticides, and herbicides. This first agricultural revolution made the people settle in harmony with nature – land, water, forest, animals, birds, and so on. The green revolution was a technical–managerial package (hybrid seeds, assured water, credit, chemical fertilisers and pesticides, and new implements/machines like tractors/harvesters) that began in the 1960s and 1970s first in the USA, then exported to Mexico, India, the Philippines, Turkey, etc. and by 1968, in total, 18 countries practised dwarf hybrid seeds of wheat. Actually the green revolution was first confined to wheat (due to the efforts of Norman Borlaug, who got the Nobel Peace Prize in 1970 for plant breeding). But the green revolution had little impact on Sub-Saharan Africa, except for higher maize yield in Zimbabwe; the focus on higher rice yield was due to the worries of the USA about the possible spread of Chinese communism after 1949. However, communist countries of China, Vietnam, and Cuba adopted hybrid seeds: In the course of history, there emerged social differentiation (class society) and, therefore, the upper class cornered the surplus. On the other hand, due to new discoveries, various fossil fuels like coal, oil, and natural gas as well as electricity power were used for machine operations for the transport, and manufacturing of goods and facilities for removing the drudgery in daily life. Thus in the 1700s CE (Common Era) in the UK and in the 1800s CE in the USA, the industrial revolution took place. Thus Newtonian science emphasised the use and exploitation of natural resources (owned differently by individuals, communities, and the State) as much as possible in order to have the optimal benefit. On a global scale, the agricultural revolution's advanced phase (green revolution) had many ecological consequences (McNeill 2007):

(a) It promoted monoculture replacing the old practice of multi-cropping biodiversity in agriculture.
(b) Monoculture, consequently, led to pest problems; even pest-resistant crops developed new pest infestation – more doses of pests led to health hazards through water sources and through fruits, vegetables, and food, reaching up to human bodies. As per WHO's estimate in 1990, pesticide poisoning killed 20,000 persons annually in the world, mostly in cotton fields as well as a million people suffered acute poisoning – two-thirds of them were agricultural workers.
(c) Huge requirements for fertilisers led to eutrophication of lakes and rivers, and the required irrigation led to the construction of big dams in India (Bhakhra Nangal in Punjab, Sardar Sarover dam in Gujarat, Tehri dam in Uttarakhand), China, Mexico, etc. As a result there was a huge displacement of the rural poor people, especially the tribals.
(d) On the one hand, it reduced the genetic and species diversity in agriculture – especially due to its confining to wheat, rice, and maize; other crops (coarse food grains, pulses, oilseeds), less responsive to nitrogen and water, but rich diets, were neglected, and, on the other hand, various indigenous varieties of paddy, wheat, and maize became extinct.
(e) It has over-used energy directly or indirectly (from using underground water through tube wells to the manufacturing of chemical fertilisers, pesticides, and

24 Environmental Sociology

herbicides, to manufacturing and operation of agricultural machinery – tractor, harvester, thresher, etc.).

(f) Due to continuous agricultural revolution in the West and Japan, they became 36 times more prosperous (in 1985 itself) than the developing countries; with a few exceptions of India, China, etc., many developing countries (especially Sub-Saharan Africa) could not become self-sufficient in food production – as per Paul Bairoch's finding, until 1981 many developing countries had been a 'net exporter' of food, but after 1981 these became a 'net importer'.

(g) It followed the path of the West (catching up) in developing countries too.

Therefore, agricultural scientist Prof M. S. Swaminathan (2007) called for an integration of agriculture, forestry, fishery and animal husbandry, integrated strategies of nutrient supply and pest control, organic recycling and water conservation, extensive tapping of sunlight, and using two tools employed by nature – 'synergy' and 'symbiosis', hence 'evergreen revolution' Goal 2 of Sustainable Development Goals (SDGs) rightly focusses on promoting sustainable agriculture.

On the other hand, the industrial revolution, too, provided various benefits to humankind, viz. more opportunities for employment for the men and women, better infrastructural facilities, better modes of transport and communication, more income to different classes and communities, more prosperity for the people, better health resulting in less mortality and more life expectancy, better and more education. However, it, along with huge population growth (four-fold during 1900–2000 CE), had many serious environmental/ecological consequences too:

(a) Environmental pollution – air, water, soil, noise, and radiation.

(b) Acid rain.

(c) Ozone layer depletion – hence ultraviolet rays of the sun causing cancer and other major diseases.

(d) Declining of the groundwater table.

(e) Culture of consumerism – 'more, faster and bigger are better' (more energy, more goods and services).

(f) Over-use and exploitation of natural resources/assets.

(g) Machines do not understand human and ecological sensitivity and inter-connectedness.

(h) The triple processes of liberalisation, privatisation, and globalisation (LPG) have accelerated the tendency of unrestrained exploitation of natural resources/assets, in the spheres of individuals (private), State, and community– the common property resources (CPRs).

(i) During 1900–2000, Gross World Product grew 14-fold, with more intensive use of energy and raw materials, resulting in more waste per unit of wealth created – the emission of carbon dioxide rose 13-fold and energy consumption increased 16-fold (McNeill 2000).

(j) Due to the massive use of technology for always-thirsty industrialism and consumerism, thousands of species of plants, fish, animals, and birds have already become extinct and the existence of many is highly threatened, or on the verge of extinction.

(k) Imperialism, being the highest stage of capitalism, used and exploited, without any restraint, the natural resources/assets of its colonies, for industrial purposes like raw materials for ship industry, aviation, rail industry, manufacturing goods, etc. As dependency theorists like A. G. Frank rightly point out, the development of the 'metropolis' (industrialised nations) led to the underdevelopment of the 'periphery' (the developing countries).

Reductionist science and technology has been practised against the flourishing of the biodiversity and it sees only the parts, not the whole, but the whole is greater than the sum total of parts, and often the associational relationship of some factors (not being cause–effect), especially related to sentiments, is quite significant for an integrated future vision. Mahatma Gandhi was against labour-replacing and ecology-damaging technology. Gautam Buddha and Mahatma Gandhi had talked of an 'economy of permanence', in harmony with nature, hence E. F. Schumacher (1973) sharpened that idea and talked of 'small is beautiful' – 'appropriate technology' easily adaptable at the local level. A modern person does not think as a part of nature; even Karl Marx fell into this error in his 'labour theory of value'. Though most of the people are better fed, clothed, and housed, Schumacher (1973) argues, this is not the 'substance' of human beings as this cannot be measured by the Gross National Product (GNP). Therefore, taking into account the ecological assets holistically, he mentions three categories of capital:

(i) Fossil fuels
(ii) Tolerance margins of nature
(iii) The human substance

Hence, he suggests new production methods in agriculture and horticulture that are 'biologically sound, build up soil fertility, and produce health, beauty and permanence'; on the other hand, in industry, there should be small-scale technology, relatively 'no violent technology', 'technology with a human face', and 'new forms of partnership between management and men, even forms of common ownership'. However, Schumacher's schema of 'small-within-large structure' does not question the capitalist mode of production. Hence he is often branded (by Lipton) as 'romantic' or 'backward-looking'; but this allegation is not fully correct because he also talks of 'intermediate technology' by improving indigenous technology or adapting modern technology as per local needs. Even then, it is correct to say that he did not question the capitalist system as such.

Realist–Constructivist Debate in the West

Sociologist Riley E. Dunlap (2010) rightly talks of three ecosystem services:

(a) The environment as a 'supply depot' provides us resources to meet 'material needs and wants'.
(b) Humans produce wastes in due course, and the environment functions as a 'waste repository'.
(c) The environment provides a 'living space' – a place to live, work, and consume.

Thus the 'ecosystems', not the environment, provide these services to humans and non-humans. Undoubtedly the wealthy (core) nations use the poor (semi-periphery

26 Environmental Sociology

and periphery) nations as 'supply depots' (taking natural resources), on the one hand, and also use them as 'waste repositories' (by dumping wastes for disposal, or locating polluting industries there or over-using global commons oceans and atmosphere). However, the West is not homogenous and undifferentiated. In North America, especially the USA, environmental sociologists empirically analysed the interaction between social and environmental phenomena – racial/ethnic, socio-economic status, and environmental exposure – hence they carried out a sociological study of environmental problems. On the other hand, in Europe, due to the influence of post-modernity, a social-constructivist view with a more idealistic orientation (in lieu of materialist) emerged. However, some scholars tried to address both symbolic and materialist dimensions of environment–society interaction. Dunlap and Catton (1979) take a realist position, while Burningham and Cooper (1999) and Latour (2004) take a constructivist position. Taking a realist view, Benton (2001, 18) criticises constructivists: 'constructivist demonstrations of the intrinsic uncertainty and politically/ normatively "constructed" character of environmental science sabotages environmental politics, and plays into the hands of powerful interest'. Hence the realists focus on how economic privilege and political power are used to ignore or suppress the scientific evidence of climate change and environmental degradation in general. Later Carolan (2005) distinguished three strata of 'nature', nature, and Nature:

(a) 'nature' (in quotes) – a socio-discursive concept signifies 'that which is not social', to refer to the natural world or human nature or human biology.
(b) nature (uncapitalised) refers to 'the nature of fields and forests, wind and sun, organisms, watersheds, landfills and DDT – involving "ubiquitous" (and obvious) overlap between the socio-cultural and bio-physical realms'.
(c) (Deep) Nature (capitalised) – 'the nature of gravity, thermodynamics, and ecosystem processes'.

Social constructivists limit their views to the (a) 'nature' – that different cultures and social sectors create and are motivated by differing images/views of the 'natural world', and hence they have different views on development and environment protection, not related to 'objective conditions'. The second stratum (b) (nature) relates to ecosystem services and disruptions, hence popular among environmental sociologists. But while the constructivists contextualise, problematise, and deconstruct the claims about ecological conditions described by the scientists, activists, and policy-makers, the realists, on the other hand, use various indicators of environmental conditions. The third stratum of deep Nature is of limited concern to sociologists. Hence, three major conclusions from these distinctions are drawn (Dunlap 2010):

(i) Realists have no problem with deconstructions of phenomena in the first stratum – primarily socio-cultural products – but do not agree when the constructivists generalise their deconstructions of cultural interpretations of 'nature' to the ecosystem services and disruptions that comprise the second stratum and conflate the two strata.
(ii) Realists criticise the constructivists' over-emphasis on problematising and relativising the evidence of ecological problems.

(iii) Realists see their focus on deconstructing both 'nature' and knowledge claims of ecological problems as reflecting a very restricted version of environmental sociology, particularly ignoring 'interactions' between socio-cultural and bio-physical phenomena.

Therefore, there are two main camps of environmental sociologists (Dunlap 2010):

(a) **Environmental agnostics**, especially in Europe, treat 'environmental matters' as symbolic/ideational/cultural phenomena examine these through a hermeneutic or interpretative approach, and take a relativistic stance towards knowledge claims about environmental conditions, and it hardly deals with materialistic aspects of ecological problems and hence avoids society–environmental interactions and represents a modern 'sociology of environmental issues'.

(b) **Environmental pragmatism**, particularly in North America, but also elsewhere, treats material aspects of environmental conditions as potential indicators of ecological problems and empirically examines how these problems are associated with social phenomena. It analyses linkages between symbolic, socio-structural, and material arenas. It takes a realist perspective, with diverse empirical approaches. This focusses on the causes and consequences of ecological problems. They focus on Carolan's second stratum, the nature of ecosystem services and disruptions, in the form of resource use, pollution, and land degradation.

However, there are attempts to merge the strengths of these two approaches, with agnostics using their rich analytical tools 'to delve more deeply into the material world and pragmatists paying greater attention to the impact of constructions, values, culture and the like' (Dunlap 2010). Nowadays the new trend is rightly 'a mix of constructivist and realist, qualitative and quantitative, micro and macro, theoretical and empirical work'.

Paul Robbins, a political ecologist, talks of the social construction of the environment. Forest, observes Paul Robbins (2012, 123), is not a 'natural' phenomenon, object, or idea, but rather a 'social' one, 'forged by convention and context, and enforced by its very taken-for grantedness'; it becomes 'political when one considers that, depending on whether this bunch of trees is considered "forest" or "degradation", significant state and international resources will be invested in its protection or its eradication'. Political ecologists, therefore, are of the view that the environment, which we take-for-granted, is actually 'constructed'. To Immanuel Kant, philosophical knowledge comes 'prior' to experience – ideas do not conform to the objects of the world, rather objects are constituted by ideas. Later Michel Foucault developed this view by arguing that assuming the universality of many concepts never existed in other times and places. To him, ideas are not powerful because they are true, rather they are true because of power. Thus 'Question Reality' was his motto. For instance, soil erosion is not a universal truth, but rather a social construction in a historical context when the colonial land management authorities, environmental bureaucracies, and other ecological elites were authorised to control other people's behaviours and property in the name of 'soil conservation'. Resistance to such efforts in colonial Africa was perceived by the colonial administrations as environmental irrationalism of the ecologically ignorant people

(Grove 1990) – thus the so-called soil erosion was a social construction that helped to secure political power. Political ecologists, therefore, argue that some environmental concepts, ideas, entities, or processes are not natural or inevitable, rather these help to secure the power of an elite community. Hence, these concepts, ideas, entities, and processes are to be 'unmasked, reinvented and changed for a better and more sustainable future', as, 'in political ecology, things are rarely what they appear' (Robbins 2012, 124). However, Robbins clarifies, the social construction of soil erosion denies the physical forces and processes that determine soil movement – a purview of soil scientists, not critical theorists.

To constructivists, categories (indigenous or scientific) may describe some commonalities in the pattern of reality, but not more correct than other possible classifications/categorisations – 'scientific' expertise gives more credibility to one arbitrary categorisation over the other. Some cultural or scientific categorisations classify palm as a tree, while others don't do so. However, constructivism is not uniform – one is 'hard' or 'radical' constructivism, while the other is 'soft' social object and social institutional constructivism.

Radical Constructivism believes that the environment is an invention of people's imagination – an experience of the world is related to us through stories, conventions, and idea systems arising from learning from other people. As per hard constructivism, social context alone conditions and determines our concepts for understanding the world and thus 'creates the world' in the process. That is, things are true because they are considered to be true by socially powerful people, because these are true on television, and because these are true in our minds. Taking a relativistic position, it argues that science, as a specific method, cannot be used for deciding disputes between different claims about what is real – all claims are arbitrary. According to Steve Woolgar (1988, 89), 'nature and reality are the by-products rather than the pre-determinants of scientific activity'. Hence environmental conflicts are seen as struggles over ideas about nature, in which the group having access and mobilising social power to create consensus about the truth prevails. However, the radical construction, first, does not consider the significance of non-human actors and processes (soil, climate, tree, etc.) in explaining outcomes. Second, though it provides space for appreciating alternative constructions of the environment by other social communities like nomadic herders or forest dwellers or religious scholars, it makes symbolic systems of humans sovereign over all other reality – thus not allowing empirical investigation in conventional environmental science.

Softer Constructivism, on the other hand, is preferred by most political ecologists. It argues that our concepts of reality are real and have force in the world, yet these provide incomplete, incorrect, biased, and false understandings of empirical reality. That is, the objective world is real and does not depend on our classification/categorisation but is filtered through subjective concepts and scientific methods which are socially conditioned. For instance, the false and socially biased category of the race does not tally with reality. But such racial experiences have quite negative effects in practice and hence need to be understood. For social institutional constructivists, wrong ideas about nature are a product of the inevitable 'socialness' of scientific communities. However, with the passage of time and through experimentation and refutation, the 'social' ideas are separated from our understanding of nature. However, it insists that only falsehoods may be explained socially, but objective facts and true understandings of nature have no social aspect. As Bruno Latour (1993, 92) rightly remarks,

> Error, beliefs, could be explained socially, but truth remained self-explanatory. It was certainly possible to explain belief in flying saucers, but not the knowledge of blackholes; we could analyse illusions of para-psychologist, but not the knowledge of psychologists; we could analyse Spencer's errors but not Darwin's certainties.

Thus social construction has many approaches to science, knowledge, and nature. Sergio Sismando (1993, cited in David Demeritt 1998) has categorised four different uses of the construction metaphor (Table 1.3).

From Table 1.3, it transpires that social constructivists are politically conservative because global environmental problems like climate change are a reality in different parts of the world. Hence, unlike the old 'Scientific forestry', 'new forestry' (Jerry Franklin 1989) talks of sustaining forest ecosystems and forest health. But the reductionism of climate change science is attached to both a moral-liberal and a rational-technocratic view of politics and science (Taylor and Buttel 1992).

In fact, science has created many environmental problems locally, nationally, and globally, but at the same time we also need the help of critical and public interest science, e.g. ecology, that sees things in totality, for their solution.

Despite the difference in some way among several constructivists, they commonly share that ideas and narratives about nature and society are mobilised in ecological struggles. Many things, not environmentally natural, are often shown to be so and vice versa. After examining the changing meaning of the concept of 'wilderness' in western history, William Cronon (1983/2003) found that it was historically contingent. Humans produce a 'natural' environment around them, while there is a presence of natural processes in non-wilderness areas like the cities – hence wilderness must be viewed as a social construction and one that bars effective conservation, thus placing humans outside of nature (exclusion of local residents). Neumann (1998) has shown that the imported ideas of Anglo-American wilderness aesthetics were imposed on African landscapes inventing environments that did not exist earlier. Calling this aesthetic natural politically led to the removal and disempowerment of local people who had worked in the creation of this very 'natural' landscape of tropical and sub-tropical Savanna that colonial and post-colonial administration preserved. One more instance recurs when environmental problems are constructed, though not existing in reality or where for international funding purposes environmental problems/events are shown as highly disastrous – this politically suits many stakeholders. Further the degradation may be overstated or understated depending on the situation. This manipulation is called the 'rhetorical' or 'tactical' approach to the construction of the environment. On the other hand, a non-conscious form of constructivism is highlighted where State administrators, local people, and international agencies hold different normative views of the environment. This view focusses less on the intentional and strategic use of ideas and narratives about nature, rather focussing more on how 'naturalisation' occurs, emphasising the social process whereby constructedness of environmental concepts and practices is forgotten (Robbins 2012).

Table 1.3 Typology of Social Constructivisms

Aspects	Common sense realism	Social object constructivism	Social institutional constructivism	Artefactual constructivism	Neo-Kantian constructivism
1. Chief tenets	Observational statements refer directly to a pre-existing, independent, and in this sense, objective reality	Taken-for-granted beliefs about reality, e.g. gender, constitute a social reality no less 'real' in its causal effects than reality itself	Science is a social construction – its institutions and social contexts of its discoveries are socially conditioned and constructed	Reality of the objects of scientific knowledge is the contingent outcome of social negotiation among heterogeneous human and non-human actors	Objects of scientific thought are given their reality by human actors alone
2. Key proponents	P. R. Gross and N. Levitt (1994)	Peter Berger and T. Luckmann (1966), J.R. Searle (1995)	R. K. Merton (1938–1970)	B. Latour (1987); D. J. Haraway (1992)	S. Woolgar (1988); H. M. Collins and T. Pinch (1993)
3. Ontology	Nature/society, subject/object, and mind/matter are ontologically distinct realms	Socially constructed reality distinct from objective facts given by nature, e.g. sex	Objective reality distinct and independent from belief about it	No absolute ontological distinction between representation and reality, nature, and society	Nature is whatever society makes of it
4. Epistemology	Truth value determined by correspondence between representation and reality	Scientific truth explained by nature; socially constructed belief is the cause of scientific falsehood	Ignorance and socially constructed bias explain belief in scientific falsehood	Ultimate truth is undecidable	Truth is what the powerful believe it to be

Source: S. Sismondo 1993 (cited in David Demeritt 1998).

Gender and Nature: Ideological/Constructionist View, with Special Reference to India

In 1974, in order to explain the human–nature relationship from a gender perspective, a French scholar, Francoise d' Eaubonne coined the term 'eco-feminism' in her work, 'Le Feminisme Ou La Mort' (1974) and called for an egalitarian, collaborative society in harmony with nature, where no group is in a dominant position. Eco-feminism seeks to explore the connections between nature and women in religion, culture, literature, art, etc. Like feminism, eco-feminism, too, has various streams like liberal eco-feminism, spiritual-cultural eco-feminism, and socialist/materialist eco-feminism. One common thread between these streams, however, is the patriarchy assigning a secondary position to both nature and woman, though other factors are also explained, with a focus on culture, economy, or ideas/ideology. When the issue of gender and environment relationship is discussed, there are mainly two broad perspectives: Ideological or Constructionist, and Realist. The mainstream eco-feminism or ideological/constructionist perspective has four major agreements (Bina Agarwal 2007):

(a) There exists a significant connection between the domination and exploitation of nature.
(b) In patriarchal thought and practice, women are seen as closer to nature and men are seen as closer to culture; since nature is seen as inferior to culture, hence women are seen as inferior to men.
(c) Since the domination of women and that of nature occurred together, women have a special responsibility 'in ending the domination of nature, in healing the alienated human and non-human nature' (to use Ynestra King's terms).
(d) Both the feminist and environmental movements stand for egalitarian, non-hierarchical systems, and hence they need to work together.
  Earlier Sherry Ortner (1974) had pointed out that:
  (a) Woman is identified with nature 'that every culture devalues, defines as being of a lower order of existence than itself'.
  (b) Man is identified with culture.
  (c) Connection between nature and women was rooted in the biological process of reproduction (i.e. menstruation, pregnancy, and birth).

However, Sherry Ortner also recognised that, like men, women also mediate between nature and culture. Later, some social anthropologists like Carol P. Mac Cormack (1980) criticised her that nature–culture divide is not universal across all cultures and terms like 'nature', 'culture', 'male', and 'female' have different connotations in different cultures and societies. Other feminist scholars like Ynestra King and Carolyn Merchant are of the view that nature–culture dichotomy is not a true one, but rather a 'patriarchal ideological construct', used to maintain gender hierarchy. Carolyn Merchant traced two contradictory images of nature in pre-modern Europe – first, nature, especially the earth, was identified with the nurturing mother. Second, nature was seen as wild and uncontrollable to render fury, violence, storms, drought, and chaos, and hence human dominance over nature was culturally sanctioned. During the sixteenth and seventeenth centuries, the scientific revolution and the growth of market-oriented culture in Europe cemented the second image. Hence Carolyn Merchant rightly suggests,

> Juxtaposing the egalitarian goals of the women's movement and environmental movement can suggest new values and social structures, based not on the domination of women and nature as resources but on the full expression of both male and female talent and on the maintenance of environmental integrity.
>
> (Merchant 1980, cited in Agarwal 2007, 319)

Thus eco-feminist discourse highlights three major points (Agarwal 2007):

(a) Important conceptual links between the 'symbolic' constructions of women and nature and the ways of 'acting' upon these
(b) The commonality between the premises and goals of the women's movement and the environmental movement
(c) An alternative vision of a more egalitarian and harmonious future society

Bina Agarwal (2007) rightly finds eco-feminists' ideological constructionist argument problematic on the following counts:

First, it considers 'woman' as a unitary category and fails to differentiate among women on e.g. class, race, ethnicity grounds. Thus it ignores other forms of domination, as much as gender.

Second, it primarily locates the domination of women and of nature in ideology and thus ignores the inter-connected material sources of such domination like economic betterment, racial supremacy, and political power.

Third, even in the sphere of ideological constructs, it is silent (with the exception of Carolyn Merchant) about 'the social, economic and political structures within which the constructs are produced and transformed'. Further it also does not address the central issue of the means by which certain dominant groups (predicated on gender, class, race, etc.) are 'able to bring about ideological shifts in their own favour and how such shifts get entrenched'.

Fourth, the eco-feminist constructionist view ignores women's lived material relationship with nature, and tracing the connection between women and nature to biology is a form of essentialism.

Needless to add that the constructionist view of nature cannot sustain because various concepts of nature, culture, gender, race, etc. are 'historically and socially constructed and vary across and within cultures and time periods'. Therefore, Bina Agarwal asserts, in order to challenge such ideological constructs, one needs a theoretical understanding of the 'political economy of ideological construction' – the interplay between conflicting discourses, and the means used to entrench views incorporated in those discourses. Further, she adds, it is necessary to analyse the basis of women's relationship with the non-human world at non-ideological levels like the works done by men and women (gender division of labour) and the gender division of property and power, and also to address the material realities in which women of various classes, castes, races, etc. are rooted might affect their responses to environmental degradation. Like Carolyn Merchant, Vandana Shiva, too, observes that violence against nature is intrinsic to the dominant industrial/development model, characterised by colonial imposition, which was in contrast to the traditional Indian cosmological view of nature as 'Prakriti' as 'activity and diversity' and as an expression of 'shakti', 'the feminine and creative principle of the cosmos' which, 'in conjunction with masculine principle (Purush) creates the world'. Thus the living, nurturing relationship between man and nature as mother earth was replaced by the conception of man as separate from and dominating over inert and passive nature. She further explains that violence against women and nature are linked not only ideologically but also materially. Women in developing countries depend on natural assets for the sustenance of themselves, their families, and their communities/ societies. Based on her study of the Chipko (hug the tree) movement, she is of the view that women in developing countries like India have a special dependence on nature and a special knowledge of nature. But their knowledge has been marginalised by the reductionist modern science that has become a 'patriarchal project' excluding women (as experts), ecology, and the holistic ways of knowing. Though going beyond the western eco-feminists, Vandana Shiva explores the links between ways of thinking about development, the process of development change, and their impact on the environment as well as those dependent on it for their livelihood, yet her argument has three flaws (Agarwal 2007):

(1) Her examples are primarily from Uttarakhand, yet her generalisations put all women from different developing countries into one category – without differentiating them on class, caste, race, and ecological zone lines. This categorisation suffers from female essentialism.

(2) She does not point out by what concrete process and institutions ideological constructions of gender and nature have changed in India, and also she does not place the co-existence of several ideological strands in India's religious diversity. Her examples of feminine principles exclusively rest on Hinduism. Even Hinduism has many sects with varying gender and nature implications; e.g. Shakti cult worships the goddess 'Shakti' as the primary deity, while Shaiva Sect gives more importance to Lord Shiva, Vaishnava Sect gives more significance to Lord Vishnu, and Arya Samaj accords primacy to Vedas over idol worship.

(3) She blames colonial rule in the developing countries, western reductionist science, and western model of development for the marginal status of women and knowledge, but earlier in the Mughal period India was substantially class and caste stratified, and it affected the access to and use of natural resources by different classes and castes. Thus Vandana Shiva ignores 'the very real forces of power, privilege and property relations' before colonialism.

Realist Alternative: Feminist Environmentalism

Bina Agarwal (2007) puts forward a realist alternative to the ideological constructionist thought of nature–human relationship and terms it 'Feminist Environmentalism', rooted in the material reality. She recognises: First, the gender division of labour, class division of labour, caste/race division of labour; second, also gender distribution of property; third, the gender distribution of power, class distribution of power, and caste/race distribution of power; and finally, the division of labour, property, and power also shaped the knowledge based on expression about nature. For example, in hill and tribal communities usually women fetch water, fodder, and fuel wood; hence they are more affected due to environmental degradation. In Jharkhand, as per our personal knowledge, only women carry head loads of fuel wood and are called '*bojharis*'. Due to regular and continuous interaction with nature as well as knowledge passed from an earlier generation (especially mothers), they have more and better knowledge of various varieties of plant species and the process of (natural) self-regeneration and self-renewal. Thus they are 'both victims of destruction of nature', and 'repositories of knowledge about nature' – in ways distinct from the men of their class. As 'victims' they resist and respond promptly to the destruction of nature, and, on the other hand, as 'repositories of knowledge', they perceive and choose what is to be done for regeneration, hence they may provide an alternative perspective to development. Thus, in Bina Agarwal's view,

> the link between women and the environment can be seen as structured by a given gender and class (caste/race) organisation of production, reproduction and distribution. Ideological constructions such as of gender, of nature and of the relationship between the two, may be seen as (interactively) a part of this structuring but not the whole of it.
>
> (2007, 324)

Thus feminist environmentalism calls for 'struggles over both resources and meanings'. Hence, on the feminist front, both notions about gender and the actual division of work and resources between genders are to be challenged and transformed. On the environmental front, both notions about the relationship between people and nature and actual methods of appropriation of nature's resources by a few persons are to be challenged and transformed. A comparative view of the two perspectives may be seen in Table 1.4.

In India, Bina Agarwal points out in detail, first, as in other developing countries, rural families have been taking various items of daily use and subsistence (foods, fibre, fruits, fuel wood, fodder, manure, water, small timber, bamboo, herbs, grasses, honey, etc.) from common property resources (CPRs) like forests, village commons as well as regional/national water bodies like rivers, streams, lakes. Since the collection of such items from such collective resource pools is time consuming and labour intensive, poor people indulge much more in their collections for everyday life. Second, regarding access to groundwater for drinking and irrigation purposes, the rich class easily affords more and deeper tube wells. Third, the poor people suffer on two grounds: Due to the degradation of land in quantity and quality, and, on the other hand, due to the increasing satisfaction (appropriation by the State) and privatisation (appropriation by a few rich persons). The former trend reduces the availability of usable land, and the latter trend enhances the inequality in the distribution of the available land. These two trends are 'primary factors underlying the class-gender effects of environment change'. In addition, she points out, the following three intermediary factors worsen the socio-economic situation:

(a) The erosion of community resource management system
(b) Population growth
(c) Technological choices in agriculture, and their associated effect on local knowledge system

Table 1.4 Comparison of Ideological/Constructionist and Realist Perspectives

Sl. No.	Theme	Constructionist Perspective	Realist Perspective
(a)	Patriarchy's dichotomy between nature and culture	Women are seen as closer to nature but inferior to men who are seen as closer to culture, unchangeable, and essentialism irreducible	Nature–culture dichotomy is false, not universal and no uniformity in meanings of 'nature', 'culture', 'male', and 'female'
(b)	Roots of domination	Domination of women and of nature is basically ideological, rooted in ideas, representations, and values/beliefs	Ideological dimension is rooted and produced in multiple material structures of domination historically and socially
(c)	Women's role in ending double domination	Women's stake in healing the alienated human and non-human nature	Both men's and women's stake in ending all forms of domination and exploitation is necessary
(d)	Forms of action and change	Link between 'symbolic' constructions of nature and women, and the ways of 'acting' upon such constructions	Through economic, social (gender), and political (power) means, dominant groups are able to bring about ideological shifts in their favour
(e)	Uniformity Vs differentiation	Women as a unitary category – no differentiation of women seen	Differentiation of women on class, caste/race, power, and privilege bases counts
(f)	Commonality of movements	Commonality between the premises and goals of women's movement and environmental movement	Feminist environmentalism goes beyond the ideological commonality principle, and links to livelihood, disasters, and displacement of local people, especially women
(g)	Orientation to change	Only patriarchy to be replaced, not concerned with redistribution of resources, nor capitalism	Transformational ecology, development, and redistribution are linked in mutually regenerative ways; both patriarchy and capitalism to be replaced
(h)	Alternative vision	Alternative vision of a more egalitarian, non-hierarchical, and harmonious future society	Alternative vision of a more socially egalitarian, ecologically just, economically prosperous, and politically participative democratic society

Source: Prepared by authors based on Agarwal (2007).

In our view, the realist perspective also includes the complementary relationship between women and men in different walks of daily life and different actions of ecological sustainability. In various environmental movements like Silent Valley, Chipko, and Narmada Bachao, both men and women genuinely participated with equal enthusiasm and willpower due to 'critical life issues'. Terming the Chipko movement solely as a women's movement (by Vandana Shiva) is to ignore the true and complex realistic narrative (Sharma 2016).

Needless to say that, in reality, the area under dense forests is declining in India though 'green cover' – sparse tree cover/green vegetation other than dense forest – is increasing in the first quarter of the twenty-first century. Again a huge area is lost due to water-logging, floods, etc., conventionally in North Bihar, Bengal, Odisha, Assam, and Eastern UP, but also due to the climate change in the recent past in Rajasthan, Gujarat, and other States. One strange phenomenon emerged in the summer of 2019 in North Bihar, especially in Darbhanga and Madhubani districts, wherein the water table declined by 10–15 feet though there used to be massive groundwater availability just at 10–15 feet depth earlier. Further, in 2022 most of North, East (Bihar, Bengal), and North-West faced drought while there was a huge flood in North-East India. All over India, due to the massive use of chemical fertilisers, pesticides/insecticides, and herbicides in various food crops, vegetables, and fruits, their residues run off into the surface water sources – destroying the fish life as well as contaminating the water for human and animal use.

One more significant issue is the narrow reductionist 'scientific' method of forest management in India since 1864. As Vandana Shiva (1987) has perceptively noted that in the reductionist worldview only the commercially profit-generating resources are valued – nature is thus seen only as of 'instrumental value' – while other resources contributing adequately to ecological sustenance, but not profit-generating, are not valued. Such a tendency, more or less, continues even today when Tendu leaves (for beedi-making) are got collected and auctioned to beedi contractors/beedi manufacturing companies by the State forest department, in Jharkhand, West Bengal, Chattisgarh, Madhya Pradesh, etc. – the local tribals and other poor people suffer the most in various processes of leaves' collection and beedi-making. It is our considered view that many natural assets like rocks, mangroves have an intrinsic value (aesthetic), too, and hence are to be respected by all humans.

All over the world, those who are absolutely deprived of and are powerless are the worst victims of pollution, environmental degradation, floods, drought, development-induced displacement, and other disasters. The case of the Ogoni ethnic group of Nigeria (Africa), being the victim of an oil spill from the oil pipeline of Shell Oil Co., is the most appropriate example (Box 1.5).

Box 1.5: Protest against Environmental Injustice Done to the Ogonis, Nigeria

Royal Duch Shell's joint ventures account for 21% plus of Nigeria's total petroleum production (6,29,000 barrels daily). There are oil spills due to 'oil bunkering' (drilling holes in oil pipelines illegally by local people), or leaving pipeline open, 'operational spills', corrosion, lack of regular maintenance of pipe and equipment, etc. The oil pipeline of Shell Oil Company in Nigeria burst badly (1992) and affected many Ogoni villages – it disrupted field drainage systems, killed a huge number of fish, gas flares fouled the air, polluted the surface water, acid rain from gas flares caused corrosion of zinc roofs of their houses, etc. Thus their life and livelihood were badly affected by such an oil spill. Hence the 3,00,000 Ogoni people (out of their total 5,00,000 population) rallied on 4 January 1993 with green twigs. Their rally was led by the famous Ogoni writer Ken Saro-Wiwa. Consequently, in the next two years, the Nigerian soldiers killed 2,000 Ogoni people and tortured and displaced thousands of them. They forced or enticed people from neighbouring regions to violently attack the Ogoni so that the repression might be coloured as ethnic clashes. Further the government army also sealed the border of Ogoniland. Ken Saro-Wiwa and eight other leaders of the protest were arrested and tortured and were charged with murder government finally executed them on 10 November 1995. Even, thereafter, for some years, the torture and killings of Ogoni leaders and activists continued. On June 15, 2001 police shot Friday Nwiido for leading a protest against 29 April 2001 oil spill that continued in Ogoni's surroundings for nine days. Later there was a peace negotiation among the Shell Oil Company, Nigerian Government, UN Environment Programme (UNEP), and the Ogoni people. In 2009, Shell Oil Company paid them $15.5 million. Again in 2015, Shell Oil Company agreed to pay $3,300 each to 15,600 fishermen for two oil spills in 2008 and 2009 – a total sum of $84 million. However, even afterwards oil continues to spill there on land and in water. During 1998–2009, 4,91,627 barrels of Shell oil spilt, averaging 41,000 barrels annually; during 2011–2018, 1,010 oil spills took place, with a loss of 1,10,535 barrels of oil. A Dutch court in 2013 ruled that Shell Oil Co. was liable for the pollution in Niger Delta for the oil spill in water. Shell Oil Co. takes about 10 days to respond to an oil spill in the water and conduct a JIV report, and for an oil spill on land, it takes more than five days.

This instance vividly shows the environmental injustice to the poor people, unequal distribution of environmental costs (bads) and benefits (goods), but local people's resistance results in the massive torture and killings of the protesters. Finally, justice is achieved only when an outside well-meaning agency (here UNEP) critically intervenes in the violence-ridden situation.

Source: M. M. Bell (2016); wikipedia.org., www.shell.co.uk

Conclusion

Both sociology and ecology take holistic perspectives, and hence the fundamental premises of environmental sociology help in the proper understanding of the relationship between environment and society. Humans, being the most progressive life form among all organisms, are to be more responsible and ethical to non-human nature (animals, birds, fish, plants and non-sentients), so that all, being parts of the whole ecosystem, flourish in a more comprehensive way, as they are inter-connected and mutually interdependent. Both over-consumption of various resources, leading to the 'culture of consumerism' in the West (led by the USA), on the one hand, and the higher rate of population growth in many developing countries, on the other, contribute significantly to the degradation of the environment. However, the former is the result of affluence, and hence may be easily avoided, while the latter is intricately associated with poverty, hunger, lack of sanitation, etc., hence linked to survival and livelihood. Unfortunately, the common property (natural) resources are being either statised (State appropriation) or privatised (private control), and thus the community as a whole suffers a lot, especially the lower classes. Due to policy defects and lack of consciousness in civil society, the people have suffered from the negative consequences of the 'green revolution', socially, economically, environmentally, and politically. Hence there is a need for an 'evergreen revolution' as Dr. M. S. Swaminathan calls for, for both intra-generational and inter-generational equity. There is also a need that environmental sociologists should give due weightage to the autonomy of non-human species by withdrawing from human dominance and hegemony over them. Finally, the realist perspective needs to be preferred to the ideological/constructivist perspective of nature.

However, now the question also arises: What are major approaches/perspectives to study environment and society? These will be deliberated upon in the next chapter.

Points for Discussion

(1) Compare various agro-climatic zones in India and illustrate the ecological costs and benefits, manifest and latent, there, with a focus on the green revolution in Punjab, Haryana, and Western UP.
(2) How have liberalisation, privatisation, and globalisation (LPG) affected the ecology in developing countries? What is their impact on energy consumption at the global level?
(3) Compare the pattern of energy consumption in developed and developing countries. What lessons do you learn from this and what are key suggestions for a better sustainable world?
(4) How does environmental sociology contribute to a better understanding of the relationship between environment and society, especially through realist versus constructivist debate?

Chapter 2

Approaches to Environment and Society

Introduction

A reality is an independent or autonomous entity that cannot be wished away by someone or a group or a formal organisation. But every human being, having a certain degree of knowledge in a particular domain, may have his or her perspective or worldview about that reality. Others may have partially or fully different perspectives or worldviews. Hence, there may be various categories of different perspectives or worldviews on the basis of core commonalities or differences. For instance, an ordinary person has a particular perspective on a cluster of trees from his everyday life experience. On the other hand, an ecologist will perceive it as an absolutely different phenomenon, treating it as a part of the natural environment having an autonomous existence as a part of the biosphere of which humans, animals, fish, and birds are also parts. An ecologist gives equal importance to human, animal, fish, and plant species; he/she sees that the seeds become plants by evolving from within with the support of soil, air, and water, and hence he/she prefers organic agriculture and horticulture to modern agriculture and horticulture (with the use of chemical pesticides and fertilisers). Further, a sociologist may think of it on a different footing – planted by some individual farmer for the individual and/or collective good in terms of preventing soil erosion or contributing to rain and providing shade, fruits, leaves, timber, etc. In the meantime a road engineer designs a road construction project passing through that cluster of trees which are considered as an obstruction. Therefore, he/she proposes for the felling of those trees at the earliest without an iota of doubt. However, on the other hand, the forest officers come to know of it and considering the various problems of climate change want to prevent the felling of the trees by bringing a stay order from a court of law. Finally, an environmental sociologist or social ecologist takes a holistic view and suggests diversion of the proposed road and thus desires to save the trees for the people and other species. The environmental sociologist or social ecologist makes an elaborate social impact assessment (SIA) of the proposed road construction project in consultation with the local community, forest officials, local administration as well as the road construction department. The environmental sociologist or social ecologist takes into account the basic needs of the present and future generations for ensuring both sustainability and inter-generational as well as intra-generational equity, (Box 2.1).

Needless to mention here that though the classical sociologist Max Weber emphasised social values and agency, he consciously reduced nature to a social construction. To put it in his own words 'culture was grounded in, even if not determined by,

DOI: 10.4324/9781003336211-2

40 Approaches to Environment and Society

nature and to take the social out of the realm of natural causality altogether was to confuse the ideal and dogmatic formulations of jurists with empirical reality' (cited in Raymond Murphy [1997, 6–7]). Later some environmental sociologists (Murphy, U. Beck, Catton, Dunlap, Devall and Sessions) clearly and emphatically argued that human activities take place within a dynamic ecosystem where a human activity is just one of many elements. For instance, U. Beck (1992, 80–81) clarifies the complex relationship between nature and society:

> nature can no longer be understood 'outside of' society, or society 'outside of' nature ... in advanced modernity, society with all its sub-systems of the economy, politics, culture and the family can no longer be understood as autonomous of nature.

Thus, according to Beck, environmental problems, produced by science, are provoking a 'risk society' and 'reflexive modernisation'; in his view, 'reflexive scientisation' ultimately demonopolises scientific knowledge and removes all the barriers and obstructions between the experts and common people. Needless to mention here that Emile Durkheim's social determinism (social facts can be explained objectively only by other social facts) developed as a reaction to biological determinism (genetic attributes), but both types of determinism bear extreme views and, therefore, had blind spots, away from the complex reality of nature–society relationships. Undoubtedly, binary thinking in both natural sciences and social sciences has been rejected, and scientific reductionism, colonial power structure, and unequal economic and ecological exchanges have been questioned by dependency theorists as well as environmental sociologists. Now it is duly recognised by environmental sociologists that every being in the ecosystem or biosphere is relatively autonomous and self-organised but at the same time is also interconnected with other beings. Further, it is emphasised that every ecological question/issue at the same time is also a political issue and every political issue is also directly or indirectly linked with ecology at the same time. For instance, in the USA in the 1930s marijuana was considered addictive and was, therefore, banned, but in the 1960s the ban was lifted and it was considered a medicine, though the natural ingredients of marijuana remained the same.

Box 2.1: Widening of Jawaharlal Nehru Marg (Bailey Road), Patna, Bihar (India), by Felling Trees

In the years 2017–2018, the road construction department, Government of Bihar, proposed the widening of Jawaharlal Nehru Marg (Bailey Road), Patna, from four lanes to eight lanes. The necessity for the widening of the road had arisen for smooth management of the traffic as a new building complex of the Bihar Museum was constructed on over 17 acres of prime land between the Secretariat and the Patna High Court to the south of the road near Hartali Chowk and, also because of the proposed construction of a road overbridge (ROB) at Sheikhpura on that road. Bihar Museum is located on the busiest road in the heart of Patna. It was partially opened to the public in August 2015 with the opening of the children's museum, the main entrance area, and an orientation area. The remaining

> galleries were opened to the public in October 2017. The estimated cost of construction was about Rs. 498.49 crores (actual cost incurred was about Rs. 600 crores), a Public Interest Litigation, being writ CWJC (Civil Writ Jurisdiction Case) No. 9939 of 2012, *Ashok Kumar v The State of Bihar* and others, was disposed of by the Patna High Court on 26 June 2015. Even though the Patna High Court held the construction of the world-class museum to be not at all in the public interest, the construction of the museum was not stalled as the project was nearing completion. By a show of a 'visible, concrete and big' project, the local people were sold the illusion of development!
>
> On both flanks of Jawaharlal Nehru Marg there were hundreds of green trees and, hence, initially the Forest Department, Bihar, was not willing to give its clearance for the felling of the trees. However, on the insistence of the higher authorities of the State Government, the Forest Department ultimately gave its clearance, and, in the dead of night, hundreds of trees were surreptitiously felled by chainsaw machines. The environmental activists were not aware of the surreptitious felling of the trees well in time to take any preventive action for saving the trees. On the other hand, the environmental sociologists or social ecologists were not consulted at all; otherwise, they would have given some alternative solution of shifting the museum elsewhere – then at least half of the trees could have been easily saved. Surprisingly, the outermost flank of a width of about one lane was lying vacant for many years and no pitch road was constructed over it.
>
> Source: Authors

Here, in order to have a proper theoretical understanding, we will discuss about various approaches to the environment and society relationships, especially human ecology, treadmill of production, ecological modernisation, ecofeminism, political ecology, and ecological Marxism.

Human Ecology

Human ecology is a new inter-disciplinary study of natural, social, and man-made (built suburb) environments. It is developed primarily by combining ecology, sociology, geography, anthropology, zoology, epidemiology, public health, and home economics.

In real life, there are various sets of interactive relationships between nature, culture, and social labour. Firstly, the 'idealists', comprising post-Marxists, post-modernists, and deconstructionists, think that culture mediates between nature and social labour. Secondly, the 'passive materialists', comprising of environmental determinists, bio-regionalists, deep ecologists, and eco-feminists, think that nature mediates between social labour and culture. Thirdly, 'active materialists' or 'eco-Marxists' (e.g. O'Connor) think that social labour mediates between culture and nature (O'Connor, cited in Sharma 2016). It may also be seen in a different and broader way. For instance, most of the biologists and ecologists emphasise that 'humans and human systems' are 'embedded in the broader webs of life in the biosphere', i.e. humans are one of many species for both biological make-up and 'ultimate dependence for food and energy provided by the earth'. On the other hand, most of the social scientists emphasise that 'humans are unique creators of technologies and socio-cultural environments that have singular power to change, manipulate, destroy, and sometimes transcend natural environmental limits' (Buttel 1986, 343).

After the industrial revolution (first in the UK in the 1780s and then in the rest of Europe, and, finally, in the USA), the second assumption about humans as an exceptional species has been the dominant view, though it is distinct from the reality of everyday life – that is, human social behaviour and action are related more to symbolic constructions of social situations than to external environments as such. As sociologists Peter Berger and Thomas Luckmann (1966) observe, people exist in natural environments, but they 'live and act' in worlds mediated and constructed by cultural symbols. What the total cultural perspective of the people, of a particular time and place, share with others is usually known as their 'world view' – Weltanschuung.

Some ecologists (like Lotka and Odum) point out more comprehensively that when various species compete for the limited energy available in the physical environment and they survive selectively, the ecosystems evolve there. If they are not obstructed, it results in a larger, more complex, and inclusive structure of species in the food chain and often in a symbiotic relationship (ranging from mutualistic to parasitic ones); similarly, when various human beings compete for control over limited resources, socio-cultural evolution takes place. There develops, in the course of time, social differentiation due to various environmental, socio-economic, and political factors. Historically speaking, there have been three major types of social exchange in various human societies (Karl Polanyi, cited in Harper and Snowden 2016):

(a) Social exchanges of reciprocity
(b) Social exchanges of redistribution
(c) Social exchanges for other goods and services
    Here I add some more prevalent types of social exchange (in India) that are as follows:
(d) *Social exchange of land* – especially in 'voluntary consolidation' of agricultural fields by local people, contrary to 'mandatory legal consolidation' imposed by a government, or for their immediate convenience like construction of houses and shops), or for public convenience like public/community common road/track/ way, or exchange of their lands mutually, at a particular time and place, through negotiation (see Box 2.2).
(e) *Social exchange of labour* – in rural areas the poor small/marginal farmers or sharecroppers lack cash (money) for hiring labour for doing various tasks, and hence needy persons in a village or villages join hands voluntarily to mutually exchange their labour for cultivation, sowing, irrigation, winnowing, harvesting, carrying head loads, thatching roofs, and other tasks. In eastern and central UP such an exchange of labour is called 'hoonr'. This has been prevalent in tribal – so-called 'simple' – societies, too, in different parts of the world. However, with the increasing tendency of monetisation ('money is a matter of functions four/medium, measure, standard, store') even in rural areas, such an exchange of labour has substantially declined since the 1990s.

How the social evolution of inter-connectedness of environment and society has taken place is a very complex process over centuries in different societies the world over. How organisms – beginning with plasma – developed into a biome passing through various evolution processes may be analysed in the next section.

Box 2.2: Social Exchange of Land for Subhash Path in Sahebganj District, Jharkhand (India)

Sahebganj District is situated about 67 km (aerial distance) to the east of Bhagalpur in the State of Jharkhand (earlier a part of the State of Bihar). It was relatively less developed and the district had hilly terrains, flood-prone areas, and plains. In the early 1990s, in its Sadar Block, and at a distance of about 3–4 km in the diara (riverine) area, Mahadeoganj Gram Sabha/Panchayat had no roads – not even a kutcha road – to connect half a dozen villages. During the floods of 1993, some villagers took the matter to the new Deputy Commissioner and District Magistrate, Sahebganj District. There were hundreds of low-lying agricultural fields with fertile soil, and the farmers, whose lands were situated in the middle way, were not ready to part with their valuable lands for building a kutcha road. There was no public fund from which lands could be acquired for connecting the villages there. As the then-Deputy Commissioner (Mr. Subhash Sharma) had committed to the local villagers to make at least a kutcha road to connect those villages at the earliest, he called the concerned Circle Officer (Anchal Adhikari) – an honest and upright officer – and directed him to bring Mr. Askaran Gope, village headman (Mukhia) of Mahadeoganj Gram Panchayat. Deputy Commissioner advised them that in a case of social exchange of land, i.e. in 'voluntary consolidation', people mutually exchange their lands for common facilities/social work, hence motivating farmers to donate a small portion (say two to three feet width) of their lands by retreating back from the proposed kutcha road. The Anchal Adhikari, after much negotiation, ultimately managed to convince the local villagers, who finally agreed to voluntarily part with a small portion of their lands for the proposed kutcha road. In the second stage, the Deputy Commissioner asked the Executive Engineer (NREP) to prepare a reasonable estimate providing for an adequate number of small bridges ('pulias' by using hume pipes) so that the excess water during the rainy season may pass through easily. He sanctioned such a kutcha road scheme under NREP and the work was completed within a few months. All the local people of half a dozen villages expressed their happiness on the completion of the kutcha road by taking the initiative to name it Subhash Path after the name of the then-Deputy Commissioner! Thus the woes of the villagers over many centuries were duly solved.

Source: Authors

Interconnectedness of Life System

The evolution of the complex interconnectedness of environment and society took place over a long period of time. Various organisms make up a species, various species make up a population, various populations make up a community, various communities make up an ecosystem, and various ecosystems make up a biome. This may be seen in Table 2.1.

44 Approaches to Environment and Society

Table 2.1 Evolution of Integrated Life System

Sl. No.	Structural Units	Definition and Examples
1.	Organism	Any individual form of life including humans, plants, birds, and animals (felix, fido, you, and me)
2.	Species	Individual organisms of the same kind (e.g. dolphins, oak trees, corn, humans)
3.	Population	A collection of organisms of the same species
4.	Community	Populations of different organisms living and interacting in an area at a particular time (e.g. interacting life forms in the Monterey Bay estuary in California, USA)
5.	Ecosystem	Communities and populations interacting with one another, and with chemical and physical factors making up an inorganic environment (e.g. lake, Amazon rain forest, High Plains grasslands in the USA)
6.	Biome	Large life and vegetation zones made of many smaller ecosystems (e.g. tropical grasslands or savannahs, northern coniferous forests)

Source: Harper and Snowden (2016).

In the 1960s, environmental economics emerged as a branch of economics wherein both 'nature' and 'human values' were taken into account, and thus, contrary to the neo-classical economists, they perceived economy as 'an open, growing, wholly dependent subsystem of a materially closed, not growing, finite ecosphere' (Harper and Snowden 2016). Two new types of problems were focussed by them:

(a) How can values ('prices') be assigned to goods that are held in common (the commons) – used by many but owned by none (atmosphere, oceans, rivers, and public space)?
(b) How can economic analysis incorporate and assign responsibility for the variety of environmental and social 'externalities', i.e. the real overhead or human costs incurred in the production process that are borne not by particular producers or consumers but by third parties, the largest social community, or the environment?

In addition, the environmental economists challenged the neo-classical economists' view that human ingenuity and technology will always overcome environmental limits and ecosystem capacities. However, environmental economists, too, are guided by the economic market tools 'in the ultimate analysis' (Harper and Snowden 2016): For instance, firstly, 'emissions trading' schemes give tradable credits to less environmentally damaging production – and these credits may be traded/auctioned to greater polluters (per unit) – e.g. 'cap' and 'trade' policies used to address global climate change; secondly, 'ecological modernisation' points out that, while modernising, firms may be more efficient by adopting an ecosystem with many feedback systems and recycling – e.g. by using wastes from one process to supply/fuel to another process ('co-generation'). Thus ecological modernists (from ecological economists, business leaders, and environmental sociologists) believe that technology per se can solve resource scarcity and pollution issues. But, in practice, such a 'business as usual' approach creates environmental problems more than solving these issues. Hence, an interdisciplinary approach is called for to address environmental problems.

Human Exceptionalism Paradigm (HEP)

Needless to say that classical sociologists like Karl Marx, Emile Durkheim, and Max Weber had rightly challenged the biological determinism, but they did not ignore the environmental factors. It is now a known fact that, in 1976, the American Sociological Association categorised a new section on environmental sociology because in the 1960s there were many environmental problems, in order to understand the changed 'sense of what is real'. According to Catton and Dunlap (1978), anthropocentrism (human in the centre) was the common thread in all prevailing sociological perspectives like functionalism, symbolic interactionism, ethno-methodology, conflict theory, Marxism. Hence, they call such a basic sociological worldview 'Human Exceptionalism Paradigm' (HEP), and due to HEP assumptions, most of the sociologists failed to 'deal meaningfully with the social implications of the ecological problems and constraints'. However, there have been two paradigms in this regard: (a) Human Exceptionalism Paradigm (HEP) and (b) New Ecological Paradigm (NEP) (Catton and Dunlap 1978, 42–48). Human Exceptionalism Paradigm (HEP) assumed four things:

(a) 'Humans are unique among the earth's creatures, for they have culture'.
(b) 'Culture can vary almost infinitely and can change much more rapidly than biological traits'.
(c) 'Thus, many human differences are socially induced rather than inborn; they can be socially altered, and inconvenient differences can be eliminated'.
(d) 'Thus, also cultural accumulation means that progress can continue without limit, making all social problems ultimately soluble'.

In fact, such an optimistic view was supported by the conception of progress in western culture, particularly American culture, maintaining that the present was better than the past, and future would be the best – that is, the 'carrying capacity' of the earth is ultimately enlargeable, and hence no question of net scarcity will ever arise due to ever progressive science and technology. In their view, in most of the sociological literature the term 'environment' refers to a society's 'symbolic environment' (cultural systems) or 'social environment' (environing social systems), or 'spatial environment', rather than the physical environment. Sociologist Daniel Bell (1973), they further add, had emphasised that the moot question before humanity was 'not subsistence, but standard of living, not biology but sociology'. Similarly, another sociologist Amos Hawley (1950) insisted that 'there are no known limits to the improvement of technology', and population pressure on non-agricultural resources was not felt in the present nor likely to be felt in the near future. Classical sociologists like Emile Durkheim had emphasised that social facts may be analysed in a natural scientific manner, and one social fact may be explained only by associating them with other social facts.

But, due to the emergence of various environmental movements, natural hazards, resource management, and 'social impact assessment' of various big projects on rivers like dams, the oil crisis in 1973 led to the evolution of environmental sociology wherein environmental variables were considered for a fruitful sociological analysis.

46 Approaches to Environment and Society

However, the conception of environment varied from 'man-made' environment to 'natural' environment – and 'human altered' environment (water, air, noise, soil, smog, etc.) in between.

New Ecological Paradigm (NEP)

Criticising the prevailing Human Exceptionalism Paradigm (HEP), Catton and Dunlap (1978, 42) drew from many environmental sociologists like Anderson (1976), Burch (1971), Buttel (1976), Catton (1976), Morrison (1976), and Schnaiberg (1972, 1975) and proposed an alternative to it, called New Ecological Paradigm (NEP), in order to take environmental variables seriously in sociological studies, with the following assumptions:

(i)   'Human beings are but one species among the many that are interdependently involved in the biotic communities that shape our social life'.
(ii) 'Intricate linkages of cause and effect and feedback in the web of nature produce many unintended consequences from purposive human action'.
(iii) 'The world is finite, so there are potent physical and biological limits constraining economic growth, social progress, and other societal phenomenon'.

Now we may have a comparison of these two paradigms – HEP and NEP – in Table 2.2.

Three major classical sociologists (Marx, Durkheim, and Weber) had some 'seminal ideas' on which environmental sociology could be grounded: Firstly, Marx's materialist view of reality included the notion of nature–society 'metabolism'. In his view, as capitalism advances, the ownership of land and other productive resources would concentrate in a few hands, and there would be class polarisation – the rich will have more political control and power. John Bellamy Foster (1999) traced Marx's 'metabolic rift theory' that sees the process of human labour as a metabolic exchange between human beings and the natural environment. Thus both human social development and

Table 2.2 A Comparison of Human Exceptionalism Paradigm (HEP) and New Ecological Paradigm (NEP)

S. No.	Aspect	HEP	NEP
I	Human existence	Unique due to culture	One of many species with inter-dependence
2	Forces shaping humans	More by culture	Both by socio-cultural forces and linkages in the web of nature
3	Restraints on human affairs	Human differences are socially induced and may be socially altered or even eliminated	A finite bio-physical environment imposes potent restraints on human affairs
4	Carrying capacity	Unlimited progress possible and all social problems finally solvable	Human inventiveness may extend carrying capacity limits, yet subject to ecological laws

Source: Based on Catton and Dunlap (1978).

nature co-evolve. In other words, human beings change nature through production processes by extracting from nature their subsistence needs and return to nature the by-products of that exchange. Their changing of nature depends on the inheritance from previous generations in both conditions of the natural environment and socio-historical conditions (tools, technology, cultural knowledge and beliefs, and institutional systems like nation-states that enforce private property rights). According to Marx, soil degradation was due to the antagonism between town and country under capitalism. Industrialisation led to a massive migration to urban areas (urbanisation was accentuated) and it caused a 'metabolic rift' (imbalance) between human beings and the natural environment – putting negative consequences on the development of both. Such migration deprived the land of human (and also animal – less people, less cattle) organic waste required for its replenishment. Further, due to the carrying away of food and fibre from rural areas to urban areas for manufacturing and consumption, their organic waste by-products were not returned to the soil for nutrient recycling. Further, the capitalists invest in technology for short-term economic gain, causing again environmental problems (soil depletion, groundwater contamination, deforestation) and social problems (displacement of rural labour, concentration of wealth and income, and health hazards) for present and future generations (Foster 1999). Thus, Marx pointed out that the technology of industrialisation may create the illusion that humans are exempt from the laws of nature and disrupt fundamental ecological exchanges simultaneously.

Treadmill of Production

The treadmill of production theory was first proposed, on the basis of environmental problems and resistance movements in the 1960s and 1970s, in 1975 by Allan Schnaiberg. Firstly, he argued that, in the USA, there was a marked shift from 'surplus' to 'scarcity' of environmental amenities.

Secondly, the profits from investments in new technologies increased substantially, but there was neither substantial expansion of employment opportunities nor the socio-economic position of industrial workers improved.

Thirdly, there was a growing rate of displacement of production workers by the new technologies developed by big corporations directly (inside industries) or through university researches indirectly (outside).

As early as 1961, then-US President Dwight D. Eisenhower recognised the growing military–industry complex with 'grave implications':

> Our toil, resources and livelihood are all involved, so is the very structure of our society. In the councils of government, we must guard against the acquisition of unwarranted influence, whether sought or unsought, by the military–industry complex. The potential for the disastrous rise of misplaced power exists and will persist.
>
> (cited in Gould et al. 2016, ii)

Due to the rise of environmental politics, by the 1970s, the US federal government established the Council on Environmental Quality, Environmental Protection Agency, and enacted the National Environmental Policy Act. Some scholars like Morrison

48 Approaches to Environment and Society

branded these as 'institutionalised environmental movement organisations'. In fact, these and other environmental organisations (NGOs) – or 'new social movements' – have been in conflict with the treadmill of production.

Then most of the environmental sociologists did realise the constraints of resources like water, oil, soil, ecosystem. For instance, distinguished sociologists Schnaiberg and Gould (1975) had spoken of the 'societal–environmental dialectic' and that 'ecological disruption is harmful to human society'; they mentioned three alternate syntheses of the dialectic:

(a) An *economic synthesis* that neglects ecological disruptions and tries to maximise growth. This is adopted by 'regressive' (inequality-enhancing) economies.

(b) A *planned (managed) synthesis* that focusses on the most perceptible environmental consequences and, hence, puts certain control mechanisms to regulate the resources and more resource-using industries; e.g. such 'non-redistributive' economy compels the industry sector to reduce pollution levels by opting for clean technology, imposing taxes and cesses on outdated technology. Thus higher prices are paid by the consumers. However, various pressures from planned scarcity may, finally, lead to economic growth synthesis. There is a likelihood of all the industrial growth-oriented people (owners and labour unions) to oppose the environmentalists (as happened in the Silent Valley movement in Kerala, South India).

(c) An *ecological synthesis* – 'substantial control over both production and effective demand for goods' in order to substantially reduce the ecological disruptions and to have a 'sustained yield' of resources – i.e. the resources should be prudently used to fulfil the needs, not the greed. Progressive (equality-enhancing) economies adopt ecological synthesis that is tantamount to 'steady-state' society by ensuring redistribution. However, it is expected that the proposed redistribution would be too difficult to be realised in practice, given various economic, political, and environmental constraints, but slower growth may result in class conflicts.

In fact, environmental sociologists, adopting NEP, question the so-called universal benefits of growth, on the one hand, and emphasise the 'costs' of growth (inequality-enhancing costs). For instance, the pollution at the location and the surroundings of factories results in health hazards for the poor people more than the upper class. In other words, it has been pointed out, time and again, by environmental sociologists, that any effort to eliminate poverty will not succeed if its environmental impact is not taken into account. To be more specific, Catton and Dunlap (1978) have emphasised that social stratification, among other aspects, would be quite substantially affected by the ecological constraints; economic growth synthesis will obviously result in social tensions, and ecological synthesis would require a steady-state society, which is a quite different goal to achieve. Thus the features of three alternative syntheses of 'societal–environmental dialectic' may be seen in Table 2.3.

The treadmill of production model was, in essence, an economic synthesis; however, the treadmill of production model was not a linear change theory, rather it was a dialectical change theory. But, as Gould et al. (2003) argue, the empirical history of the USA and global political-economic scene since 1980 has been 'only weakly so' – treadmill of production has expanded exponentially, not linearly. Treadmill of production's

Table 2.3 A Comparison of Three Alternative Syntheses of Societal–Environmental Dialectic

Sl. No.	Alternative Synthesis	Features and Implications
I	Economic synthesis	(a) Maximise the growth – higher growth is better for the economy and society
		(b) Trickle-down effect – the bigger cake will help in poverty alleviation
		(c) Ignores ecological disruption
2	Planned (managed) synthesis	(a) Economic growth is managed and regulated by control mechanisms – more resource-using industries are regulated
		(b) Non-reducible economy, though liberal to the poor
		(c) Partly takes into account ecological disruption
3	Ecological synthesis	(a) State's substantial control over production and effective demand for goods, without regard to issues of profitability and wages/employment
		(b) Substantial yield of resources, to fulfil the needs, and not the greed, of the people
		(c) To reduce ecological disruption substantially
		(d) Similar to the conception of sustainable development

Source: Based on Schnaiberg and Gould (1975).

main goal is growing return on the investment; US investors and managers prefer to invest abroad due to lower wages and often lower environmental protection. This way US investors convince environmentalists based on the logic of reduction of production and of pollution. Though in the twenty-first century the rivers and streams in the US are cleaner than earlier, there has been a destruction of the habitat due to mining, logging, and intensive agriculture since 1980, both in the USA and US-investor locales abroad.

It has the following features:

(a) It points out the increasing capital – intensification of production – more capital, more production.
(b) It highlights that new technologies substituted a higher level of energy and a wider range of chemicals for human labour – thus labour displacement to a large extent. On the other hand, in order to reduce the costs of new technology, production was to be increased substantially, and thus there was more demand for natural resources, and, hence, more waste and more chemical toxicity of wastes.
(c) It also emphasises that the producers and investors in the post -World War II period saw the natural environment as merely surplus, easily available, and exploitable without limit.
(d) It adds that more exploitation of natural resources meant more profit for investors/producers, and that meant presumably a better economy and society (though not so in reality). Many times such an illusion was created among the local people by misleading them and practising environmental racism and classicism (see Box 2.3). Thus treadmill mode points out the significance of power, social inequality, and conflict underlying environmental behaviour – not only people's thought but also actual happenings in nature.

50 Approaches to Environment and Society

(e) It has influenced the anti-corporate globalisation movement that emphasises macro-structural processes, class conflict, corporate power, inequality, and environment. The southern environmental movements have rightly integrated livelihood issues and ecological struggles; thus it indirectly used the treadmill analytical framework in opposing the corporate power market mechanisms, and international financial institutions. Anti-corporate globalisation movements converged various movements like organised labour movements, social justice movements, human rights movements, southern environmental movements.

(f) It emphasises that environmental politics is driven by both social and ecological factors and their limitations. In fact, exporting 'hazardous chemical wastes and the transfer of toxic technologies' to developing countries has serious implications for the latter's health and ecology. Undoubtedly the loss of biodiversity in developing countries is primarily because of extractive industries (mining and oil) and oil refineries by the developed countries. Unfortunately, in the present age of globalisation there is 'global treadmill of production', due to the close linkage of the State, capital, and the environment, as Islam (2015, 144) rightly observes,

> ecological resources are increasingly converted into profits via trans-national organisation of production and market exchange. Consequently, with an intensification of production processes driven by competition and drive to increase profits in the capitalist world economy, there has been a rising amount of both resource withdrawals from and toxic additions to the environment.

Due to the globalisation of capital, to a large scale, various Trans-National Corporations (TNCs) have out-sourced various processes of production or established new plants in developing countries to tap cheap labour, cheap raw materials (primarily natural resources like land, water, minerals), cheap electricity, and free infrastructures like roads, thereby earning huge profits which, finally, go back to the (developed) country of origin. In India, for instance, various State Governments compete out of the way – often secretly – with one another in providing more and more amenities to the MNCs! Further, the reallocation of unskilled jobs from the developed countries and cheap labour (both skilled and unskilled) from developing countries has actually created a New International Division of Labour (NIDL) that finally led to 'world factory', on the one hand, and, consequently, the dumping of huge amounts of toxic by-products and various waste materials (including electronic wastes) into developing countries like India. There exists an 'unequal ecological exchange' (Islam et al. 2015), because demands of more production from developed 'core' countries led to the dumping of more wastes in, and environmental pollution of, the 'periphery' (developing countries). This unequal ecological exchange is again accentuated and accelerated by the processes of liberalisation, privatisation, and globalisation (LPG).

(g) It also points out that the root cause of the environmental problem is the power of the elite institutions to construct reality. Now, environmental activists have become aware of 'the problematic alliance between corporations and the State' that has negative consequences for democratic governance in general and environmental protection in particular. The treadmill model provides environmental movement activists to ask relevant critical questions, and thus a 'diagnostic frame'; e.g. the traditional mechanisms fail to understand 'the anti-ecological logic of capital', on the one hand, and ignore the role of 'social inequality' in creating both 'treadmill support and ecological decline'. The notion of green technology does not take into account power relations.

(h) It forcefully argues for a steady-state economy and, therefore, State intervention is certainly required for the progressive redistribution and growth deceleration by dismantling the free market system(s), because the citizens will not normally accept the low or no growth trajectory required to protect the ecosystems. This ultimately threatens the privileged economic class, especially the big corporations which usually fund the environmental movements in the North. Hence, it argues, a political conflict with the ruling elites, transnational financial institutions, transnational corporations, and States is inevitable in order to have a major goal of socio-ecological sustainability. Nowadays treadmill of production theory and environmental justice movements have been converging – the environmental justice movement usually considers the macro-structural analysis and, on the other hand, the treadmill of production theory now considers cultural and institutional racial discrimination.

(i) In essence, the treadmill of production theory synthesised both changes in the forces of production and the relations of production (Marx's concept of mode of production) and, further, integrated such changes with the disruption of the ecosystem due to the large scale and form of production. Obviously it focusses on production rather than consumption because the process of production takes place before consumption. The treadmill model further contextualises the consumer decisions within the 'material parameters of their political economic contexts'; consumer choices are guided by the following three factors (Gould, Pellow, and Schnaiberg 2003):

   (i) The constraints of specific production decisions
   (ii) Specific prior economic distribution decisions
   (iii) A specific distribution of policy and decision-making power – the ability to influence the choices of the less powerful people (which is obscured in neo-classical economics)

In fact, the alternative forms of production are decided by a small coterie of powerful treadmill elites, not the consumers themselves. In practice, producers' access to natural resources and ecosystem waste depends on the following five factors in a stratified and politicised society (Gould, Pellow, and Schnaiberg 2003):

(i) Producers' assessment of marketability
(ii) Producers' access to capital
(iii) Producers' access to labour
(iv) Producers' assessment of potential liability
(v) Producers' assessment of profitability

However, the decisions of producers are influenced/regulated by the State and negotiations with the labour force, and hence the treadmill of production model, first, focusses on the role of common individuals as 'citizens' (in the polity) and 'workers', rather than as consumers. Second, this model emphasises collective actions (of NGOs and social movements) over individual actions or choices.

52 Approaches to Environment and Society

Here presumption is that more democratic control of production is likely to ultimately solve the social and ecological problems, not the attempts to control consumer choice of one or another product.

In fact, the treadmill metaphor symbolised a decline in the efficiency of the production system – post-World War II it was favourable to investors for huge profits, but each unit of the ecosystem involved in the production system created less support and was unfavourable to US workers – thus there was a higher rate of depletion of the ecosystem (resource extraction) and ecosystem pollution (using ecosystems as dump sites).

Box 2.3: Cases of Environmental Racism and Classism (Inequality)

First: Operation Silver Shovel was a phenomenon of hatred, racism, and inequality in the city of Chicago during the mid-1990s; tons of construction waste were racially, illegally, and unethically dumped in the Latino and African American neighbourhoods by the connivance of some construction companies' white owners, waste dumpers owned by the whites, and some bad Latino and African American political leaders who took bribes for such a heinous criminal activity (Pellow 2002). This was a case of both environmental racism and classicism.

Second: On numerous Native American reservations, tribal leaders took bribes in order to allow nuclear waste and other 'locally unwanted land uses' to be located despite collective opposition by members of the tribe (LaDuke 1999). This was a case of both environmental racism and classicism.

Third: In the home-based high-tech sweatshops of Silicon Valley, Vietnamese immigrant entrepreneurs exploited the people of their own ethnic community in the name of 'profit and the American Dream' (Hossfeld). This was a typical case of environmental classism and intra-racial exploitation.

Source: Gould, Pellow, and Schnaiberg (2003)

Criticism: The treadmill of production model may be criticised on the following grounds:

(a) The treadmill of production approach is capable of analysing the basic problem of how the big industrial corporations connive with different government agencies or departments (individually or collectively) to promote highly hazardous treadmill (wherein wealth goes to private pockets usually) leading to huge environmental degradations and social inequalities on a large scale. And this injustice is fostered by capitalism and its many arms. But this model fails to suggest any reasonable and easily workable alternative for their remedies.

(b) This model primarily focusses on production, and as such the supply side of the economy fosters more and more consumption, and hence a culture of consumerism emerges as a vicious cycle. The large corporate industries also promote consumerism through 'hidden persuaders' (advertisements) by creating false illusions in two ways: Happiness is wrongly equated with the possession of 'more' (quantity), 'bigger' (size), 'faster' (speed), and 'newer' (new product or new technology) commodities. Further there are alluring sale offers like 'buy one, get one free', 'buy five litre of edible oil, get 0.5 litre free', 'up to 50% rebate', 'clearance sale'. Thus the creation of hyper-consumerism, too, influences the size, form, and direction of production by transnational companies, but the treadmill model does not look at the consumption issue.

(c) The treadmill model has not been widely adopted in practice by mainstream environmental movements. However, for this, partly the treadmill model is responsible, and partly the mainstream environmental movements in the USA that somehow adopted pro-treadmill values in the 1980s and 1990s – this facilitated their access to policy-makers and funding from various private foundations and donors.

(d) 'Voluntary simplicity' and 'green consumerism', the individual consumer choice efforts, are not recognised at all by the treadmill model; though these efforts cannot substitute collective ecological action and democratic governance (as desired by the treadmill model), these efforts may have the shape of collective action in due course and thereafter may take collective ecological action with the required vision for structural change as 'a thousand miles journey begins with the first step'! For instance, in India, Anna Hazare's 'voluntary simplicity' began at first at an individual level and later turned into collective actions, and thus his entire Ralegan Siddhi (Maharashtra) was changed into full greenery (tree plantation on a vast scale – social forestry), water sustainability (with large scale rain water conservation), sustainable dairy production (from scarcity to sufficiency and supply), and agricultural sustainability (by pursuing organic farming). Hence, some scholars like Arthur Mol (1995) have branded the treadmill of production model as an example of deterministic neo-Marxist environmental sociology.

(e) As Hooks and Smith (2004) clearly point out with evidence, the treadmill model has not focussed much on the phenomenon of materialism, violence, and coercion in US environmental politics and history; they find the US military–industrial complex as a treadmill institution but driven basically by 'geopolitical' and 'social/racial' motives rather than industrial-capitalist influences (as treadmill scholars suggest). They distinguish between two major institutions – market and polity – that mould residential location, especially for Native American Indians. In six States in the USA, they find a positive correlation between Native American counties (in total there are 332 Native Americans' counties) and US military hazardous sites, resulting in what Daniel Brook terms as 'environmental genocide'. This fact of class and racial inequality is enforced by capitalism, coercive State, and military–industrial complex. Hence Hooks and Smith (2004) call it the treadmill of destruction (Table 2.4). In fact they identified the core conditions under which the value of the treadmill of production approach is the highest. Coercive and racist State policies constrain residential choice. During the apartheid, South Africa located black communities downstream of polluting industries and badly managed waste fill sites. In the USA, slavery and segregation shaped environmental inequalities in Louisiana (where during the hurricane in the recent past no State help was provided to the blacks); from New Orleans to Baton Rouge, known as the 'chemical corridor', large chemical factories dump hazardous chemicals into the air, water, and land – located in former plantation lands surrounded by blacks' counties. For the treadmill of production approach economic competition for profit and market share explains the increase of human influence on the

54 Approaches to Environment and Society

Table 2.4 Treadmills of Production and Destruction: Core Conditions for Understanding the Distribution of Environmental Inequality

Production of Environmental 'Bads'		
Spatial Distribution of People Relative to Environmental 'Bads'	Capitalism/commercial	Militarism/geopolitical
Market (choice as constrained by the ability to purchase)	Treadmill of production	Military–industrial complex
Polity (coercion prominent, often on the basis of race or ethnicity)	Coercive polity	Treadmill of destruction

Source: Hooks and Smith (2004).

environment. But, on the other hand, in the treadmill of destruction approach arms race and geopolitical competition enforce the environmental impact of militarism. Further coercive policy in a housing market determines who lives closest to the hazardous sites. Regarding Native Americans in the USA both militarism-generated environmental dangers and coercive policy of the State determine the location of reservations (the lands not desired by the whites). Hooks and Smith (2004) also point out that the Indian Removal Act of 1830 resulted in ethnic cleansing – Native Americans residing in all States, east of the Mississippi River, were forcibly displaced to western territories by 1850; a treaty with the Cherokee Nation (concentrated in Georgia and North Carolina) was broken and they were forced to move to Oklahoma. On 29 December 1890, over 300 Native Americans (mostly unarmed women and children) were killed by the US military because they participated in the Ghost Dance ritual – hoping for Indian resurgence, but it frightened white settlers and soldiers. In 1940, the US Government seized 3,42,000 acres of Pure Ridge Reservation in South Dakota for a bombing range to train World War II pilots; it forced 125 Oglala and Sioux Indians to sell their farms and ranches for 3 cents an acre. Thus it was treadmill of destruction.

Ecological Modernisation

Ecological or eco-modernisation perspective implies that an economy benefits from various actions of environmentalism. It was proposed in the 1980s and 1990s by various scholars like Arthur Mol, Gert Spaargaren, U. Beck, Joseph Huber, Udo E. Simons, Martin Janicke, Arthur H. Rosenfeld, Rene Kemp, Donald Huisingh, and Ernst Ulrich Von Weizsacker who were of the view that various dimensions of production processes have been increasingly grounded in both ecological and economic criteria. It argues that due to the new industrial revolution restructuring of production along ecological lines is necessary. One of the fundamental assumptions of eco-modernisation is an environmental adaptation of economic growth and industrial development, so that there may be innovative structural changes. It is argued that productive use of natural resources ('environmental productivity') and environmental media (water, soil, air, ecosystems) may possibly be a source of future growth just like labour productivity and capital productivity. It encompasses an increase in resource and energy efficiency, on the one hand, and structural changes in the product and process like: (a) sustainable supply chain management, (b) environmental management, (c) clean and green technologies, (d) substitution of hazardous substances, and (e) better product design for environmental stability, on the other hand.

Thus various changes in the arenas of production processes would (a) reduce the emissions, (b) reduce the quantity of resource turnover, and (c) change the structure of 'industrial metabolism'. This is, therefore, called 'greening of industrial ecosystems'.

Janicke (1990), Mol (1995 and 1997), Spaargaren (2000), and Giddens (1998) were the major proponents of ecological modernisation theory, highlighting two specific points (Buttel 2000):

(a) Both political processes and practices are specifically critical in enabling ecological phenomena to be 'moved into the modernisation process' (Mol 1995); hence, ecological modernisation theory has to be a theory of politics and State.
(b) Ecological modernisation has close links with 'embedded autonomy', civil society, and State–society synergy theories in political sociology.
     Ecological modernisation is used in four different ways by different scholars (Buttel 2000):
     (i)   Arthur Mol, Gert Spaargaren, Joseph Huber, and Martin Janicke were 'core (objective) ecological modernisation' theorists (North American and British perspectives).
     (ii) Political-discursive and social-constructionist perspective (M. Hajer 1995) – it is a category to analyse the dominant discourses of the environmental policy areas of the developed industrialised nations – in contrast to the core perspective mentioned above may dilute the political desire for environmental reforms by obscuring the State's ability to bring better environment.
     (iii) It means strategic environmental management, industrial ecology, or eco-restructuring – thus primarily private sector actions lead to minimising pollution and wastes, and enhancing efficiency (e.g. Hawken 1993, Ayres 1978, Anderson 1976).
     (iv) Any environmental policy innovation or environmental improvement; e.g. Murphy (1997) speaks of State policies which make the internationalisation of environmental externalities feasible.

At the fag end of the twentieth century, Arthur Mol (1999) differentiated between '*first generation ecological modernisation thought*' (the 1980s and early 1990s), especially by German and Dutch social scientists, and '*second generation eco-modernisation thought*' (in the late 1990s). The former views that capitalist liberalist democracy has the institutional capacity to reform its effect on the natural environment and modernisation would improve it. The latter pinpoints, on the other hand, the particular socio-political processes, and further modernisation leads to (or obstructs) beneficial environmental outcomes. However, Buttel (2000) thinks that ecological modernisation was an effective response to many situations or socio-ecological thoughts in the 1990s:

(a) That significant decrease in the use of fossil fuel, checks on tropical forest damage and loss of biodiversity, regulation of the industry, decentralisation, and localisation were necessary; this understanding somehow led to the beginning of radical environmental movements in North Europe.

(b) The notions of sustainability and sustainable development were constructed in the context of primary renewable sectors in the South, and hence ecological modernisation was considered to address the 'transformative sectors' of metropolitan regions of the developed industrialised nations.

(c) There was North American dominance and bias in environmental sociology by over-theorising the intrinsic tendency to environmental disruption and degradation, and hence there was hardly any space left for environmental improvement. Hence, an idealistic view of environmental movements as a panacea was called for. Hence, most of the scholars started evaluating eco-modernisation in the sense of the third and fourth uses mentioned earlier.

However, according to Buttel (2000), the first meaning of ecological modernisation – a distributive social theory – proposes that most of the environmental problems of the twentieth and twenty-first centuries are caused/will be caused by the modernisation and industrialisation, but their solutions lie in *more modernisation* and 'super-industrialisation'. That is, capitalism is seen as quite flexible institutionally to allow movement in the direction of 'sustainable capitalism' and its inherent competition among capitals may be used to have eco-efficiency (by preventing pollution) within the production process, and finally in the consumption processes too (as per Spaargaren 1996).

Ecological modernisation is in fact a critical response to radical environmentalism (or 'counter-modernity'); as Mol (1995, 48) observes: 'the role of environmental movement will shift from that of a critical commentator outside societal developments to that of a critical – and still independent – participant in developments aimed at ecological transformation'. To Mol and Spaargaren, ecological modernisation has parallels in other concepts like 'long cycle' (Schumpeter, Kondratieff), 'dis-embedding' (Karl Polanyi), and 'four dimensions of modernity' (Anthony Giddens), but it has the closest link with 'reflexive modernisation' and 'risk society' (Ulrich Beck). While the Netherlands (Mol–Spaargaren's interest) is a highly centralised society, Germany (Beck's interest) is far less than that, and both have parliamentary democracy and environmental ideologies which are well established in their political cultures. Further, these three scholars recognise the changing role of the State – towards less bureaucratisation and more decentralisation. They also argue that the problems of industrialisation and science can be solved only through 'more modernisation, industrialisation and science'.

But, on the other hand, there are some differences between them. Firstly, while, Mol and Spaargaren do not give much significance to new social movements or radical environmental groups in realising environmental modernisation processes, Beck's '*reflexive modernisation*' theory accords a significant role to new social movements and sub-politics in the restructuring of the State and political discourses.

Secondly, the anti-nuclear and anti-biotechnology protests have been supported by Ulrich Beck, while Mol and Spaargaren do not go beyond incremental environmental reform.

Thirdly, Beck's concept of 'risk society' moves to the arena of identity politics and extra-scientific policy-making, whereas Mol and Spaargaren stick to environmental improvement.

Fourthly, Beck distinguishes strongly the 'risk society' from 'industrial society', while the core ecological modernisation theory of Mol and Spaargaren argues that eco-efficiency benefits may be attained without radical structural changes in State and civil society.

Finally, though Beck's *reflexive modernisation* got much prominence in Europe in the 1990s, it could not get acceptance and recognition in North America, and hence the ecological modernisation thought did not take much interest in Beck's 'reflexive modernisation' (Buttel 2000).

Mol (1995) makes two fundamental points for the 'modernisation' of the State and the transfer of more tasks to the market:

(a) Transformation of the State's environmental policy is essential:

> from curative and reactive to preventive, from exclusive to participatory policy-making, from centralised to decentralised wherever possible, and from domineering over-regulated environmental policy to a policy (creating) favourable conditions and contexts for environmentally sound practices and behaviour on the part of producers and consumers.

Hence, the State, as a facilitator, will obviously widen the scope of civil laws in the environmental policy and focus on economic mechanisms and change in the management strategy through collective self-obligations.

(b) 'Transfer of responsibilities, incentives, and tasks from the State to the market', as the market is considered a more efficient and effective mechanism (than the State) to tackle environmental problems; the State ensures adequate conditions and stimulates social 'self-regulation' through economic mechanisms or through citizens' groups, environmental NGOs and consumer organisations.

Thus Mol's above-mentioned points reflect the neo-liberal, neo-classical approach in economics and political science, by reducing the role and arena of the State and widening the role and arena of the market in the socio-economic and political life of the people. He also argues for advocacy–coalition type relationship among State officials, corporate managers, and environmental NGOs in realising ecological modernisation processes. This is compatible with Evans' (1995, 1996) notions of 'embedded autonomy' and 'State–society synergy'. This form of embedded autonomy (Weberian notion of State) is a State structure that integrates 'corporate coherence', and connectedness of, and social ties between State agencies and officials, and various groups in civil society. However, before Evans, Martin Janicke (1990) – a German founder of ecological modernisation – talked of 'State failure', hence a need for closer State–society ties, especially in solving environmental problems. Thus Evans' neo-Weberian notion of 'embedded autonomy' is consistent with the ecological modernisation theory.

Scholars like Buttel consider the ecological modernisation perspective 'more useful than sustainable development as a macro-framework' for the environmental problems of metropolitan transformative industry in the North.

Methods for Using Green Technology in Eco-Modernisation

Green technology or eco-technology, associated with ecological modernisation, is an applied system that attempts to use eco-friendly technology to have minimum

58   Approaches to Environment and Society

ecological disruption (MED) by designing, using, and managing all the processes of production, exchange, distribution, marketing, and consumption in a more sustainable way; for instance, production of bio-fertilisers in lieu of chemical fertilisers for practising 'organic farming', or production of less energy consuming air conditioners or generators with symbols of more stars using less energy. Here two dimensions of 'ecology of technics' and 'technics of ecology' are positively integrated in order to reduce the damage to the ecosystems. It ensures minimum ecological disruption through the following methods:

(a) By increasing the efficiency in the selection and use of raw materials and the sources of energy – e.g. preferring renewable sources of energy (solar, wind) to non-renewable sources of energy like fossil fuel (coal, oil, and natural gas)
(b) By applying the control mechanisms of the impact of eco-technology on ecosystems
(c) By devising new and better cleaning processes and thereby green products, and ensuring their constant improvement
(d) By engaging eco-technology savvy personnel to handle various tasks
(e) By ensuring transparent and green modes of marketing
(f) By ensuring an environmental management system (EMS) in all the sectors of production, marketing, services, etc.
(g) By ensuring awareness generation among the people about the use of green technology and minimum ecological disruption
(h) By ensuring the provision of dynamic corporate social responsibility (CSR) with a focus on popularising green technology
(i) To contribute to the reduction of ecological footprint in some way or another

Conditions for Eco-modernisation

The following conditions are to be ensured for eco-modernisation:

(a) To change various laws, national and international, in order to follow industrial ecological principles
(b) To integrate environmental costs into the accounting system
(c) To develop industrial metabolism design to change the behaviour of industrial and service firms (sectors)
(d) To reduce industrial wastes
(e) To recycle industrial wastes
(f) To ensure environmental audit
(g) To focus on the relationship between trade and the environment
(h) To change the consumers' attitude through consciousness-raising
(i) To make the life-cycle analysis of industrial products
(j) Most significantly, to integrate economics in accordance with ethics and environment

Christoff (1996, cited in Roger Hildingsson (2007)) has classified ecological modernisation into two broad categories – weak and strong – having six different dimensions that may be seen in Table 2.5.

Approaches to Environment and Society   59

Table 2.5 Weak Versus Strong Ecological Modernisation

Sl. No.	Aspects	*Weak Ecological Modernisation*	*Strong Ecological Modernisation*
1.	Perspective	Economistic	Ecological
2.	Degree of broadness	Technological (narrow)	Institutional/systemic (broad)
3.	Value of nature	Instrumental	Communicative (intrinsic)
4.	Orientation of participation	Technocratic/neo-corporatist (closed)	Deliberative democratic (open)
5.	Worldview and range	National (euro-centric/westernised)	International (globalised fairly)
6.	Degree of diversity	Unitary (hegemonic)	Diversifying

Source: Modified after Christoff (1996, cited in Hildingsson [2007]).

Criticism: Buttel (2000) admits that the eco-modernisation perspective has the following flaws:

(a) It suffers from (North) Euro-centricity (both theoretical tenets and empirical examples are from North Europe).
(b) Excessive focus on transformative industry.
(c) Uncritical acceptance of transformative potentials of modern capitalism.
(d) Pre-occupation with efficiency and pollution control vis-à-vis cumulative resource consumption and its impact on the environment.
(e) Fundamental questions raised about modernisation have not been addressed from an eco-modernisation perspective.
(f) Eco-modernisation's view that radical environmentalism is not directly responsible for many environmental benefits, in North Europe and elsewhere, is not correct because the radical ecological groups have raised issues (usually ignored by mainstream environmental groups), like toxics and chemicals, highlighted by the radical 'environmental justice' groups.

Therefore, Tim S. Gray (1997) terms ecological modernisation as a right-wing and reformist political ideology. But Arthur Mol argues to distinguish between ecological modernisation as a normative and prescriptive programme for change and its status as a theory of social change. Ultimately we may say that the ecological modernisation perspective stands for a superficial and incremental change (not fundamental change), in the existing capitalist economy and society.

Modernisation Theory

Since the theory of ecological modernisation came out of the mainstream theory of modernisation by adding an ecological dimension to it, it is quite appropriate that we should also discuss the basic tenets of modernisation. Talcott Parsons, Eisenstadt, W. W. Rostow, Lerner, and other sociologists propounded modernisation theory with varying focus on one or another aspect. Talcott Parsons (1951) talked of a social

60 Approaches to Environment and Society

Table 2.6 Pattern Variables in Traditional and Modern Societies

Sl. No.	Traditional Society	Modern Society
(a)	Affective (kinship)	Affective neutrality
(b)	Collective orientation (unit of life)	Self-orientation (individualism)
(c)	Particularism (range of worldview)	Universalism
(d)	Ascription (of works)	Achievement
(e)	Diffuseness (of role)	Specificity

Source: Based on Talcott Parsons (1951).

system (social interactions) getting patterned and institutionalised, and then institutionalised patterns become social systems and 'pattern variables' (culturally shaped). A social system is the key sub-system in order to hold other sub-systems together. Through two mechanisms of socialisation and social control, human personality is integrated into the social system. He dichotomised pattern variables for traditional and modern societies and indicated that people from traditional societies should shift to modern society by adopting the pattern variables of the latter. This may be seen in Table 2.6.

Thus Talcott Parsons desired that the people of traditional societies (developing countries) should adopt the pattern variables of modern society (affective – neutrality, self-orientation, universalism, achievement, and specificity) by discarding their own affective, collective – orientation, particularism, ascription and diffuseness. Singh (1973) talks of the 'modernisation of Indian tradition', while Gupta (2000) talks of 'mistaken modernity': While the former focusses on distinctive features of Indian tradition like collectivity and transcendence, the latter emphasises the universal features of modernity like citizenship and secularism, but India has not achieved such features of modernity.

In 1960, W. W. Rostow wrote a book entitled, 'Stages of Growth: A Non-communist Manifesto' – thus it was written purposively to refute the theoretical foundations of Marx's theories of capital accumulation, class contradiction, social transformation, etc. He propounded five stages of economic growth and argued that all the human societies in the world would pass through such stages in a sequential manner, sooner or later. These stages of growth are listed in Table 2.7.

The various societies passing through different stages of growth all over the world may be seen in Figure 2.1.

But, the modernisation theory may be criticised on the following grounds, Sharma (2016):

(a) Various societies put in the category of pre-Newtonian science had a wide range of differences in their economies, technology, values, policies, and political order.
(b) A developing country may have one modern sector (manufacturing) along with another traditional sector (agriculture).
(c) All developed societies cannot be kept in the category of the 'age of mass consumption' – only the USA seems to be in that category.

Approaches to Environment and Society   61

Table 2.7 Features of Societies in Various Stages of Growth

Sl. No.	Society's Stage of Growth	Socio-economic and Political Features
1.	Traditional society	(a) Primarily self-sufficient agricultural economy (b) Rigid ascriptive social structure based on kinship (c) Pre-Newtonian science and technology adopted
2.	Pre-conditions of take-off society	(a) Due to an impulse from outside, agriculture is augmented by an increase in trade, service, and mining industry (b) Less self-sufficient and localised agrarian economy due to growth in trade and communication (c) Due to new scientific ideas, the natural world is no more seen as given
3.	Take-off society	(a) Investment rises to a minimum of 10% of national income (b) One or more sub-sectors of manufacturing take a leading role by adopting modern technology (c) Socio-political conditions are reshaped to suit a high rate of growth reforms implemented
4.	Drive to maturity (society)	(a) Modern science and technology is extended to all sectors of the economy, and consolidation done (b) Rate of investment is above 10% of the national income (c) Political reforms continue (d) Economy stands on its own feet in the international market

5.	Age of high mass consumption society	(a) Energy consumption per head is generally very high
		(b) Natural resources are considered infinite, hence maximum exploitation of natural resources
		(c) One of three strategic choices is adopted
		  (i) Wealth may be concentrated on individual consumption (as in the USA).
		  (ii) It may be channelled into a welfare State (as in West Europe).
		  (iii) It may be used to build up global power (as in the former USSR).

Source: Based on Rostow (1960).

(d) It ignores the fact that some developing societies had indigenous (non-western) industrialisation, but colonial rule by the western imperialist powers de-industrialised them (e.g. India) through the introduction of reductionist Newtonian science and technology and draining of raw materials to the empire at cheaper rates.

(e) It ignores the fact that even a modern society like Japan practises the ascription criteria (family background, age, and sex) in giving promotions to employees appointed on the basis of achievement criteria of skills and education.

(f) It confused pre-industrial and non-industrial paths of development and wrongly put the two together.

(g) Even modern technology may, directly or indirectly, contribute to the resurgence of old traditional religious/ethnic values (e.g. Islamic revolution in Iran, Sikh separatist movement for Khalistan in Punjab, India, linked with the prosperity due to the green revolution).

(h) It suffers from the 'tragedy of psychologism' blaming the peoples of developing countries as fatalist, passive, other-worldly, superstitious, and backward-looking.

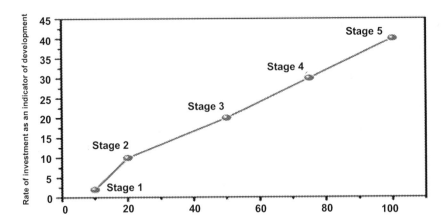

Figure 2.1 Different societies in different stages of growth. Percentage of use of natural resources/energy over time. NB: Not on scale, only illustrative. Source: Prepared by the authors based on Rostow (1960)

(i) It created a mad race for over-use and exploitation of natural resources; it ignores the combination of people's entitlement (basic needs, not greed) and ecological sustainability; traditional society did not over-use such resources.
(j) As the proponents of the dependency theory have rightly pointed out, the capitalist development of the metropolitan developed countries led to the 'under-development' of the developing countries in Asia, Africa, and Latin America.

Thus most of the criticisms of the mainstream modernisation theory are, more or less, also applicable to the eco-modernisation theory currently in vogue in the field of environmental sociology. In fact, there is a new exceptionalism in the form of ecological modernisation theory, resurrecting many dimensions of the human exceptionalist model (that assumes that humans are free from natural constraints) of post-world war II modernisation theory; on the other hand, there is a growing 'planetary rift' due to the crossing of planetary boundary (Foster 2012). Now we may discuss ecofeminism as a distinct perspective on environment/ecology.

Ecofeminism

Ecofeminism developed in different parts of the world in different ways. The term 'ecofeminism' was used for the first time by Francoise D'Eaubonne in 1980. In fact, ecofeminism grew out of various feminist, peace, and ecological movements. For instance, there was a huge protest against the establishment of a nuclear power plant at the Three Mile Islands, and, consequently, a large number of women in the USA participated in the first conference of ecofeminism in Amherst in March 1980. Therein the interconnectedness of militarisation, feminism, ecology, and healing was duly explored for the first time.

One of the organisers of the conference, Ynestra King, perceptively remarked:

> Ecofeminism is about connectedness and wholeness of theory and practice. It asserts special strength and integrity of every living thing ... we see the devastation of the earth and her beings by the corporate warriors, and the threat of nuclear annihilation by the military warriors, as feminist concerns.

Thus they expand it as the stages of exploitation of women, nature, and lower classes in both North and South, and developing countries have passed or are passing through stages of aggression, conquest, possession, and control.

However, earlier French scholar Simone de Beauvoir in her classic work, 'The Second Sex' (1949), argued quite boldly that 'a woman is not born but made'. That is, by birth, there is only a 'biological difference' (sex) between male and female in terms of some body parts, genitals, chromosomes, and hormones, but there is a 'social construct' (gender) through socialisation and other processes which are discriminatory against women. Simone de Beauvoir makes a strong point that, just like non-human animals, women have been more occupied with the 'regeneration' and 'repetition' of life; on the other hand, men have been free to search for ways of transcending life by 'remodelling', 'reshaping', and 'recreating' the future through technology and symbols. In addition, especially from the ecological point of view, she also contends that women's activities are usually perishable and involve 'lower level' transformations of nature; on the other hand, men's activities are more lasting, involving major transformations of nature. Women are restrained in different arenas of rural life by families, neighbourhoods, communities, civil society as well as the State (Table 2.8), though the scene is changing fast due to new technology, new products, and progressive policies and attitudes (mindsets).

Maria Mies and Vandana Shiva (1993/2014), in their work 'Ecofeminism', show a strong contrast between the decline of passive consumerism and the vitality of self-sufficient and autonomous livelihood economies: Subsistence. Maria Mies emphasises that the developed industrialised nations create public fear of terrorism in order to justify self-interested foreign interventions; on the other hand, Vandana Shiva proves that in India the imposition of a 'structural adjustment' programme (with free trade and deregulation) led to disorganisation and stress, resulting into manifold increase in attacks on women in some communities during the period 1993–2011. Both financial and environmental crises, according to them, are gendered, and inter-connectedness between different spheres, lives, and entities shows why some people are still treated like animals, why women do 65% of the world's work for 10% of its wages, why the earth is manipulated as a weapon of war, various species are lost, water stress is increasing, the soil is losing organic integrity, etc. Firstly, as Vandana Shiva (1993/2014), in the joint work with Maria Mies, perceptively remarks:

> We are in the midst of an epic contest ... between the rights of Mother Earth and the rights of mother corporations and militarised states using obsolete worldviews and paradigms to accelerate the war against the planet and people. The contest is between the laws of Gaia and the laws of market and warfare ... between war against Planet Earth and peace with it.

(1993/2014 Preface)

64 Approaches to Environment and Society

Table 2.8 Society's Restrictions on Women in Rural Areas in India

Sl. No.	Public Sphere of Life	Type of Restrictions
1.	Social	(a) Not to wear shorts or 'revealing' dresses, or male dresses (b) Not to remain outside the home after sunset, and not to move alone outside (c) Not to go to study at places far away from home (d) Not to climb on the tree (e) Not to give 'Mukhagni' (fire) to the parents and other members of the family (f) Not to enter certain temples (e.g. Sabarimala, Kerala), synagogues, mosques, tombs (e.g. Haji Ali Dargah, Mumbai), 'dakhmas' (e.g. the Parsees Towers of Silence, Mumbai), etc.
2.	Economic	(a) Not to plough the fields (b) Not to make thatch (roofs by leaves/grasses) (c) Not to weave cots (d) Not to work as a skilled 'Raj mistri' (mason), but permitted to work as an unskilled labourer (e) Not to expend, of her own volition, the wages earned by her, but to expend it under the control of the family
3.	Political	(a) Not to contest elections, or at least, not to contest elections on important seats (e.g. in Meghalaya) (b) Not to act independently after being elected on seats reserved for women in local bodies – Panchayati Raj Institutions (PRIs) – dominance of husband and sons – informal 'Pradhan Pati', 'Mukhia Pati', and 'Sarpanch Pati', if the village head is a woman!

Source: Authors.

Secondly, the economic growth model discounts women's contribution to the economy: 'production for sustenance is counted as non-production. The transformation of value into disvalue, labour into non-labour, and knowledge into non-knowledge … the patriarchal construct of GDP … which commentators have started to call the "gross domestic problem"'. But there is now a 'paradigm shift' to co-operation, relational and holistic view, working as 'co-creators and co-producers with the Earth'.

Maria Mies, too, states therein that, firstly, Hiroshima, Nagasaki, Chernobyl, and Fukushima (nuclear disaster in 2011) are 'guilt names for a system which promises a better life for all but ends in killing life itself'. Secondly, she adds, after the 9/11 terrorist attack in 2001, the violence against women and children is a 'normal' side effect, 'collateral damage', and nowadays children are trained in computer games 'to fix a target and kill an enemy' – hence, 'fight against virtual enemies in virtual wars'. Thirdly, though mothers are the beginning of human life ('arche' in Greek), reproductive and genetic technology has produced 'motherless children'. This is new violence against women. Fourthly, New Economy, under liberalisation, privatisation, and globalisation, has exploited young women as the cheapest labour, e.g. in the textile sector in Bangladesh the wages for women are the lowest, and they work in unsafe and unhealthy working conditions, with the insecurity of jobs and long working hours. Fifthly, after the incident of 9/11 in 2001, 'communism' as an enemy was replaced by

'Terrorism and Islam'. Hence, the security system in the USA and NATO countries was tightened to the extent of spying on everyone and waging wars against Islamic countries like Afghanistan and Iraq, and one of many reasons for such wars was to liberate women from backward Islamic traditions like wearing 'hijab' (head scarf). But usually the enemies do not emancipate women, only brutalise and humiliate them. Finally, poverty has come back to southern Europe – Greece, Spain, Italy, and Cyprus are deeply indebted to banks like Deutsche Bank; now resources (like oil, coal, gas, iron) have been exhausted, and there is destruction of water bodies, soil air, forests, and climate.

Mies and Shiva also think that the capitalist patriarchal world system 'emerged, is built upon and maintains itself through colonisation of women, of "foreign" peoples and their lands; and of nature which is gradually destroying'. Modernisation and development processes were thus responsible for environmental degradation. Further, to them science and technology are not gender-neutral. Lois Gibbs opposed the toxic waste dumping in Love Canal.

They further opine that modern civilisation's cosmology dichotomises reality and one part is seen as superior to the other – man–woman, man–nature, global–local, production–consumption, etc. Feminists in general and eco-feminists, in particular, have strongly questioned this irrational dichotomisation. Such a European project of modernity by colonising (whiteman's burden) is replaced by eco-feminists' notions of 'reweaving the world', 'healing the wounds', and reconnecting and inter-connecting the 'web' – thus a holistic cosmology.

Many eco-feminists like Irene Stoehr and Angelika Birk have questioned the socialist conception of 'emancipation from nature', implying independence from, and dominance over, natural processes by rationality; socialists argue for a historic march from the 'realm of necessity' (the realm of nature) to the 'realm of freedom' (real 'human' realm) – by transforming nature into a 'second nature' or culture. Hence, eco-feminists have criticised such dominance over nature, including human, and female nature that is responsible for the destruction of nature. Today many forms of environmental pollution and degradation are linked to household technology which claimed to emancipate women from drudgery. Hence, many eco-feminists, especially Maria Mies and Vandana Shiva (1993/2014), talk of freedom and happiness within the limits of nature and call it subsistence perspective – as nature's subsistence potential, in all its aspects, should be nurtured.

Vandana Shiva (1993/2014) also argues that the so-called economic reforms have led to

> the subversion of democracy and privatisation of government ... corporate-driven reforms create a convergence of economic and political power, a deepening of inequalities, and a growing separation of the political class from the will of the people they are supposed to represent.

Mies and Shiva brand 'the culture of rape' 'a social externality of economic reforms'. On the other hand, humans are responsible for the extinction of 75% of agricultural biodiversity due to industrial farming.

Mies and Shiva also emphasise there that nowadays 'global' means 'global domination of local and particular interests, by means of subsuming the multiple diversities of economies, cultures and of nature under the control of a few multinational companies'. Thus the so-called 'global' does not represent the universal human interests of all societies. They further criticise the cultural relativists who wrongly justify many regressive traditions like dowry, female genital mutilation, caste system as expressions of particular people – or emphasise 'difference' so much that commonalities may not be seen. They remind us of European liberalism and individualism which are rooted in many negative factors like colonialism, destruction of common natural property resources, privatisation, and commoditisation for profit. Global capital is using the 'ethnic' label for the tourism industry. Hence, they perceptively remark: 'To find a way out of cultural relativism, it is necessary to look not only for differences but for diversities and interconnectedness among women, among men and women, among human beings and other life forms, worldwide'. They also clarify that human needs, not human rights, of all are to be fulfilled, and distinction be made between 'basic needs' and 'higher needs' (freedom, knowledge, etc.). They also consider 'spirituality' as the female sensuality of love (not other-worldly). They eliminate the opposition between spirit and matter, transcendence and immanence. It is the rediscovery of the sacredness of all life forms, hence to be respected. This should be celebrated in rituals, dances, and songs. However, leftist men and women criticise such spirituality as a part of ecofeminism.

Maria Mies, in the joint work with Shiva (1993/2014), puts forth seven methodological guidelines for feminist research:

(a) 'Conscious personality' with identified research objects, not a value-free (neutral) research towards objects; current science and technology is fundamentally 'military science and technology'.
(b) View from below (not view from above) – having both scientific and ethical–political dimensions.
(c) Active participation in actions, movements, and struggles for women's emancipation – research as an integral part of such struggles. It rejects Weber's view of separating science and politics (praxis) as it is not in the interest of women's liberation.
(d) Change of status quo becomes the starting point for a scientific investigation. 'If you want to know a thing, you must change it'.
(e) The research process must become a process of 'conscientisation' both for 'research subjects' and 'research objects' (women as target groups), by following Paulo Friere, who thinks a study of an oppressive reality is carried out by the objects of oppression, not by the experts.
(f) Going beyond Paulo Friere, the collective conscientisation of women through a problem-posing methodology must be accompanied by the study of women's individual and social harmony.

Streams of Ecofeminism

Ecofeminism is not uniform; rather it has various streams focussing on one or another dimension of women and non-human nature. For instance, Marxist or socialist ecofeminism focusses on economic deprivation and desires structural changes in the economy, while liberal streams focus only on formal equality and liberty (equated with accumulation) and desire only incremental and functional changes within the system. On the other hand, post-modernism challenges grand theories like Marxism and liberalism and believes in 'personal is political' and gives due importance to culture. For liberal and socialist perspectives Newtonian science and technology are emancipatory, while for post-modernist and subsistence perspectives it is reductionist, is one-dimensional, and ignores ethical, local, and indigenous knowledge systems. Post-modernist perspective questions grand narratives (meta-theories) like liberalism and Marxism (socialist perspective). The subsistence perspective (of Mies and Shiva) questions 'patriarchal capitalism' (both patriarchy and capitalism), and also the materialism of the Marxist perspective. Except for the neo-classical perspective that has little concern for biodiversity, others respect biodiversity and cultural diversity; however, the post-modernist and subsistence perspectives also have a concern for linguistic diversity. All four perspectives question various sorts of dualism – man–woman, man–nature, etc.; however, the subsistence perspective goes a step further by challenging the dualism of immanent and transcendence, of spirit and matter. Neo-classical/neo-liberal perspective and socialist perspective favour a shift from traditionalism to modernisation, but post-modernist and subsistence perspectives reject this view – post-modernist perspective respects local, specific, 'difference', while the subsistence perspective stands for ethical spirituality and respects the sacredness of all life forms. A comparative analysis of four major streams of ecofeminism may be seen in Table 2.9.

Thus Maria Mies and Vandana Shiva (1993/2014) theorised ecofeminism by arguing that the marginalisation of women has strong links, and substantial similarities, with ecological destruction. They have criticised neo-liberal economic theory, the myth of 'catching-up' development following the capitalist path of the West, the foundation of modern science and technology, ignoring ethics in biotechnology and reproductive technology, culture of consumerism, etc. They provide a new critique of modernity as 'capitalist patriarchy'. New social movements, especially ecological ones, have questioned the death of nature and women together. The local rural women have come out of their homes and hearth to protest against such patriarchal capitalism – colonisation of the developing countries, women, and nature – and also for waging a struggle for the regeneration and renewal of the earth.

Ecofeminism is thus an umbrella perspective with a plurality of theoretical dimensions and core assumptions.

Core Assumptions of Ecofeminism

(a) There have existed historical as well as contemporary inter-connections and linkages between nature's oppression and women's oppression.
(b) While studying the problems of women, one needs to study the ecological problems; similarly, while studying the ecological problems, one needs to study women's problems – otherwise attempts for a separate study will result in partial and fragmented conclusions.
(c) For their marginalisation, certain myths of human–nature dualism, man–woman dualism, etc. have been prevalent and responsible in most of the cultures in the world; on the other hand, the conception of 'mother earth' was relegated to the margin due to the capitalist mode of production.

Table 2.9 A Comparison of Various Streams of Ecofeminism

Aspects / Streams	Equality and Freedom	Challenging Dualisms (Man–Nature, Man–Woman)	View of Nature	View of Economy (Capitalism)	View of Science and Technology	View of Culture	Diversity	View of State, Market, and Civil Society	Principle of Cosmology
Neo-liberal, neo-classical perspective	Equating accumulation with liberty and happiness	No	Humans superior to nature (instrumental value only), proper valuation of natural resources needed	Conforming to capitalism – private gain, no limits to growth	Progressive – emancipating	Promotes a culture of consumerism	Little concern	State as a facilitator (green tax) market-oriented, less scope for civil society	For the modernist world, catching up development (westernisation)
Socialist (Marxist) perspective	From the realm of necessity (nature) to the realm of freedom	Yes	Both – instrumental value of nature to fulfil basic needs; in socialist society intrinsic value of nature (inorganic body of humanity)	To transform capitalism into socialism (redistributive justice)	Progressive – emancipating	Questions the culture of consumerism – labour mediates between nature and culture	Broad concern for biodiversity and cultural diversity	Regulated market, State's primary role	Modernisation for a socially just society
Post-modernist perspective	Freedom of equality for the silenced voices	Yes	Intrinsic value of nature	Against industrialisation, not against capitalism	Science as reductionist – for a plurality of perspectives, no single truth	Regional narratives, language games, and heterogeneous discourses – culture mediates between nature and labour	For bio-cultural and linguistic diversity	Less State-regulated market, more civil society (not privatisation)	Modernisation and Marxism suspected – no universal values; local and specific difference respected
Subsistence perspective	Equality, liberty, justice for all life forms – freedom and happiness within the limits of nature	Yes	Intrinsic value of nature and interconnectedness of humans and non-humans	Against capitalism	Reductionist – to be replaced by a plurality of perspectives and respect for local indigenous knowledge	Against the violent culture of rape, war, consumerism, and destruction of nature	For bio-cultural and linguistic diversity	State, to ensure welfare and justice, regulated market, more civil society (not privatisation)	For spirituality – respecting sacredness of all life forms; modernisation of the West is questioned

Source: Authors.

(d) Modern science and technology has resulted in reductionism, and it has ignored ethics and spiritualism that have substantial bearings on everyday life experiences of human beings in different societies.

(e) The nature, form, and direction of modern industrialisation have been one-dimensional, highly exploitative of natural resources violently, and it has also marginalised women and colonised the developing countries – actually the development of the North essentially under-developed the South.

(f) The western notions of knowledge being 'objective', and the knower being 'rational' and 'detached', and of non-human nature seen as a passive object of study, have been questioned here.

(g) There are also linguistic connections between the oppression of women and nature – the sexist–naturist language represents women, animals, and land as inferior to male-oriented culture.

(h) It is concerned with the sense of self – as 'deep ecology' does – and not the formal value only.

(i) Women's identification with non-human nature is easier than men's on two different grounds:

  (1) The 'biological' argument holds that women and nature are similar due to their physical bodies (e.g. ovulation, menstruation, pregnancy, delivery, and suckling by the baby);

  (2) Gender oppression argument contends that the similarity between nature and women is in terms of the oppression of both by men due to patriarchy or 'patriarchal capitalism'.

(j) 'Ecological feminists' are both street fighters and 'philosophers' (Ariel Salleh) – thus they combine theory and practice (reality in life situations). Most of the eco-feminists have been activist scholars (e.g. Ynestra King, Maria Mies, Vandana Shiva).

Criticism: The following criticism of the ecofeminism is note-worthy:

(1) Ecofeminism's attempt to stretch the analogy of women and nature too far is not supported by logic and fact. For instance, due to the use of contraceptives and other family planning devices birth has been prevented, planned/regulated, and controlled – thus women have been liberated from the burden of bearing children to a large extent, 'other things remaining the same'. But such is not the case with the non-human nature (plants, birds, and other animals).

(2) The emancipation of women due to better and transformative gender policies and programmes of action need not be emancipatory for non-human nature. For instance, more women's work participation rate may put more pressure on natural resources (land, water, forests, mining minerals, etc.), hence ultimately anti-nature. Similarly emancipation of nature may not be emancipatory, in many cases, for women. For example, imposing restrictions on the collection of forest produce as tendu leaves or hunting may be helpful for the regeneration of various plants and wildlife, yet it may be quite negative for women (especially for tribal women) who collect fuel wood, fruits, foods, flowers, fibres, twigs, and leaves from forests.

(3) Often some eco-feminists, due to over-enthusiasm or wrong perception, both ideologically and empirically, bring local social–ecological movements under

the rubric of eco-feminist movement. For instance, Vandana Shiva brands the 'Chipko movement' in Garhwal (State of Uttarakhand, India) in the 1970s as an eco-feminist movement, though, it had primarily male leadership in all three strands: Crusading Gandhians (Sunderlal Bahuguna), appropriate technologists (C. P. Bhatt), and Marxists (Uttarakhand Sangharsh Vahini). Yet there were some major women activists, too, like Gaura Devi who hugged the trees in order to prevent the trees from being cut by the contractor's labourers, coincidentally when the male leaders had gone to the district headquarters to receive compensation for their lands which had been acquired long back by the State (Guha 1989, Sharma 2016).

(4) Eco-feminists often put undue emphasis on gender inequality and accord relatively less significance to the prevailing connected economic and political factors. For instance, in various environmental movements against dam construction projects, both men and women victims of the local area protest, on the one hand, while often 'the political class', representing various major political parties, jointly and separately, support such projects in the name of 'development' (to bring irrigation to fields of the rich farmers' cash crops and drinking water to urban areas, to prevent the floods, to generate energy for towns and industries, and to attract tourism), on the other.

Political Ecology

Political ecology is actually a trans-disciplinary field, emerging mainly from political science, geography, anthropology, environmental history, sociology, and ecology. It tries to analyse 'political forces at work in environmental access, management and transformation ... to demonstrate the way that politics is inevitable ecologically and that ecology is inherently political' (Robbins 2012, 3). Thus it sheds light on environmental dynamism, issues of social equity, and sustainability. Robbins, for instance, goes beyond 'the political ideology as a body of knowledge' and shows political ecology as 'somethings people do', or a 'community of practice' united around a 'certain kind of text'. It explores the cases of problems like starvation, soil erosion, decline in biodiversity, health crisis, and unequal power relations in access to natural resources and also suggests less exploitative and more sustainable ways of addressing livelihood and other related issues. However, there are various definitions of political ecology emphasising one factor/aspect or another – e.g. political economy, formal political institutions, environmental change, or narratives about change. In contrast to political ecology (wherein both politics and ecology matter substantially), there are two major 'apolitical ecology' approaches: 'eco-scarcity' and 'modernisation'.

Eco-Scarcity Argument

In the early eighteenth century in Western Europe for the major problems of social–ecological crisis, the explanation given by some scholars was the burgeoning population growth, the absolute number of people in the world. Drawing heavily from Thomas Malthus' theory of population growth, who argued that the human population grows geometrically (1, 2, 4, 8, 16, 32, etc.), while food production grows arithmetically

Approaches to Environment and Society 71

(1, 2, 3, 4, 5, 6, etc.), these scholars argue that the exponential growth of population results in the failure of the environmental system to support them; hence, the consequence is a major structural crisis for both humans (whose number falls due to mortality because of starvation and diseases) and nature (because natural resources are over-used beyond its capacity of self-generation and renewal). Paul Ehrlich's (1968) 'population bomb' and Club of Rome's 'Limits to Growth' (Meadows et al. 1972) share this argument of eco-scarcity and warn the developing countries as these have higher growth rates of the population as well as their absolute number is high. The problem becomes compounded in the poor countries which are blessed with valuable and scarce environmental assets like rain forests in the Amazon and Asia or wildlife in African countries, especially Kenya. This argument has been criticised on three empirical grounds:

(a) At the world scale, food production's growth rate has surpassed the population growth rate as history shows.
(b) Instead of higher population growth in the poor developing countries, the high consumption pattern in developed countries, especially in the USA, is more responsible for more use of natural resources than the people in the developing countries.
(c) Humans are not just passive objects; rather they are creative and active subjects, and new skilled human power changes the whole scenario of scarcity. As Julian Simon (1981) calls humans 'the ultimate resource', due to resource scarcity prices rise and it ultimately creates an incentive for people to discover new substitutes, or ration and recycle the old ones (e.g. in the case of copper).

Thus the apolitical approach of over-population does not hold good, and it is implicitly political.

If we compare the USA, representing the developed world, and India, representing the developing world, we find that though India is three-and-a-half times larger than the USA in terms of population, per capita consumption of food, water, and energy, and per head carbon emission in the USA is many times more than that in India.

Modernisation Argument

The second major apolitical approach to ecology is the modernisation argument that primarily holds that most of the environmental problems arise due to the adoption of obsolete and unscientific inefficient techniques, methods, and processes of conservation and management of natural resources. Hence, the modernisation principle pleads for open markets and better and more efficient technologies for both higher economic growth (more profit) and, simultaneously, ensuring the conservation of nature (less carbon emission) – thus a win–win situation will be there. This approach suggests three principles and policies (Robbins 2012):

(a) Western technology (and its accessories) must be diffused more all over the world, especially the global South.
(b) Both corporate firms and individuals need to be connected to larger markets and these be given more property controls over natural resources.

72 Approaches to Environment and Society

(c) For the conservation of biodiversity and wilderness, the benefits of efficiencies need to be realised by institutionalising valuation; in the open market, all environmental goods must be duly priced.

The modernisation approach has been criticised on the following grounds (Robbins 2012; Sharma 2015):

(a) Technology in developing countries has not always resulted in benefits; for instance, the green revolution has brought negative consequences like desertification of soil, decline in the groundwater table, salinisation of groundwater, resistance to pesticides, use of chemical fertilisers and pesticides leading to diseases like cancer.
(b) Transfer of technology from, and origin of scientific knowledge in, the North to South assumes a paternalistic colonial attitude, and, on the other hand, most of the technologies transferred to the South in practice are obsolete, hence less beneficial to the developing countries.
(c) There is nothing like a free and fair market system nationally and globally because of the uneven distribution of information, adverse terms of trade against the developing countries, and protectionism allowed for the developed countries. Globalisation has remained selective and discriminatory to developing countries whose products are not given reasonable prices and access to the world market.
(d) 'Polluter pays' principle has not ultimately resulted in ecological conservation and justice due to the purchase of carbon, and not reducing the high consumption rate, by the developed countries.
(e) The trickle-down effect has not been fruitful to the deprived classes, races, and women.
(f) Like the eco-scarcity argument, this modernisation argument, too, claims to be apolitical – but in reality both are political and ignore political-economic forces that work at different levels in real-life situations.

On the other hand, political ecologists like Bryant and Bailey (1997, 28–29) rightly consider that ecological conditions and changes are the products of the prevailing political process, and hence three assumptions are linked while seeing any environmental problem:

(a) 'Costs and benefits associated with environmental change are for the most part distributed among actors unequally' (winners and losers).
(b) Which 'reinforces or reduces existing social and economic inequalities' (hidden costs).
(c) Which holds 'political implications in terms of the altered power of actors in relation to other actors' (uneven power relations).

Thus, observes Paul Robbins (2012), political ecology puts forth Jekyll and Hyde persona:

(i) Explaining what is wrong with the dominant accounts of environmental change.

(ii) And, simultaneously, exploring alternatives, adaptations, and creative human action to solve the problem of environmental exploitation – as a 'hatchet' (to take dangerous accounts) and a 'seed' (to grow into new socio-ecologies) linking of environmental change to political-economic marginalisation came out first in the 1970s and 1980s by applying 'dependency theory' to the environmental crisis.

However, much earlier, Peter Alexeyevich Kropotkin (1842–1921) was an anarchist activist–observer of nature, rather an 'early political ecologist'. He had travelled 800 miles of mountains in the Russian Far East and appreciated the constructive works of the local people, and collective mutual co-operation and organisation between individuals. The political ecology, in fact, addresses the following diverse issues (Robbins 2012):

(a) Possibility for community collective action (e.g. ecological movements)
(b) Role of human labour in environmental metabolism
(c) Risk taking and risk aversion in human behaviour
(d) Diversity of environmental perceptions
(e) Causes and effects of political corruption
(f) Relationship between knowledge and power

Paul Robbins has put forward five theses of political ecology that may be seen in Table 2.10.

Thus Paul Robbins raises five big questions and theses of political ecology. Firstly, 'State development intervention' and 'increasing integration in regional and global markets' result in more poverty and over-exploitation of natural resources. Further, due to the enclosure of community natural resources by the State or private firms there is unsustainable traditional collective resource management and a decline in equity of resource distribution.

Secondly, new efforts to preserve 'nature', local livelihood, production, and socio-political organisations are disabled by the State and global interests. Further, for controlling natural resources, the State questions the benign traditional local production practices as unsustainable.

Thirdly, due to more scarcity resulting from enclosure or appropriation by the State, private firms or elites amplify group conflicts on gender, class, and/or ethnicity lines. On the other hand, environmental problems become 'socialised' when the dominating groups succeed in controlling the common natural resources at the cost of the deprived others.

Fourthly, due to the institutionalised and power-laden environmental management regimes, new environmental actions, and behaviours or rules emerge and these provide a new opportunity for representation by the people politically. That is, aggrieved local people start an environmental movement with a new kind of political action by connecting on gender, class, and ethnicity grounds.

Fifthly, hegemonic institutions and individuals (environmental ministries, MNCs, corrupt forest officers, etc.) have undue influence by controlling new connections and transformations, leading to negative consequences. Consequently resistance arises from 'traditional, alternate or progressive' human/non-human alliances marginalised by such efforts, along class, gender, and ethnicity lines.

74 Approaches to Environment and Society

Table 2.10 Paul Robbins' Five Theses of Political Ecology

Sl. No.	Theses	What Is Explained	Relevance
1	Degradation and marginalisation	Environmental conditions (degradation) – why do environmental systems change?	Shown in a larger political and economic context – capital accumulation is viewed as benign, but the poor are blamed for degradation.
2	Conservation and control	What are conservation outcomes (especially failure)?	Four consequential types of degradation: (a) Loss of natural productivity (b) Loss of biodiversity (c) Loss of usefulness (d) Creating or shifting risk ecology Environmental conservation efforts are shown to have negative effects – failure.
3	Environmental conflict and exclusion	Who have access to natural resources and who are excluded (from natural resources)?	Environmental conflicts are shown as part of a larger class, race, and gender struggles and vice versa.
4	Environmental subjects and identity	Why do social identities of people/ groups change (especially new or emerging ones)?	Political identities and social struggles are shown to be linked to livelihood and environmental activity.
5	Political objects and actors	What are socio-political conditions (especially deeply structured ones)?	Political and economic systems are shown to be affected by the non-human actors with which they are intertwined.

Source: Paul Robbins (2012).

Cultural Ecology

Cultural ecology has seen human–environment issues eco-systemically – humans as a part of the larger ecosystem, controlled by universal forces for subsistence. However, at first, some light needs to be thrown on the 'hazards' research approach that focussed on rational management and amelioration of risk (calculable likelihood of problematic outcomes of human actions and decisions). It arose in the political urban activism (by settlement house workers, mostly women) concerned with environmental health and welfare – due to waste, water, air, and toxins. Women's National Rivers and Harbours Congress, Women's Forestry Commissions, and Federation of Women's Clubs in entire America joined the progressive mayors, unions, and municipal bodies for food safety and urban infrastructure reforms. This informal 'sewer socialism' by activist researchers (Alice Hamilton, Florence Kelly) did a street-level analysis of environmental hazards, connecting typhoid with plumbing, toxins/machinery, work place injury, and deaths. Further, Jane Addams, with her social workers, systematically established the connection between municipal garbage collection in Chicago's wards and local death rates (Bailes 1985, Darnovsky 1992, Jane Addams 1910, cited in Robbins 2012). In 1945, this pragmatic and apolitical hazard research approach was formalised by Gilbert White who opined that the traditional way to handle flood (by building more engineered structures) is costlier, irrational, and not sensitive to humans, and hence better land use planning and change in people's behaviour would mitigate such problems better. He emphasised not to differentiate between 'natural' and 'social' while seeing the environment as a hazard – a flood is a hybrid human–environment problem. Later Robert Kates and Ian Burton claimed that the environment is really becoming more hazardous as a result of human development; i.e. the present economic-political systems actually enhance the risk of natural events (Burton et al. 1993). However, with the passage of time its activist research was lost, and, on the other hand, it could not address bigger issues like global warming, ozone depletion, acid rain, nuclear waste. In addition, it restricted to individual choice, free markets, and rational regulation and thus ignored the important questions of political economy, role of capital in agrarian development, role of power relations, vulnerability of the poor people, as well as control of legislative processes (see Box 2.4, Robbins 2012).

Box 2.4: Limits of Hazards Research Approach

The hazards research approach is limited because here individuals see a hazard, then analyse the alternatives, and finally adjust rationally their behaviour. Here 'irrationality' results due to one or more of four factors: (a) cognitive bias, (b) wilful ignorance, (c) faulty perceptions, and (d) personal and social–psychological problems. Hence, Ben Wisner and Maureen Fordham rightly criticise this approach:

'How are populations made more vulnerable to these hazards by war, by government policies, by misguided development projects? What about the spiking incidence of domestic violence after hurricane Andrew in Florida and the Red River floods, both in the USA? What about the fact that 40 percent of all deaths from tornadoes in the USA occur in mobile homes – inhabited by low-income people?'

Source: www.radixonline.org, cited in Robbins (2012, 35)

As transpires from the above, contrary to the hazards research approach, cultural ecology proposed universal and generalised rules of human–environment interaction. But this approach, too, like the hazards research approach, did not take into account the change in the political economy. However, cultural ecology focussed on the exploration of culture on and within the environment. Carl Ortwin Sauer established cultural landscape studies in the field of Geography at the University of California at Berkeley – focussing on the adaptation of culture over time, diffusion of cultural trains, and intersection between cultures. He saw the ecological degradation in western civilisation while he appreciated indigenous practices and agro-ecology vis-à-vis modern green revolution. He restricted his study to pre-modern cultural and economic contexts, not modern economy–environment relationships. But he established an empirical tradition of field work in geography.

Later, some anthropologists and geographers took a positivist scientific study of culture, contrary to Sauer and other historicist cultural ecologists. Julian Steward

(1972) focussed on cross-cultural comparison. He criticised the historicist notion of history creating culture because it could not explain the common, global, cultural types. Disagreeing with environmental determinists, he claims 'human interaction with nature through substance and work' is the determinant and directing influence of the environment on the socio-cultural order. He talked of a 'culture core' at the centre of all human societies – social, political, and religious patterns closely connected, and secondary features are determined, by and large, by purely cultural–historical factors – by 'random innovations or diffusion' – and they give 'appearance of outward distinctiveness to cultures with similar cores'. Therefore, cultural ecology focusses on the features 'most closely involved in the utilisation of environment in culturally prescribed ways' (Steward 1972, 37). According to him, semi-arid ecosystems do not 'determine' the structure of a society in a simple way – as all hunting–gathering societies in semi-arid ecosystems have similar production problems, and hence common social structural solutions may be proposed. He talked of multi-linear evolution of culture, subsistence technology, and environmental constraints. Steward's cultural ecology is primarily concerned with subsistence producers in developing societies. This concern for production and division of labour, at the centre of the 'culture core' takes a materialist view of culture, quite close to Karl Marx's 'mode of production'. Cultural ecologists, through empirical studies in various developing countries, show that small peasant's subsistence farming is more efficient with far less ecological costs than modern farming (green revolution) with costly fossil fuel, soil erosion, desertification, salinisation of soil, and health hazards due to chemical fertilisers and pesticides, etc. Cultural ecology shows that local people's ecological knowledge and adaptive capacity in its application is quite appreciable for sustainability. However, it has been criticised on the following grounds, conceptually and practically (Robbins 2012):

(1) Excesses of the logic of adaptation lead to reductionist conclusions, suffering from a teleological flaw – if people do it, it must be adaptive (Trimbur and Watts 1976). For instance, the Aztec human sacrifice tradition of Mexico in the pre-Columbian period was explained as an adaptation to protein deficiencies. But the alternate sources of protein could be found in pulses – hence such reduction is not acceptable both logically and socially.
(2) Its 'neo-functionalism' crudely used the notion of 'carrying capacity'.
(3) Neo-functional cultural materialism (of Marvin Harris, 1966) is not empirically proved; for example, Marvin Harris simplistically argued that the cow became sacred in India because of its ecological value in terms of protein provision and agricultural traction power.
(4) Its one argument neutralises and legitimises contingent social behaviours and practices. For instance, native Americans of Bennett's (1969) Northern Plainsmen are said to fill an 'adaptive niche' by living at the edge of subsistence, scavenging at the periphery of the larger economic and ecological system – implying this status is natural, and not the consequence of land seizure, discrimination, political marginalisation, and long term exploitation.

Now we may see a case study of a failed dairy development project in Odisha State (in India) in the late 1970s and 1980s that had disastrous effects on ecological diversity (Box 2.5).

Approaches to Environment and Society   77

Box 2.5: Failure of Dairy Development Project in Odisha

The Government of India, in 1978, in close association with some NGOs, attempted to modernise the cattle breeding system among the poor people in Odisha that was highly affected by drought and hence poverty, malnutrition, and hunger prevailed there. Rural poor people were given free cows impregnated with Jersey semen for the reproduction of high milk yielding she calves. Further, each one was given one acre of land for growing fodder. Again they were given the costs of labour involved in tending cows, growing fodder, etc. In 12 years by 1990, there were neither hybrid animals nor traditional indigenous breed animals in the concerned villages. Rather Odisha's indigenous Khariar bull became biologically extinct. The lands assigned for growing fodder were used for growing food crops. It had three major consequences:

(1) Such a 'modern development' project led to the depletion of the Odisha region's environmental resources and damage to its biodiversity. The local climate did not suit the American bull, while the local Khariar bull breed was eliminated in the name of purity of the western bull breeding system. The local climatic conditions also did not suit the fodder species.

(2) The local people became poorer – small peasants became low-wage labourers.

(3) Though it intended to reduce out-migration, it increased that enormously.

Thus the development administration wrongly dismissed local traditions of dairy (breeding and milk yielding). But the question remains – who benefitted from foreign animal and semen sales, international consultancy fees, and salaries of officials, etc.? How were small peasants reduced to labourers? And development for whom and at whose cost? These are thus the issues of political ecology.

Source: Based on P. Sainath (1996) cited in Robbins (2012)

Collective or common property theory is obviously a variant of political ecology theory because it is grounded in collective or common property resources, mainly natural resources like land (including traditional paths), water sources, forests, groves, fisheries (in rivers, seas), pastures, gardens. These resources cannot be divided into mini or micro-individual units of ownership due to their temporal and spatial variability. However, these common resources have their informal or formal rules of use, management, maintenance, and control, primarily with the help of the local indigenous knowledge system. But Garret Hardin (1968) extensively talked of the 'tragedy of commons' due to its over-use by one and all, resulting in degradation; hence, he suggested that either the State should own and control it or it should be privatised. Robert Wade (1987) classifies Hardin's hypothesis with three choices for use of the commons:

(a) 'Everyone else abides by the rules while the individual enjoys unrestrained access' (he 'free rides' or 'shirks').
(b) 'Everyone, including himself, follows the rule' ('co-operates').

78 Approaches to Environment and Society

(c) 'No one follows the rule; he follows the rule while no one else does' (he is 'suckered').

The second alternative is the most desirable, but Hardin sticks to the first ('free ride') or third ('suckered') in his hypothesis in the European context. However, in various developing countries commons are conserved, used, and managed with practical rules – e.g. fisheries in Turkey, pastures in India and Morocco, and forests in Japan and Madagascar. In Chapter 9, 'tragedy of commons' thesis has been explained and criticised in this book. Some scholars have rightly shown that commons are not unowned (legally 'res nullius'), but rather commonly held property (Ciriacy-Wantrup and Bishop 1975). Further, Ostrom (1990, 1992) has also clearly shown that in a society the failure of collective resource management is the failure of the prevailing rules; hence, there is a need for new and better rules about use, sanction, control, maintenance, redistribution, inclusion, and exclusion, and not the division of commons into small private entities. Thus degradation of commons is not necessarily the consequence of collective ownership – hence it fits well with the theory of political ecology.

In addition, two concepts of 'moral economy' and 'everyday resistance' (E. P. Thompson 1978, James C. Scott 1976, 1985) are also relevant here. The first concept, the 'moral economy' of peasants, argues that small peasants face various kinds of subsistence risks that create and promote some mode of mutual help and minimum exploitation. Thus they redistribute surplus more appropriately and protect themselves in bad crop years/seasons when they take help from their community members, other villages, and relatives in the village or outside; in good crop years/seasons, they help them in a similar way. Such mutual help and co-operation may take the following forms in Asia in general, and India in particular:

(a) Helps in kindness, by providing seeds, fertilisers, water from private tanks/wells/ boring tube wells and other inputs
(b) Helps in cash (credits/loans) on favourable terms
(c) Share and exchange of labour – avoiding the use of labour with higher wages
(d) Suitable share-cropping – 'adhia', 'adhbatai', 'bataidari', ' bargadari', etc.

Thus, in everyday life, many peasants accept some kind of tolerable extraction as 'moral subsistence obligations'. However, when the socio-economic situation crosses the tolerance limit, the peasants start protest movements against the authorities. But small peasants cannot start the armed struggle against the local landlords; hence, they start 'everyday resistance' (James C. Scott 1985), through slander, back talk, work slow-down, and pilfering. This concept is rooted in Antonio Gramsci's concept of 'hegemony' (the ability of the elite to have spontaneous consent of the deprived population through the use and control of culture, opinion formation, and ideology). It is called the 'ideological State apparatus' (press, education, culture). On the other hand, the State uses 'coercive State apparatuses' (army, police, etc.) for direct and strong control over the people. Political ecologists thus use the notions of 'moral economy' and 'everyday resistance' in their studies because due to the negative effects of liberalisation and market dominance on rural grasslands in India, co-operative and mutual help declines, as moral economy dissolves, leading to the over-grazing of the range (Jodha 1985, 1987).

However, James C. Scott has been criticised on the following grounds:

(a) 'Top down' view of ideological control
(b) Ignoring the gender aspect – especially extraction of female labour value in peasants' households
(c) Risk-centred view of producer logic

Anyway, the two concepts of James Scott's find support from the empirical evidence in different parts of developing countries. Another concept relevant to political ecology is 'environmental justice' that argues that the disempowered communities, especially the coloured ones, and the urban poor, are disproportionately and unevenly located in 'ghettoes' around hazardous places with the dumping of nuclear wastes, garbage incinerators, lead smelters, and so on. A report of United Church of Christ's Commission for Racial Justice in 1987 and Robert Bullard's (1990) 'Dumping in Dixie' exposed racial and classist discrimination in the urban housing system because the rich and privileged white people control the city's metabolism and expose their externalities on the coloured people. Various empirical evidence shows that black minorities are denied environmental 'goods' (not planting trees/grasses) and over-loaded with environmental 'bads' (wastes) (Box 2.6).

Another associated notion of political ecology is feminist development studies that traced in the 1970s and 1980s that human–environment relationship and processes are 'gendered', i.e. men and women have different access and control over ecosystems due to their different socio-cultural roles. Women's marginal status stemmed from the very development institutions, especially in forestry, farming water development, etc. (Shiva 1988). Such gendered social reality is reflected in three ways (Robbins 2012):

(a) Women's survival skills are based on a different knowledge of environmental systems and processes than that of men. For instance, women in rural Africa reject the government's tree plantation scheme, because the selected plant species will not fit in with the local ecology. Further, their ecological knowledge and priorities are ignored by government agencies. Hence, conflicts may occur.
(b) Women's right to access ecological resources differ from men's. For instance, Gambian women traditionally control the produce of all their agricultural labour on private plots but owe the harvest of communal village lands to the household. Development schemes for agricultural intensification may reduce the resources claimed by women, though enhancing their labour burden.
(c) Women's social motivation for taking environmental action differs from men's. For instance, women in Malaysia took action against their agricultural employers regarding their exploitation, but men did not do so. Similarly, in Harlem in New York City (USA), women of colour took action against the dumping of hazardous wastes' processing there.

Further, more cropping intensity and extension demands more female labour, and hence their burden for fuel, fodder, etc. increases substantially. Obviously hidden costs of agricultural change are often reflected in women's time, health, autonomy, and drudgery. Hence, the linkage of female labour, nature, and social power (resistance against it) may be explained by political ecology.

Thus it clearly transpires that there are insights from various sub-streams/notions like environmental justice, post-colonial studies, feminist development studies, moral economy, everyday resistance by the peasants, and green materialism that could be integrated for proper political ecology perspective. Further, in the late twentieth century and early twenty-first century, many global challenges, like climate change, natural disasters, have arisen, most of which are man-made and call for the investigation into the capitalist mode of production, and the off-shoots of industrialism and consumerism well operated by MNCs. This could be realised by duly applying political ecology because most of the global and national environmental problems are interconnected. Paul Robbins (2012) has clarified that political ecology is (a) neither a theory nor a method, rather it is a 'community of practice'; (b) its constituency operates in the borderlines between analysis and action, and between social practice and environmental resources, impacts, and changes; (c) it 'simultaneously constructs and deconstructs', criticises, and defends, it listens and argues; and (d) it is a large constituency, heterogeneous, and differentially invested in theory and practice, divided in its aim and scope.

Political ecologies' four characteristics are as follows:

(a) 'Track winners and losers to understand the persistent structures of winning and losing'.
(b) 'Are narrated using human–non-human dialectics'.
(c) 'Start from, or end in, a contradiction'.
(d) 'Simultaneously make claims about the state of nature and claims about the state of nature' (Robbins 2012, 87).

According to Robbins, political ecology is a tradition that 'dismantles other accounts' (wielding its intellectual hatchet) while 'making space for, and nurturing, other possibilities' (planting intellectual and practical seeds) – both hatchet and seed. As a critical historiography, deconstruction, and myth-busting research, it is a hatchet, 'cutting and pruning away the stories, methods and policies that create pernicious social and ecological outcomes'. On the other hand, as seed, it is 'equity and sustainability' – it is progressive (not reactive or retrospective), and it plants the seeds for reclaiming and asserting alternative ways of managing water.

However, Robbins admits the limitations of political ecology regarding global issues like, firstly, the dumping of wastes/scraps in developing countries; for instance, during the years 1997–2007 an estimated 500 million computers in the USA became outdated, hence the dumping of more than a billion pounds of lead, two billion pounds of cadmium, and six billion pounds of plastics at different locations in Asia and Africa.

Secondly, despite the increase in the plantation of forest from 124 million hectares to 187 million hectares in the world in 1990, total forest cover is decreasing due to deforestation; the new forest planters 'control the character, diversity, and habitats (and their absence) of the global forest' (Robbins 2012, 252).

Thirdly, the global corporate media is dominating the imagery, views, practices, and environmental imaginary of people and communities in the world; for instance, Discovery Communications (owning Discovery Channel, Animal Planet, and Learning Channel) made $372 million in the second quarter of 2010 alone.

Fourthly, migrant workers exposed to agricultural chemical pesticides experience daily fatigue, nausea, muscle pains, and cramps and have the risk of developing leukaemia.

Fifthly, during the twentieth century 90% of global agricultural crop diversity was lost, even when transgenic maize has been introduced in central Mexico (the original place of growing maize), hence the serious risk of genetic decline. Though at the global level there is food surplus, a large number of people, including children, die of malnutrition and starvation every year.

Finally, every US citizen consumes 1,600 litres of gasoline annually, though some petroleum exporting countries like Nigeria are still the poorest in the world.

But political ecology, observes Robbins, is not capable of addressing the problems of global climate change or urban food and health. Yet the insights of political ecology and its theoretical traction are necessary to understand such problems rooted in power and privilege, dialectical relationships between humans and the world, and the contradictions of political economies.

Criticism

(a) The political ecology approach has not duly recognised the global ecological issues – e.g. dumping of hazardous electronic and other wastes by the developed 'metropolis' in the 'peripheral' developing countries.
(b) It boasts of addressing the real political, economic, and ecological problems as well as the constructionist views of such problems, and hence it is a too broad and general approach to properly address specific local ecological issues in their concrete forms.

Box 2.6: Maya Culture Identity and Ecology: 'Decolonising Development'

The great Maya culture and civilisation in Belize (Central America) was ruined by the Spanish imperialist colonisers in the fifteenth century. It had shifted agro-forestry agriculture ('milpa') that supported its communities and maintained the civilisation. On the other hand, it also provided Achilles' heel from over-intensification. However, Joel Wainwright's *Decolonizing Development: Colonial Power and the Maya* (2008) traces the historical–political ecology of Maya as the product of colonial and State writing and thinking. For instance, the well-meaning recording and mapping of Maya culture, territory, agriculture, and ecology have resulted in the extensive enclosure of Maya lands, conversion of community resources into private property rights, and control of State authorities. Consequently the loss of land, autonomy, and resources of indigenous people there – thus creating a Maya identity has been an element of Maya marginalisation. He argues that even well-meaning participants in modern development are necessarily tied with territorial and economic logics, ultimately supporting capitalism and the State at the cost of the local people – as development becomes a trap: 'we cannot want development' because we reach the contradictions, facing head on 'aporias' (Jacques Derrida and Gayatri Chakravorty Spivak use it to denote spaces or passages that are 'non-passages' that become 'the experience of the impossible' and produce 'radical doubt'). Joel Wainwright worked with Julian Cho, an activist leader of the Maya movement, but the latter was killed in December 1998. He rightly observes,

> I came to see the necessity of the analysis of the colonial roots of the current crisis, on one hand (which led to the archives) and the de-structuring of capitalism qua development ... There were the conditions that created the problematic of Decolonising Development.

Source: Joel Wainwright (2008), cited in Paul Robbins (2012)

Ecological Marxism

Ecological Marxism is historically traced from the ideas of Karl Marx and Friedrich Engels. To them, firstly, the degradation of the environment was the primary consequence of capitalism. Hence, the politics of environment was linked to the politics of industrialisation, capital accumulation, and class struggle. Marx's famous statement is relevant here: 'It is not the consciousness that determines the existence, rather existence determines the consciousness'. Such existence is made up of the natural environment and economic mode of production. In fact, the way people interact with natural objects provides the 'base' upon which politics, culture, law, etc., are founded. The society changes as the forces of production and relations of production change. Secondly, the capitalist mode of production needs the extraction of surpluses from nature and labour. The intensification of production results in contradictions that need to be resolved either by increasing growth or increasing the exploitation of workers and natural resources. Thus environmental degradation is inevitable in capitalism. Marxists have expanded common property theory by linking the process of accumulation with the encroachment and dismantling of traditional commons. No doubt, in the Marxist theory of the mode of production, human beings are shown as 'Promethean', capable of more and more exploitation of natural resources as the economies develop.

Among the ecological Marxists or eco-Marxists, there are two broad schools: Orthodox eco-Marxism (Charles Tolman, Howard Parsons) and humanist eco-Marxism (Andre Gorz, David Lee, O'Connor).

(A) **Orthodox Eco-Marxism**

The arguments put forward by the orthodox eco-Marxists are as follows:

(a) History is viewed as a progressive dialectical struggle from the primitive to the advanced stage, resulting in more domestication of non-human nature through labour and technology.

(b) Environmental problems are traced to exploitative capitalism; hence, their solution needs a revolutionary transformation of the relations of nature.

(c) The preservation of wilderness makes no sense unless it has some instrumental value for humans. As Howard Parsons observes, 'it is hard to know what could be meant by nature 'in itself', either in Kantian sense or in the sense of discrete reality entirely independent of our cognition and action' (cited in Sharma 2016, 69).

(d) Orthodox eco-Marxists see the concern for the welfare of animals as a 'displacement of human concern' that is restricted by the privileged class position of its advocates.

(e) Since private capital accumulation gives rise to resource depletion, pollution, urbanisation, and occupational hazards, private and inequitable mastery over nature under capitalism must be replaced with the public and socially equitable mastery of nature under communism. Thus Marx and Engels were the pioneers of 'human welfare ecology'.

(f) Orthodox eco-Marxists are of the view that the working class is the agent of change, both social and environmental.

(B) **Humanist Eco-Marxism**

On the other hand, the humanist eco-Marxists' arguments are as follows:

(a) In Marx's early writings, there was no human–nature dichotomy, rather Marx was concerned with overcoming alienation between human beings and their work, and between humans and their external nature. According to David Lee (1980), Karl Marx was not anthropocentric because he considered nature as humanity's 'inorganic body', unlike the capitalist notion of nature as an alien 'other' to be exploited for profit. David Lee suggests the replacement of the dichotomy between nature and humanity by 'a rational, humane, environmentally un-alienated social order' (1980, 11).

(b) Friedrich Engels argues that the more one understands the laws of nature and the consequences of human actions, the more humans will come to 'know themselves to be one with nature', and there is no inherent 'contradiction between mind and matter, man and nature, soul and body'. Hence, we cannot rule over nature. Marx's active materialism was both ontological and epistemological, 'a unilateral dependence of social on biological (and physical being) and the emergence of the former from the latter' and in the sense of 'the independent existence and trans-factual activity of at least some of the objects of scientific thought' (Foster 2001, 6). His critical materialism belonged to the 'process of natural history'.

(c) In Economic and Philosophical Manuscripts of 1844, Marx referred to the progressive 'humanisation' of nature and the 'naturalisation' of humanity. He recognised the interdependence of nature and humans. According to Karl Marx,

> Man lives on nature means that nature is his body, with which he must remain in continuous interchange if he is not to die. That man's physical and spiritual life is linked to nature means that nature is linked to itself, for man is a part of nature.
>
> (cited in Sharma 2016, 65)

(d) Karl Marx was highly concerned with the antagonistic division of town and country due to industrialisation on capitalist lines.

O'Connor (1996) once perceptively remarked, 'If you think, as I do, that social labour mediates culture and nature (which in turn, mediates back in various ways), you are an eco-Marxist, An active materialist' (cited in Sharma 2016, 72).

Needless to mention here that the history of ecological Marxism may be traced to the thoughts of three scholars from different arenas: Epicurus (materialism), Charles Darwin (natural selection, evolution), and Justus Freiherr von Liebig (agricultural chemist's circulation of soil nutrients and its relation to animal metabolism). The relevance of materialism to the development of ecology may be clearly understood in four 'informal laws' of ecology by Barry Commoner (Foster, 2001):

(a) Everything is connected with everything else.
(b) Everything must go somewhere.
(c) Nature knows best – evolution knows best (with levels of contingency at every stage).
(d) Nothing comes from nothing.

First, second, and fourth 'informal laws' were the main principles of Epicurus' physics. Here it is duly emphasised that humans should be cautious in making fundamental ecological changes – introducing chemicals into the environment may result in major damage. The dialectical materialist approach of ecological Marxism recognises that organisms in general do not simply adapt to their environment: They also affect that environment in many ways and, by affecting so, change it also – thus a reciprocal relationship exists. Therefore, the ecological community and its environment should be seen as a dialectical whole. As Richard Levins and Richard Lewontin ('The Dialectical Biologist') clarify it (cited in Foster 2001, 16):

> Ecology must cope with interdependence and relative autonomy, with similarity and difference, with the general and the particular, with chance and necessity, with equilibrium and change, with continuity and discontinuity, with contradictory processes. It must become increasingly self-conscious of its own philosophy, and that philosophy will be effective to the extent that it becomes not only materialist but dialectical.

Further, dependency theory was developed from mainstream Marxism in the late 1960s and 1970s in Latin America. It pointed out that the marginalisation and underdevelopment of peripheral countries in Latin America, Asia, and Africa was the result of the adverse terms of trade against the colonialised countries who were forced to produce primary products, not the more valuable industrial and craft goods. In India, the textile industry was shunted aside by the colonial authorities (in the metropolis) who demanded and wanted cheap raw materials like cotton from India, but not any competition in finished goods, for textile mills in Manchester. Needless to say that the British rule forced the farmers in Bengal and Bihar to cultivate indigo in at least 3 out of 20 *Katthas* ('*Tinkathia*') on the most fertile lands. Even now that unequal exchange continues as transnational corporations (TNCs) control technology, payments of interests, royalties, and profit. Following Marxism, the dependency theory holds that the capital accumulation needs the exploitation of both nature and labour. By exploiting nature's bounties (including capital) and not investing or investing only nominally in the restoration of the damaged natural environment, the capitalists, especially TNCs squeeze surplus from there. For intensive (capitalist) agriculture, the fertility of the soil is damaged due to over-use. So is the case of the destruction of the forests, especially for building sleepers for the tracks of railways, ships for the navy, and also 'developing' of the hill stations.

A variant of ecological Marxism is 'green materialism' (but not a uniform category) that argues like the launching of the labour movement due to labour exploitation, the response to the exploitation of natural resources should be the launching of an environmental movement (O'Connor 1996). O'Connor's group has developed a school of 'political economy of nature' (Ted Benton, James O'Connor, John Bellamy Foster, Bakker, Bridge, Luke, etc.) – especially exploring the politics of water resources, mining, and training the environmental technocrats. Needless to say that much of the research work in political ecology takes three basic principles of ecological Marxism or dialectical materialist theory of ecology (Robbins 2012):

(a) Social and cultural relationships are rooted in economic interactions among people and between people and non-human objects and systems.
(b) Exogenous imposition of unsustainable extractive regimes of accumulation results in environmental and social stress.
(c) Production for the world market leads to contradictions and dependencies.

According to Roy Bhaskar, rational philosophical materialism is a complex worldview that consists of three types (Foster 2001):

(1) Ontological materialism – it asserts the unilateral dependence of social upon biological (and physical) beings, and the emergence of the former from the latter.
(2) Epistemological materialism – it points out the independent existence of transfactual (causal and law-like) activity of at least some of the objects of scientific thought.
(3) Practical materialism – it points out the constitutive role of human transformative agency in the reproduction of social forms.

Marx's dialectical materialist view of history was 'practical materialism' – man's relations to nature were practical, i.e. established by action. However, his materialist conception of nature and science used both 'ontological materialism' and 'epistemological materialism'. But his conception of nature was not a rigid, mechanical determinism.

Criticism: The following points are notable here:

(a) Many post-Marxist green theorists like R. Bahro, John Clark, Bookchin, and Carl Boggs have questioned the 'production ideology' of Marxism (even that of orthodox eco-Marxism), and the inherent conservatism of the western labour movement. Bahro rightly argues that the western standard of living, which the people of developing countries aspire for, is not achievable.
(b) It over-emphasises the relations of production at the cost of forces of production – including natural resources.
(c) It accepts the industrial technology and 'instrumental reason' to solve all problems including those of ecological limit that has its own limitations – even the distorted industrialisation in Eastern Europe was not substantially different from the mainstream capitalism in the USA, Canada, and Western Europe, as Bahro points out (Sharma 2016).

86 Approaches to Environment and Society

(d) He was influenced by the English classical labour theory of value which undergirds his conception of man as 'homo faber' (as remarked by Hwa Yol Jung [Sharma 2016]).

(e) Marx was 'speciesist' (like racist), radically disconnecting human beings from animals and taking sides with humans over animals.

(f) Marx also finally failed to address the exploitation of nature (ignoring to incorporate it into his value theory) and rather adopted a 'Promethean' (pro-technological, anti-ecological) view (Foster 2001).

However, Karl Marx desired in his early writings (Economic and Philosophic Manuscripts of 1844) to remove all kinds of alienation (between human beings, between humans and non-human nature, alienation from oneself, and alienation from species, etc.), to remove the soil's nutritional loss and degradation of land, as well as to bridge the antagonistic division between city and rural areas, but this would be possible in his idea of socialism after people's struggle (Box 2.7).

Thus various perspectives/approaches to the environment and society highlight the range and depth of different dimensions of the interaction and inter-connection between humans and the environment – the latter is not merely a social construction but a real entity (surroundings) linked with health and livelihood of the local people in a complex way.

Box 2.7: Victory of People and Defeat of the Mining Giant Vedanta: Environment, Health, and Livelihood

On 22 May 2018 tens of thousands of protesters, under the leadership of 67 years old Prof. Fatima Babu (and her group 'Veeranganai' – meaning valourous women) took to the streets of Thoothukudi (Tamil Nadu) against the expansion of Anil Agarwal's mining giant Vedanta Resources' subsidary Sterlite's 4,00,000-ton annual capacity smelter (that began production in 1997). Unfortunately the State police fired and killed 13 protesters! The UN also condemned the 'excessive and disproportionate' use of force by the police and sought an inquiry. National Human Rights Commission (New Delhi) conducted an inquiry and lauded govt. for paying compensation 'adequately' to victims' families! They were paid a compensation of Rs. 2 million ($19,000) only, so the victims' families termed it meagre. However, within a week of this incident, the Tamil Nadu govt. shut it down in May 2018. Sterlite appealed against it in Madras High Court that rejected it in August 2020 and did not allow the re-opening of the copper plant as it caused 'widespread environmental degradation, destroying farmlands, flouting local laws, and subjecting thousands to detrimental health effects'. Various experts found sulphur dioxide emitted from smelters and power plants affecting the human respiratory system and killing the vegetation in its surroundings. Tirunelveli Medical College and Hospital found in a survey of 80,000 persons in a 5-km radius of a copper smelter that, first, 14% had respiratory diseases, significantly higher than the State average, 'attributable to air pollution due to the presence of gases or a mixture of gases and particulate matter'. Second, there was also a high incidence of nervous system disorders, brain tumours, cancer, and menstrual disorders among women. Third, there was also groundwater contamination around 5 km of the plant-iron content in the groundwater of two villages near the plant was 17 and 20 times the permissible level of less than 1 mg per litre (as per Bureau of Indian Standards). Since 1997 Tamil Nadu Pollution Control Board ordered its closure four times (1997, 1998, 2010, and 2013) for excessive emission of sulphur dioxide and in 2013 Supreme Court of India imposed a fine of Rs. 1 billion for environmental restoration (due to adverse effects from its major gas leak on 20 March 2013).

Sterlite had faced strong resistance earlier, too, in Gujarat, Goa, and Maharashtra as well as in Zambia (to Vedanta Resources). But then-Chief Minister of Tamil Nadu Ms. J. Jayalalithaa saw it as a 'dream project in the process of industrialising the state' at the time of laying its foundation on 31 October 1994. Even then the fishermen had opposed the 8-km long waste water pipeline from the plant to sea due to effluents decreasing fish and threatening their livelihoods. Similarly, the local farmers had opposed the diversion of 10% of the city's water supply from the Thamirabarani river to the copper smelter. Due to resistance to the first consignment of copper ore from Australia from entering the harbour, Sterlite abandoned the waste water pipeline project. There Christian fisherpersons are referred to as Fernandos, while Hindu fisherpersons are called Paravars – but all united then and there. She (Fatima Babu) united fisherfolk and Nadars (trading community), divided by Vedanta earlier. Her Christian religion was used against her, but she was firm. She founded the Anti-Sterlite People's Movement to bring different social and political groups together. The trigger for the final phase of protest in 2018 was clearances given by Central and State authorities for doubling the copper smelter's capacity from 1,200 to 2,400 tons daily! But there was no public hearing as per law; hence, Madras High Court agreed with the activists. The company engaged in a massive construction work during December 2016–October 2017, but villagers nearby started organising secret meetings; even Kumaraveddiyapuram villagers met Fatima Babu, though this village had supported Vedanta earlier. Within a day 5,000 persons signed against the plant. At her call, the traders shut down their shops on 24 March 2018 and joined the fisher folk in a large number. It lasted for 10 days and on 22 May 2018 State police killed 13 persons. Then Tamil Nadu govt. was forced to shut it for the fifth and final time (after a week). People blamed slag dumps (iron and gypsum, a by-product of smelting copper, toxic groundwater, and sulphur dioxide fumes) for poisoning their fields. Tamil Nadu Pollution Control Board submitted in Madras High Court in 2019 blaming Vedanta for dumping 6 million tons of slag in 20 locations across the city and its surroundings, and choking rain water run offs to river Uppar (along Thoothukudi) leading to flooding. In Veerapandianpuram, a local well's water turned green and poisonous, while the CEO of the local company claimed a 'Zero effluent' facility there! But Fatima Babu wanted Vedanta not only to leave the place but also must pay for all the damages done to health, environment, and loss of livelihood. The appeal by Vedanta is pending in the Supreme Court (Shankar 2021). Thus this struggle was not for purely environment but also involved livelihood and health issues together.

Source: K. Shankar (2021)

Conclusion

Integrated life system has evolved over a long period of time from an organism, through species, population, community, and ecosystem, to a biome. Hence any human being or group should not think that he/she is a better species and hence may dominate over other species like animals, birds, fish, and plants. In fact, the anthropocentric view (dominance of humans over other species) has been, to a large extent, responsible for the damage to ecology locally, nationally, regionally, and globally. Hence there is a need to take a holistic view of the planet (earth) for correct understanding. Various approaches like human ecology, treadmill of production, ecological modernisation (an offshoot of modernisation theory), ecofeminism, political ecology, cultural ecology, and ecological Marxism have been discussed here, with their positive and negative aspects, on the grounds of logic, fact, and eco-ethics. Hence these approaches are tested on the bedrock of social context (time and place). Undoubtedly, the ecological equilibrium is nowadays highly disturbed due to the prevailing capitalist system (wherein the natural resources are extracted and exploited to the maximum extent, surpassing the self-generation and self-renewal capacity of nature), and the individual tendency of greed dominating over the genuine basic needs. Mahatma Gandhi had rightly remarked, 'the earth has everything to fulfill everybody's need, but not anybody's greed'. Unfortunately, various State apparatuses, business-industry corporates, and most of the citizens are not paying heed to this. Ecofeminism has rightly underlined the gender perspective in ecology. But ecofeminism is not unified, rather it has many streams. However, some ecofeminists have taken an extreme view of essentialism and have ignored the political economy as well as the positive role played by men in the genuine struggle for women's emancipation – socially, culturally, economically, and politically. The political ecology approach emphasises that 'politics is inevitable ecologically' and 'ecology is inherently political'. That is, the question of ecology is also a political question. As far as ecological Marxism is concerned, it has two streams of orthodox eco-Marxists and humanist eco-Marxists. Obviously, the stream of humanist eco-Marxists is more relevant in the context of ecology as a whole. The contradictions in the early writings (Economic and Philosophical Manuscripts of 1844) and later writings of Marx ('The Capital') are quite visible – the former is inclined towards a holistic social–ecological perspective, while the latter is inclined towards an anthropocentric view of nature. However, the dependency theory is also quite relevant in the humanist eco-Marxist stream, as well as a new school of neo-Marxists who call for 'green materialism'. Thus, from various approaches it is crystal clear that all of these approaches have some positive and negative points. Hence a holistic view of social ecology is the need of the hour that takes into account the eco-centric or deep ecology perspective combined with the political economy perspective for the correct understanding of the environment, development, and society. Further, it has to be emphasised, finally, that the 'local' is to be given priority over the 'national', and 'national' is to be given priority over the 'global', at least in the context of the developing countries; otherwise, in view of the grim reality of 'unequal ecological exchange' and 'unequal economic exchange' at the global level, the 'underdevelopment' of the developing countries will get accentuated. Already there is capitalisation and corporatisation of various natural resources by global capitalism under the garb of liberalisation, privatisation, and globalisation. Thinking globally and acting locally should be our motto of everyday life.

Here the question arises: How can development be sustained for what is sustainable development (SD)? This will be attempted in the next chapter.

Points for Discussion

(1) Compare two major approaches to environment and society; which approach is more suitable in the Indian context and why?

(2) What is the inter-connectedness of the life system? Compare HEP and NEP.

(3) How does the environment affect the poor? Which institution is responsible for it – State, market, or civil society?

(4) What is the relevance of W. W. Rostow's stages of growth theory in the context of developing countries' environment, economy, and politics?

(5) In the context of developing countries like India, which approach is the most suitable in the current situation? And why?

(6) How do humanist eco-Marxists differ from the orthodox eco-Marxists?

(7) Discuss various streams of ecofeminism; which stream is the best and why?

Chapter 3

Sustainable Development
Concept, Principles, and Practice

Introduction

Sustainable development has two concepts intertwined: 'sustainability' is possible generally in close proximity and harmony with nature, and 'development' is generally a product of culture by taming or controlling nature. Before addressing the main issue both in range and in depth, let us discuss ecology from a holistic and historical perspective, in the Indian context in particular, and in the international context in general. Perhaps the oldest Greek mythology mentions Gaia who was the primordial mother earth or goddess earth. Homer's 'Iliad' mentions Gaia to whom people sacrificed black sheep and declared oaths to invoke her. Greek historian Hesiod wrote about the union of Gaia and chaos leading to the birth of Uranus; later Gaia and Uranus birthed Oceanus, the Titans, the Giants, and the rest of the world. Greek goddess Gaia is thus seen as an all-nourishing and all-producing mother. In Roman mythology, she was called Terra or Tellus, personifying the earth. Tellus Mater or Terra Mater means 'mother earth' in Latin. So in the modern period, earth Scientists see the earth as 'a complex living organism' (Madeleine 2019). The Jain religion talks of compassion for all living entities – humans, animals, birds, fish, and plants. Lord Buddha got enlightenment under a boe (peepal – Ficus religiosa) tree. Semitic religions –Christianity, Islam, and Judaism – have relatively less concern for ecology than Hinduism and Jainism. Practising sacredness in various parts of ecology goes beyond the realm of romanticism. First, in ancient India, 'Aranyani', or 'Van Devi' (the goddess of forests), was worshipped with due reverence.

Second, various trees – parts of the forest and symbolising forest too – were worshipped and this tradition of animistic worship still continues – more so in rural areas. For instance, Peepal (Ficus religiosa), Bel or Vilva (Aegle marmelos) bar or Vat or Bargad (Ficus benghalensis), Neem (Azadi rachtaindica), Shami (Acacia ferruginia), Amla, and Banana plants are worshipped as Hindus continue to believe these trees/ plants as the abodes of their different gods and goddesses. Similarly various flowers are presented to these gods or goddesses on different sacred occasions. Various Vedic hymns and cantos express their obeisance and peace to the earth, plants, water, rivers, fire, space, etc.

Third, the gods and goddesses use various animals as their 'Vahans' (transporting assistants); e.g. goddess Durga uses a lion, Lord Ganesha uses a mouse, Lord Vishnu uses an eagle (Garuda), Lord Shiva uses a bull (Nandi), Ma Saraswati uses a goose (Hans), Lord Indra uses an elephant, etc. Among all animals, the cow is worshipped by almost all the Hindus – as most sacred, hence called 'Gomata' (mother cow)!

DOI: 10.4324/9781003336211-3

Fourth, rivers, lakes, and ponds are also worshipped and people salute Ganga, Yamuna, Godavari, Narmada, and other rivers in their daily life. In Sanskrit, there is a saying: '*Gange! tav darshnat muktih*' (O Ganga! one gets salvation just from your view).

Fifth, the earth is called *Dharati Mata* (mother earth) (like 'Gaia' among the Greeks) and one Sanskrit hymn speaks of '*Vasudhaiva Kutumbakam*' (the earth is a family) as the philosophy of the liberal people and 'this is mine and that is yours' (some kind of the 'us' versus 'them') is considered the characteristic of the lowly persons. Thus there are five mothers to many Indians, especially to Hindus' – birth-giving (human) mother', 'mother earth', 'mother cow', 'guru mata' (teacher's wife), and 'mother tongue' (matribhasa).

Sixth, even some of the ten directions ('*Dishas*') are considered sacred for various purposes like worship, opening for the main gate of the house or temple, Yajna, marriage rites – especially North-East, East, and North are regarded highly sacred and auspicious (in that order) from the viewpoint of Hindu Vaastu!

Finally, even *Shanti Mantra* (peace hymn) is concerned with fire, air, water, space, plants, earth, herbs, gods, etc.

Thus the 'social ecological' perspective is ingrained in Indian culture. Though there used to be the sacrifice of animals at the altar of some gods and goddesses (like Kali) among Hindus, yet it has declined considerably and new practices are as follows:

(a) To offer the fruits/vegetables like a gourd, pumpkin, lemon, or coconut with vermillion, red small scarf
(b) To offer a he-goat symbolically by slightly cutting a part of his one ear to take a few drops of blood in the name of goddess Kali and leaving the goat scot free
(c) To sacrifice the he-goat actually and eating his meat as '*prasadam*' by a minority of some people – as Maithils do in North Bihar

History shows that as soon as people became more educated and started settled agriculture, they distanced themselves from the forests both physically and socially. In addition, such people also began condemning the tribals (and other forest dwellers) who had usually their temporary Kachcha houses (built from wood, mud mortar, etc.) in the midst of the forests. The most celebrated Indus Valley civilisation, wherein traces of various animals like tigers, rhinoceros, elephants, and bullocks were found as per its seals, collapsed mainly due to environmental degradation and climate change. Many other civilisations and established cities/towns, too, collapsed due to climate change, environmental degradation, drought, or tsunami; this may be seen in Table 3.1.

Further in ancient India, we clearly find in the Tamil Sangam literature (during the start of the common era, CE) five ecozones, signifying ecological differences or varieties, and these were linked to culture in some way or other – littoral, wetland, pastoral tracks, dry zone, and hilly backwoods. Such ecozones' association with various occupations is quite notable (Thapar 2007), as seen in Table 3.2.

In other parts of India, too, there have been different categories of landscape, fit for different occupations.

On the other hand, various Sanskrit texts like *Vedas* distinguish between '*grama*' ('settlement') and '*aranya*' (forests). Grama is shown as an orderly, disciplined,

92 Sustainable Development

Table 3.1 Collapse of Civilisations/Cities due to Climate Change/Environmental Degradation

Sl. No.	Climate Change/Environmental Degradation	Death/Collapse of Civilisations/Cities
1.	Decreasing soil fertility causing food scarcity (4500–1900 BCE)	Sumer (Mesopotamia, now in Iraq and Kuwait) civilisation (Chaleolithic and early Bronze Age)
2.	Repeated floods, earthquakes, change of course of Indus river/climate change (3300–1300 BCE)	Indus Valley Civilisation (Bronze Age civilisation)
3.	Recurring droughts and volcanoes causing mass out-migration (3000–1900 BCE)	Disappearance of Saraswati river (from Haryana to Allahabad, UP)
4.	Volcano eruption for two days on Mount Vesuvius in 79 CE and leaking of carbon dioxide from earth	Towns of Pompeii and Herculaneum destroyed – thousands of people buried; after 1,500 years excavated and now a UNESCO World Heritage site
5.	Tsunami and earthquake in Helike (Greece) in 373 BCE	Destroyed and submerged the city-State of Helike almost fully, rediscovered in 2001
6.	Five million years ago Mediterranean sea was sealed off from the Atlantic ocean – rate of evaporation exceeded the incoming flow of waters	Mediterranean sea dried up
7.	Drought in old Egypt – loss of crops, repeated larger sand storms, and famine	People and pyramids wiped out in Egypt
8.	Drought and climate change	Collapse of Bronze age empires, Hittite empire, early Greek city-States
9.	Drought and steady climate change (800–1000 CE)	Fertile crescent in the Middle East (between Tigris and Euphrates rivers) lost
10.	Mega-drought (one of the main reasons)	Maya Civilisation (Central America and South Mexico) collapsed
11.	Wasteful irrigation methods, subsidy, and over-production	Drying of Aral Sea (a big lake in Central Asia)

Source: Authors.

Table 3.2 Different Occupations in Various Ecozones in Tamil Nadu (India)

Sl. No.	Eco Zone	Occupations
1.	Coastal area	Fishing, making of salt
2.	Wetlands	Rice growing
3.	Pastoral tracts	Animal husbandry and shifting cultivation
4.	Dry zone	Animal husbandry
5.	Backwoods	Hunting and gathering

Source: Prepared after R. Thapar (2007).

known, predictable, and civilised location; here Vedic rituals were performed, agriculture was practised, the exchange of commodities (barter) took place, government functionaries worked, and arts and cultures of the elites were performed and practised. On the contrary, *aranya* was considered disorderly, unknown, unpredictable, and inhabited by predators and strange persons, usually thought to be 'savage'

Table 3.3 Contrasting Perceptions of People Residing in Forests and Settlements

Sl. No.	Aspects	Aranya (Forests)	Grama (Settlements)
1.	Orderliness	Disorderly	Orderly
2.	Discipline	Indisciplined	Disciplined
3.	Knowingness	Unknown	Known
4.	Occupation	Gathering fruits, hunting wild animals	Agriculture and trade
5.	Predictability	Unpredictable	Predictable
6.	Civilisation	Inhabited by the savage and strangers	Inhabited by the civilised people
7.	Power and status	Powerless and lower in social hierarchy	Powerful (by controlling nature) and higher in social hierarchy
8.	View of location	Place of darkness with spirits	Place for performing Vedic rituals/ sacrifices

Source: Prepared after Thapar (2007).

(Thapar 2007). Such confrontational perceptions of the *grama* and *aranya* may be seen in Table 3.3.

Thus it is crystal clear that for the *grama*, the forest was needed to be tamed and controlled, while for the tribals/forest dwellers it has been a natural habitat. However, as an exception, some selected trees/mangroves are worshipped as abodes of some gods or goddesses. Further sacred mangroves were maintained by three categories:

(a) By the people of a city (e.g. Vaishali, Kushinara, Champa)
(b) Others by a monastery
(c) Still others by the wider community (who lived on the edge of a forest)

In India, there has existed another dichotomy between *Prakriti* (nature) and *sanskriti* (culture) – the letter created by humans, hence artificial. Thus forest is natural, while settlement is artificially created (Thapar 2007); consequently, Sanskriti was almost equalled with civilisation (material culture). During the colonial period in India, the forests were cut for railway sleepers and ships in a large number. Similarly for canals and canal colonies, too, various forests were cleared. Hence the holistic understanding of nature and culture was fully lost during the colonial rule due to the British attempts to take maximum revenue out of forest:

> In the formidable alliance of the politician, the contractor, the bureaucrat and the industrialist, there is little hope for the forest and the people of the forest, for they are not permitted even to denounce, let alone stop, the desecration.
>
> (Thapar 2007, 41)

However, there has also existed historically an alternative narrative of forests in India. For instance, '*Aranyani*' (the goddess of forests) was worshipped with full reverence.

Further, in the ancient period in India, there was also an established tradition of practical penance by the learned *Rishis/Munis* who were living in 'Ashramas' in the midst of forests and doing their penance (e.g. Kanva Rishi, Valmiki Rishi, Vishwamitra); the kings and princes of that period used to visit them with high humility to gain specialised knowledge skill or training as well as to get boons for their personal life and/or for the polity as a whole. In addition, some Rishis used to run highly professional '*gurukulams*' (residential schools) to teach and train the selected pupils in different fields of knowledge and skills. Thus we are of the firm view that forest (*aranya*) was not viewed always, and everywhere as a negative force or inferior natural system.

Interestingly the worship of a tree is generally associated with the cult of fertility, especially among pastoralists, peasants, tribals, and also some urban peoples of lower order. Sociologically speaking, at the beginning of human society, the first religion was animistic religion wherein people were worshipping plants and animals – as was practised by the people in the Rigvedic period too and is still practised by many tribal communities; e.g. the tribals of Jharkhand worship 'Sarna' (sacred groves) even today and have '*gotras*' (clans) named after various animals (e.g. Kachchhap – tortoise – is one of many clans). This signifies some kind of integration with nature – showing respect to, not domination over, nature. In Jharkhand State, many tribal villages were also named after plants and animals; for instance, in the erstwhile Palamu district (now Medininagar) one village is named 'Kunjar' (elephant) and there used to be elephants. It is notable that Hanuman (monkey god), the special devotee of Lord Rama, is worshipped by many Hindus – thus the animistic feature of tribals is accommodated here too. The above-mentioned narrative indicates holistic sustainability in India. After the 4th century CE, in the Gupta period, the forest resources of tribals and other forest dwellers were appropriated and they were made marginalised communities. Needless to say that with the growth of population, trade and commerce forests were cut down; similarly for *ashramas* and monasteries, too, forests were cleared. Even globally humans played havoc with the ecology in different ways, especially the empires (in the name of 'scientific management of forests' in 1864 in India). As Fairfield Osborn (1948) in his book 'Our Plundered Planet' recorded how the 'cradle of civilisation' in the Middle East became a desert with the passage of time, how Greece and Turkey were deforested, and how, in the recent past, the destruction of the American prairies led to the Dust Bowl.

Concept of Sustainable Development

In 1972, the UN Conference on Human Environment was held in Stockholm (Sweden) wherein heads of 113 countries and over 250 (green) NGOs participated. Shrimati Indira Gandhi, the then-Prime Minister of India, too, participated there and observed that poverty is the biggest pollutant in developing countries like India. Hence development is the need of the hour. Then the UN General Assembly decided to establish the UN Environment Programme (UNEP). This was the first occasion when global environmental problems were addressed at an inter-governmental forum for initiating corrective measures globally. This conference sensitised the scientists, politicians, bureaucrats, NGOs, and people, in general, towards various environmental problems. Later, in 1983, the UN General Assembly decided to establish World Commission on Environment and Development (WCED) under the chairpersonship of Gro Harlem

Brundtland (the then-Prime Minister of Norway). The task assigned to WCED was to chalk out the strategies, devices, and action plans as to how economic development and environmental safety could be pursued simultaneously, nationally, and globally. WCED came out with a detailed report called 'Our Common Future' (1987) with its central idea that a high level of economic development should be framed fully compatible with more environmental safety. Why was it the intent of the Brundtland Commission? The following reasons perhaps were duly considered for taking this stand:

(a) Since poverty is usually considered as causing various global environmental problems, it was accordingly assumed that poverty may be reduced only through economic development that, in turn, may solve the environmental problems.
(b) The developing countries had pointed out during some international conferences (1971 and 1972) and at a session of the UN General Assembly (1983) about their primary concern for eliminating national poverty (environmental pollution and degradation were considered a small price for more economic prosperity), and hence not addressing the major issue of development would have resulted into non-cooperation from the developing countries.
(c) Most of the developed countries were used to a high level of standard of living and the massive use of energy per head in their personal and professional lives, and hence they were not willing, personally, collectively, professionally, and nationally, to cut down the rate of growth (production) and consumption in real sense.

The Brundtland Commission's Report 'Our Common Future' (1987), that indulged in a 900-day international exercise and analysed and synthesised various views/suggestions from government representatives, scientists, industrialists, NGOs, and the common people the world over, had the following broad scope, and contents, covering three major arenas: Common concerns, common challenges, and common endeavours (Table 3.4):

Hence WCED asked for a 'global agenda for change', as the UN General Assembly had called for:

(i) Proposing long-term environmental strategies for achieving sustainable development by the year 2000 and beyond
(ii) Recommending ways and concerns for greater cooperation among developing countries and between countries at different stages of economic and social development, and leading to the achievement of common and mutually supportive objectives taking account of interrelationships between people, resources, environment, and development
(iii) Considering ways and means by which the international community can deal more effectively with environmental concerns
(iv) Helping define long-term environmental issues and appropriate efforts for protecting and enhancing the environment and aspirational goals for the world community

Thus 'Our Common Future' duly recognised that human resource development in the form of poverty reduction, gender equity, and wealth redistribution was very crucial for making strategies for ecological conservation. On the other hand, it, too, recognised

96 Sustainable Development

Table 3.4 Broad Scope and Contents of Brundtland Report (UN)

Broad Scope	Contents	
Part I. Common concerns	1.	A threatened future
	2.	Towards sustainable development
	3.	The role of the international economy
Part II. Common challenges	4.	Population and human resources
	5.	Food security: Sustaining the potential
	6.	Species and ecosystems: Resources for development
	7.	Energy: Choices for environment and development
	8.	Industry: Producing more with less
	9.	The urban challenge
Part III. Common endeavours	10.	Managing the commons
	11.	Peace, security, development and the environment
	12.	Towards common action: Proposals for institutional and legal change

Source: WCED, 'Our Common Future' (1987).

the environmental limits to growth. Thus it tried to suggest various ways and means for the proper balancing between the economy and ecology, and hence it used the term 'sustainable' (to symbolise the environment) as an adjective before the term 'development' (noun). This is its strength (less in the real sense) as well as limit or weakness (more in the real sense).

Brundtland Commission Report defined sustainable development (SD) in the following words: 'to meet the needs of the present (generation) without compromising the ability of the future generations to meet their own needs'. However, in this brief definition, all the salient features and principles of sustainable development are not covered. Hence further clarification is very much required. However, at first, it is clarified that WCED first met in October 1984 and published its report ('Our Common Future') after 900 days, later in April 1987. It recognised the inseparability of the environment ('where we live') and development (what we all do to improve our life), and also emphasised three components of SD: Economic growth, environmental stewardship (need-based prudent use of natural resources), and social inclusion (intra- and inter-generational equity).

During 900 days of WCED's deliberations following six major international incidents, with wider environmental consequences, took place:

(a) Due to the drought crisis in Africa, 36 million people were affected and about 1 million were killed.
(b) The Bhopal gas tragedy occurred in December 1984 due to gas leakage from Union Carbide Co. (pesticide factory), killing several thousands of people (in total 16,000) and it also blinded/injured or affected 6,00,000 persons.
(c) Liquid gas tanks exploded in Mexico city, killing 1,000 persons and thousands became homeless.
(d) Chernobyl nuclear reactor explosion in April 1986 in the former USSR (Ukraine) had fall-outs in other countries, especially Europe, increasing the risks of future human cancers.

(e) Agricultural chemicals, solvents, and mercury flowed into the Rhine river during a warehouse fire in Switzerland, killing millions of fish, and threatening drinking water in Federal Republic of Germany and the Netherlands.
(f) An estimated 60 million people died of diarrhoeal diseases related to unsafe drinking water and malnutrition – most of the victims were children.

WCED had also noted that during 1970s twice as many people suffered annually from 'natural' disasters as during 1960s – mostly droughts and floods; first, 18.5 million people were affected by drought annually in 1960s, and 24.4 million were affected in 1970s; second, from floods, 5.2 million were affected annually in 1960s and 15.4 million were affected in 1970s; third, the number of victims of cyclones and earthquakes, too, increased substantially as most of the poor people built unsafe houses; fourth, globally military expenditures was about $1 trillion a year and the arms race 'preempts resources that might be used more productively to diminish the security threats created by environmental conflict and the resentments that are fuelled by widespread poverty'; fifth, 'a world in which poverty is endemic, will always be prone to ecological and other catastrophes'. Later, the United Nations Development Programme (UNDP) in Human Development Report 2003 defined 'environmental sustainability' as 'achieving sustainable development patterns and preserving the productive capacity of natural ecosystems for future generations' (Human Development Report, 2003, 123). Peter Bakker (2021), President and CEO of the World Business Council for Sustainable Development (WBCSD), goes beyond the optimisation of financial capital, to nature and people also; businesses should be held accountable for both financial and non-financial results by taking four steps:

(a) Every company should integrate environmental and social risks into its financial risk management system.
(b) How decisions are made – to see these risks or to see opportunities for new types of solutions.
(c) To disclose its results transparently (by including environmental, social, and governance into the accounting rules).
(d) ESG's performance disclosure will change capital allocation models, and evaluation models used by investors to select an investment.

In his words, 'Products should be circular by design. The flow of waste into the environment is ended and nature is restored'. That is, the system of production should also ensure the regeneration of resources, as only 17% of e-waste globally is being collected. The World Circular Economy Forum is taking initiatives in this direction. However, even this does not question and come out of the exploitative world capitalist system.

Principles of Sustainable Development (SD)

To our understanding, SD has ten principles for its operationalisation:

(1) There is no watertight option between environment and development – rather such a form of development wherein the future generations' ability to meet their

needs is not reduced/compromised. Thus *inter-generational equity* is focussed. This motivates the present generation to use natural resources prudently.

(2) The *needs* of the present and future generations are emphasised – that is, *need, not greed*, should be the ultimate criterion for the use of various natural resources. This implies 'limitations' imposed by the present technology and social organisation on the environmental resources, on the one hand, and 'the ability of the biosphere to absorb the effects of human activities', on the other. In WCED's words, 'the Earth is one but the world is not'!

(3) Widespread *poverty is not inevitable*; it is usually prone to 'ecological and other catastrophes', hence to be tackled with a new kind of development that addresses the issue of poverty directly. Further food security should address the questions of distribution because hunger arises often from a lack of purchasing power, not from the unavailability of food. Land reforms and integrated rural development will increase the purchasing power of, and work opportunities for, vulnerable groups.

(4) SD is 'not a fixed state of harmony', but rather a '*process of change* in which the exploration of resources, the direction of investments, the orientations of technological development, and institutional change are made consistent with future as well as present needs'. On the one hand, the diversity of species is necessary for the normal functioning of ecosystems and the biosphere as a whole, but, on the other hand, 'there are also moral, ethical, cultural, aesthetic, and purely scientific reasons for conserving wild beings'.

(5) Ultimately SD must rest on a strong '*political will*' – that is, unless the local, State, or national government has a strong will power to change the nature and direction of the development (or rather the economy), it would not be sustainable in the long run; the political system should ensure that the poor get a fair share of the resources required to sustain the growth (i.e. equity be ensured).

(6) All systems of production and business firms should adopt *better and safer environmental management processes and procedures* and should not threat the lands, rivers, seas/oceans, or atmosphere as a free dumping ground for the waste materials; the need for informational cooperation to manage economic and ecological interdependence, and within nations Environment Ministry and environmental agencies be given more power to cope with the effects of unsustainable development. There are thus no technical or military solutions to 'environmental insecurity'.

(7) In future, people should prioritise *quality of life*, not the higher material standard of living, measured mainly in money or energy terms; hence, affluent people should adopt life style 'within the planet's ecological means', and there is also a need for population stabilisation for environmental sustainability.

(8) There should also be intra-generational equity (inclusion) – that is, in the present generation all social classes/castes/races/genders should have equal access and affordability to use natural resources.

(9) To ensure efficiency (more output per unit of input) by saving economic costs, labour materials, energy and/or time in the processes of production, distribution, and consumption. A safe, environmentally sound, and economically viable energy pathway is imperative, and developing countries will need assistance for this.

Sustainable Development   99

(10) To ensure techniques of the seven Rs: Reduce, replace, recover, reuse, recycle, remediate, and restore:

   (a) *Reduce* – to add some equipment/device in the production/distribution operation/ consumption system(s) to reduce harmful wastes (e.g. pollution due to carbon emission is to be reduced)

   (b) *Replace* – replace/substitute a harmful input/output with an unharmful/less harmful one – cotton/paper bag in place of a plastic bag.

   (c) *Recover* – convert wastes into resources (heat, electricity, compost, fuel, etc.) through thermal or biological means.

   (d) *Reuse* – one item/product may be used again without changing its form e.g. tooth brush, after using for six months or so, may be reused for cleaning a small box, drawer, window grills, etc.

   (e) *Recycle* – used items are collected, reprocessed/recycled as new products – saving materials, money, and energy.

   (f) *Remediate* – to clean the effects of pollution on the environment-remediation technique is used to clean the contaminated soil/water due to chemical wastes/ oil, etc.

   (g) *Restore* – to return structure, function, or particular organisms of a degraded landscape, forest mountain slope (landslide), or wetland to its original state. It is a costly, time-taking, and labour-intensive technique.

Undoubtedly, these seven Rs techniques are quite useful in our daily life, but these demand a new thought process, new technology, finance, time, labour, and other inputs. As the famous scientist Albert Einstein rightly remarked: 'We cannot solve our problems with the same thinking we used when we created them'.

Rio Declaration (1992)

Later, in the same direction, the UN Conference on Environment and Development (UNCED), known as Earth Summit, was held on 5–18 June 1992 in Rio de Janeiro. It was attended by 178 nations, with 118 heads of State or government, 8,000 official delegates, about 1,400 NGOs represented by 3,000 observers, 9,000 journalists, and about 15,000–20,000 visitors. Both the developed North and under-developed South raised various issues, often conflicting ones. For instance, Malaysia asked the former, especially the USA: How should the costs of environmental restoration and future protection in the developing world be shared? Who was more to be blamed for the rise in pollution and drawdown of natural resources: The rich nations because of their profligate consumption? Or the poorer nations because of their population explosion? Which environmental issues deserved immediate priority: Global environmental issues such as climate change and biodiversity? Or livelihood issues such as access to freshwater, desertification, and food security? Thus economic and political contradictions concerning the environment were deliberated there, and the final reply to these difficult questions was obviously 'all of the above'. Rio Declaration finally elaborated 27 principles, as guidelines (not binding) to govern the future environmental decision-making, based on various political–economic compromises among nations of competing positions and interests – North and South were competing for dominance. Out of these 27 principles, 11 more significant may be seen in Table 3.5.

100 Sustainable Development

Table 3.5 Main Selected Principles of Rio Declaration (1992)

Principle No.	Details of Selected Principles
1.	Human beings at the centre of concerns for SD; 'entitled to a healthy and productive life in harmony with nature'.
2.	States have 'the sovereign right to exploit their own resources pursuant to their non-environmental and developmental policies', but to ensure that their activities do not cause damage to the environment of other States.
4.	In order to achieve SD, 'environmental protection shall constitute an integral part of the development process, and cannot be considered in isolation from it'.
7.	State shall cooperate in the spirit of global partnership to conserve, protect, and restore the health and integrity of the earth's ecosystem. In view of the different contributions to global environmental degradation, States have *common but differentiated responsibilities*. Developed countries should acknowledge this responsibility.
10.	Environmental issues are best handled with the participation of all concerned citizens, at the relevant level. States shall facilitate and encourage public awareness and participation by making information widely available. Effective access to judicial and administrative proceedings, including redress and remedy, shall be provided.
14.	States should effectively cooperate to discourage or prevent the relocation and transfer of other States of any activities and substances that cause severe environmental degradation or are found to be harmful to human health.
15.	The precautionary approach shall be widely applied by States according to their capabilities. Where there are threats of serious or irreversible damage, lack of full scientific certainty shall not be used as a reason for postponing cost-effective measures to prevent environmental degradation.
16.	National authorities should endeavour to promote the internalisation of environmental costs and the use of economic instruments, taking into account the approach that the polluter should, in principle, bear the cost of pollution, with due regard to the public interest and without distorting international trade and investment.
17.	Environmental impact assessment, as a national instrument, shall be undertaken for proposed activities that are likely to have a significant adverse impact on the environment and are subject to a decision of a competent national authority.
20.	Women have a vital role in environmental management and development. Their full participation is, therefore, essential to achieve SD.
22.	Indigenous people/communities and other local communities have a vital role in environmental management and development because of their knowledge and traditional practices. States should recognise and duly support their identity, culture, and interest and enable their effective participation in the achievement of SD.

Source: UNCED (1992).

Rio declaration also approved 'Agenda 21', a detailed non-binding blueprint (115 programmes are as in 800 pages) for putting sustainable development into practice (Box 3.1).

Box 3.1: Broad Areas of Agenda 21 (UN)

Agenda 21 has four broad areas:

1. **Social and economic development:** Highlighting international cooperation and assistance, poverty, reduction, over-consumption population trends, health, human settlements, and policy-making for SD.

2. **Conservation and management of resources for development:** Addressing issues of energy use, integrated land resource use, deforestation, desertification and drought, mountain ecosystems, agricultural needs and rural development, biodiversity, biotechnology, oceans, fresh waters, toxic chemicals, and hazardous and radioactive wastes.
3. **Strengthening the role of major groups:** Focussing on actors other than the government: Women, youth, indigenous peoples, NGOs, business and industry, scientists, communities, workers, trade unions, and farmers.
4. **Means of implementation:** Addressing how international and national support should be organised including a transfer to South of financial resources and environment-friendly technology, building capacity through technical assistance, environmental education and scientific information, creating better environmental databases to bridge the data gaps between nations, and improving international environmental organisations, coordination, and legal processes.

Source: Speth and Haas (2006)

Two trends were notable there (Speth and Haas 2006): (a) Linking of poverty alleviation and official development assistance (ODA) with a goal of environmental protection – Earth Summit argued for doubling of ODA by developed North to South to support the implementation of Agenda 21; and (b) linking of the local and the global–local actions in different countries are to be supported by international organisations, governments, and businesses, as a part of a collective response.

Criticism of Sustainable Development

Following criticisms are made against the concept of sustainable development:

(a) It stands for an incremental, formal, functional, and reformative change only; it does not stand for fundamental/structural change of the economy – i.e. capitalism which is the root cause of various social, economic, and ecological problems locally, nationally, and globally. It criticises industrialism, but not the capitalism of which industrialism is an off-shoot.

(b) It is an attempt to offer a 'metafix' (Lele's term cited in Cohen & Kennedy 2000) trying to unite green activists, conservationists, poor small farmers in developing countries, development-oriented governments, and large corporations. Hence it fails to provide a substantive solution to any group. The proponents of SD claim to reduce poverty by reducing environmental problems through more economic growth, but it has not happened historically in most of the developing countries. In fact, this model is silent on the introduction of a new policy for the redistribution of resources, income, etc. in favour of the deprived people.

(c) The proponents of SD have created a new breed of elite global 'eco-crats' (Sachs' term 1993) who have hijacked the green agenda from the more radical groups. Such eco-crats do not regard the biosphere as a fragile heritage, to be protected for future generations, but rather simply as a 'commercial asset in danger'.

102   Sustainable Development

(d) It does not strongly attack the culture of consumerism (promoting more use of energy per capita) that is prevalent in developed countries, not to say of the roots of the environmental problems lying in the capitalism.

(e) Sustainable Development Goal (SDG) 15 talks of sustainably managing forests, combating desertification, *halting and reversing land degradation*, etc. – more to improve the quality of land (from degraded status to upgraded status). However, in most of the developing countries like India, land issue has the following multiple dimensions that need to be taken into account for sustainable development :

   (i) Status of land as a source of livelihood varies according to the patterns of use, possession, and ownership, e.g. a peasant, an absentee landlord (land as a status symbol), a rich farmer (using paid labour), a share cropper, a contract farmer, and a poor labourer have different types of interest in a piece of land – hence the nature, extent, and direction of the improvement of degraded land depend substantially on this real aspect.

   (ii) A poor peasant, especially a female peasant, often sees a small piece of land as social security for his/her daughter's marriage, and hence the degraded land is sold on a priority basis (instead of investment for its upgradation).

   (iii) Land also symbolises the culture of a community, especially the tribals; e.g. '*Sarna*' (Sacred groves in Jharkhand) land is collectively owned and local people do not interfere in changing its status, even if it has lost all the trees/groves (that stood on it earlier).

   (iv) Among the poor peasants, the land is also considered as an 'emergency security' during calamities, e.g. floods, drought, fire, cyclone, and hence they do not want to invest in land upgradation.

However, SD does not take into account these four aspects of land in developing countries like India.

Thus, the major criticism of the SD may be seen in Table 3.6.

Table 3.6 Criticism of Sustainable Development

Aspects	Specific Features
a) Kind of change pursued	a) Only for incremental, formal, functional, and reformative change, not for fundamental/structural change
b) Stakeholders	b) 'Meta fix' – loosely encompassing the opposite camps like green activists, conservationists, poor small peasants, governments and large corporations – no ideological unity among opponent stakeholders
c) Proponents	c) Elite eco-crats hijacking the agenda from the more radical ecological groups
d) Views on consumerism and economy	d) Does not question the prevailing culture of consumerism and capitalist system of economy
e) Views on land	e) Does not see multiple aspects of land like culture, gender, terms of practical use, security in an emergency, but rather sees land's physical degradation aspect only

Source: Authors.

Perspectives on Sustainable Development

Two scholars J. Clapp and P. Danvergne (2011) have analysed following four perspectives or worldviews on the interactions between environment, social, and economic processes in a societal context:

(a) Bio-environmentalists
(b) Institutionalists
(c) Market liberals
(d) Social greens

Bio-environmentalists

According to them, human beings are self-centred and dominate over other species of the earth. They are also of the view that humans use, consume, and exploit various natural resources like water, land, forests, minerals much more than the 'carrying capacity' of the earth – beyond its capacity of self-regeneration and self-renewal. Hence three factors are leading to global environmental crises: (i) Infinite economic growth, (ii) the culture of consumerism, and (iii) high population growth. These three factors will ultimately lead to the scarcity of natural resources, at least to be faced by the next generations, resulting in more disasters (both natural and man-made), more socio-economic and ecological conflicts, and finally to more wars between different nations. It is predicted that the Third World War will take place over the issue of critical natural resources like oil or water. World Wide Fund for Nature (WWF), World Watch Institute (Washington DC), Thomas Robert Malthus ('Principle of Population'), biologist Paul Ehrlich ('Population Bomb'), and Herman Edward Daly (ecological economist in the USA who talked of 'steady state economy', by imposing permanent government restrictions on all resource use) represent this group.

Market Liberals

They mostly hail from neo-classical/neo-liberal economics and believe in a free market which will create competition and equal opportunity, leading to the reasonable allocation of resources to one and all. According to them, environmental degradation and other problems are caused by the restrictions imposed on the market, and hence there is market failure and policy failure (e.g. no clear property or access rules or no strong institutions to enforce them). They do not recognise major structural environmental problems caused by the prevailing economy; on the other hand, they also think that natural resources are not a constraint for economic development. Nevertheless, environmental pollution, poverty, and social inequality are just temporary problems, according to them, and could be duly tackled in the long run. They obviously do not recognise the culture of consumerism and high growth for mass production as problems. Various international organisations like World Bank, World Trade Organisation (WTO), World Business Council for Sustainable Development (WBCSD), the US magazine, The Economist, and World Economic Forum (WEF) usually represent this group.

Institutionalists

They primarily hail from the fields of political science and international relations and believe that unsustainability often comes from the lack of organisation or from institutions which do not support genuinely sustainable choices. They are concerned with both the scarcity of natural resources and social inequality, but they do not question and reject the discriminatory global economic and political system. However, they do recognise that the States need strong institutions to cross the boundary of self-interest and opt for collective interests/norms. For instance, there should be strong rules and regulations for the use of commons (grazing land, ponds for fishery, forests, etc.). They give due significance to economic growth, technology, international trade, and foreign investment. UNEP (United Nations Environment Programme), Gro Harlem Brundtland (the then-Chairperson of the World Commission for Environment and Development), and Maurice Strong (Canadian diplomat and founder of Earth Charter) represent this group.

Social Greens

They point out that the root cause of social and environmental problems lies in the inequality within (class inequality) and between countries (North versus South due to unequal ecological exchange and adverse terms of trade against the developing countries). In other words, the rich people dominate over the poor people due to unequal access to natural resources (the poor have less access to these resources), and unequal exposure to environmental pollution (the poor have more exposure to pollution). Like bio-environmentalists, the radical social greens, too, think of the existence of global ecological crises, but while the bio-environmentalists consider the faster population growth as its primary cause, the radical greens, on the other hand, consider the over-consumption (high rate of per capita energy consumption) in the developed countries, as well as the physical limits to growth all over the world, not the high population growth, are its primary cause. Thus, the radical social greens consider the environmental resources as a matter of fairness and equity within and between countries because the poor people have a small ecological footprint *vis-a-vis* the large ecological footprint of the rich people. Thus these radical social greens are of the view that the just and fair distribution of assets will certainly reduce the population growth rate, especially in developing countries. Thus these radical social greens, influenced by the dialectical materialist theory of Karl Marx (eco-Marxists), blame capitalism and its offshoots like trans-national corporations (TNCs) which have a monopoly over the capital, hence political power, too, over the poor social classes and poor developing countries. Radical journal *The Ecologist*, International Forum on Globalisation (IFG), Third World Network (TWN), and World Social Forum (WSF) represent this radical group who finally put forward an alternative development paradigm – thus 'an alternative world is possible'.

Thus, four perspectives on interactions between social, economic, and environmental processes may be seen in Table 3.7.

After deliberating about the four perspectives on social, economic, and environmental processes, it is now imperative to discuss the holistic sustainability framework (HSF), as we propose here.

Table 3.7 Four Perspectives on Interactions Between Social, Economic, and Environmental Processes

Sl. No.	Perspectives	Interaction Processes					
		Earth's Physical– Biological Limits	Humans as Self-centred or for Collective Norms	View on Social Inequality/ Poverty	Fast Population Growth	Scarcity of Resources	Proponents/ Supporters
1.	Bio-environmentalists	Yes	Self-centred humans	Not concerned	Major problem	Yes, due to population growth	WWF, World Watch Institute, P. Ehrlich, Thomas R. Malthus, H. E. Daly
2.	Market liberals	No	Lacking clear property/ access rules	Poverty and inequality cause environmental degradation in short run but removed in long run through free market correction	A part problem	Not concerned	World Bank, WTO, WBCSD, The Economist, World Economic Forum
3.	Institutionalists	Yes	Strong institutions needed to stick to collective norms	Much concerned	A part problem	Much concerned	UNEP, Gro Harlem Brundtland (Chair of WCED), Maurice Strong (Earth Charter)
4.	Social greens	Yes	The rich exploit the poor due to unequal access to natural resources	Yes, inequality within and between countries – domination of the poor by the rich	Not a problem, a just and fair redistribution of assets will reduce population growth	Over-consumption stresses global environment	John B. Foster and other eco-Marxists, The Ecologist, IFG, TWN, WSF

Source: Prepared after J. Clapp and P. Danvergne (2011).

106 Sustainable Development

Table 3.8 Multiple Conditions for Holistic Sustainability Framework

Social Conditions	Economic Conditions	Political Conditions	Environmental Conditions
1. Intra-generational equity 2. Inter-generational equity 3. Quality education for all 4. Health for all 5. Cultural diversity	1. Food security 2. Adequate livelihood for all 3. Economy to adjust to ecology 4. Self-reliance 5. Priority on agricultural sustainability 6. Green budget	1. Freedom from oppression 2. Freedom of participation in public affairs 3. Pro-people good governance 4. Eco-democracy (earth democracy) 5. Pro-active ombudsman (Lokpal)	1. Ecosystem's inter-dependence between humans and non-humans 2. Biodiversity 3. Clean environment for all 4. Optimal use of renewable green energy (in lieu of fossil fuel) 5. Need-based prudent use of natural resources (green transport, green park, green housing, green/organic fertilisers)

Source: Authors.

Holistic Sustainability Framework

To our understanding, there are four types of conditions necessary for a holistic sustainability framework (HSF):

(a) *Social conditions* – intra-generational equity, inter-generational equity, quality education for all, health for all, and cultural diversity
(b) *Economic conditions* – food security, adequate livelihood to all people, economy to adjust with ecology, self-reliance, priority on agricultural sustainability, and green budget
(c) *Political conditions* – freedom from oppression, freedom of participation in public affairs, pro-people good governance, eco-democracy, pro-active ombudsman
(d) *Environmental conditions* – ecosystem's interdependence between all human–non-human entities, biodiversity, clean environment, optimal use of renewable energy, need-based prudent use of natural resources

This may be seen in Table 3.8.

Millennium Development Goals

As mentioned earlier, most of the developing countries, including India, have been reiterating since the UN Conference held in Stockholm in 1972 that poverty is one of the major causes of environmental degradation in the global South. Hence, before the next millennium, eight millennium development goals (MDGs) were declared collectively

Sustainable Development 107

Table 3.9 Millennium Development Goals (MDGs) (UN)

Sl. No.	MDGs	Target
1.	Eradicate extreme hunger and poverty	Halving the proportion of people living on less than $1 daily and halving malnutrition
2.	Achieve universal primary education	Ensuring that all children are able to complete primary education
3.	Promote gender equality and empower women	Eliminating gender disparity in primary and secondary schooling, preferably by 2005, and not later than 2015
4.	Reduce child mortality	Cutting under-five death rate by two-thirds
5.	Improve maternal health	Reducing the maternal mortality rate by three-fourth
6.	Combat HIV/AIDS, Malaria, and other diseases	Halting and beginning to reverse HIV/AIDS and other killer diseases
7.	Ensure environmental stability	Cutting by half the proportion of people without sustainable access to safe drinking water and sanitation
8.	Develop a global partnership for development	Reforming aid and trade, with special treatment for the poorest countries

Source: Prepared after UN MDGs.

by 189 countries of the world under the aegis of the United Nations in the Millennium Summit in September 2000. These goals (basically human rights) were targeted to be realised by all the developing countries by 2015. These goals and their targets are mentioned in Table 3.9 before discussing in detail, the SDGs for a chronological understanding of the major issues:

Thus it is crystal clear that MDG 1 is directly related to eradicating extreme hunger and poverty. MDG 7 is for ensuring environmental stability by reducing by half the proportion of people without sustainable access to safe drinking water and sanitation. MDGs 2 and 3 are related to ensuring universal primary education and also secondary education with gender equity respectively. MDGs 4, 5, and 6 are related to improving health and removing diseases. Finally, MDG 8 is for a global partnership for development, by reforming development aid and trade (as the terms of world trade are against the interests of the developing countries). However, most of these MDGs and their associated targets could not be achieved by 2015. Hence, the United Nations adopted on 25 September 2015 various sustainable development goals (SDGs) for *all countries* of North and South to be achieved by 2030.

Sustainable Development Goals (SDGs)

If we look at the comparison between MDGs and SDGs, the first major difference was that all the 8 MDGs were meant for the developing countries of the South only, while all the 17 SDGs are meant for all the countries of the North and South.

Second, while 8 MDGs were sectorally targeted to only social (health and education), economic (hunger and poverty), environmental (drinking water and sanitation), and development aid and trade specifically, on the other hand, the SDGs are holistic with 17 goals, 169 targets, and 300 plus indicators!

108   Sustainable Development

Third, in SDGs for the first time new goals and associated targets were fixed which had not been included in the earlier MDGs. For instance, sustainable agriculture (Goal 2), affordable, reliable, sustainable, and modern energy to all (Goal 7), inclusive and sustainable economic growth, employment and decent work for all (Goal 8), promoting sustainable industrialisation (Goal 9), reducing inequality within and among countries (Goal 10), making cities inclusive, safe, resilient, and sustainable (Goal 11), ensuring sustainable production and consumption patterns (Goal 12), combating climate change and its impact (Goal 13), conserving and sustainably using oceans, seas, and marine resources (Goal 14), sustainably managing forests, combating desertification, halt and reverse land degradation, and halt biodiversity loss (Goal 15), promoting just, peaceful, and inclusive societies (Goal 16), and finally revitalising the global partnership for sustainable development (Goal 17).

Fourth, in MDGs poverty was defined as living on less than $1 a day, while in SDGs it was changed to $1.25 daily.

Finally, these SD goals and associated targets are to be realised by 2030 by all countries, while the period of MDGs ended in 2015.

All SDGs and selected targets may be seen in Table 3.10.

Critical Appraisal of SDGs

Thus these sustainable development goals (SDGs) are to be achieved with the efforts of all stakeholders like governments (public sector), corporate (private sector), NGOs, civil society (community-based organisations), and also the people at large by 2030. However, though SDGs are wider and more comprehensive in scope and direction than the MDGs, these, too, suffer from neo-liberal considerations and constraints of 'management' of natural resources as of 'instrumental value' only, not of 'intrinsic value'. That is why SDG 9 talks of building 'resilient infrastructure, promote sustainable industrialisation and foster innovation', and thus it focusses on infrastructures and industries which usually destroy, on a huge scale, lands, forests, water bodies as well as minerals, on the one hand, and pollute the environment massively, on the other.

Similarly SDG 11 talks of making cities 'inclusive, safe resilient and sustainable' – but does not question the very conception of cities and the very process of urbanisation that is responsible, to a large scale, for environmental damages and substantial pollution due to huge emission from numerous vehicles, construction activities, modern electronic gadgets, cutting of trees, and encroachment of water bodies. The very conception of urbanisation has created a huge hiatus between towns and rural areas all over the world. Further the notion of 'smart cities' is also cementing the socio-economic, political, and cultural gaps – cities as 'haves' and villages as 'have-nots'! For instance, at the time of the Independence of India, its national capital city Delhi had more than 800 ponds, lakes, and other water bodies. But, by the third decade of the twenty-first century, most of them will have disappeared due to public and private encroachments, if no drastic measures are taken. So is the case of other metropolitan cities like Mumbai, Chennai, Bengaluru, Hyderabad, Kolkata, Lucknow, Allahabad, Kanpur, Patna, Jaipur, and so on, leading to the massive choking of rainwater, causing huge floods (resulting in loss of humans and properties) in the first two decades of the twenty-first century. Further, SDG 12 talks of ensuring 'sustainable production and consumption patterns'; actually *mass production* and *consumerism* are salient by-products of capitalism, but SDGs do

Sustainable Development   109

Table 3.10 Sustainable Development Goals and Selected Targets (2015–2030)

SDGs	Selected Targets
1. End poverty in all its forms everywhere	a) By 2030, eradicate extreme poverty for all people everywhere (living on less than $1.25 a day) b) By 2030, reduce at least half the proportion of men, women, and children of all ages living in poverty in all its dimensions
2. End hunger, achieve food security and improved nutrition, and promote sustainable agriculture	a) By 2030, end hunger and ensure access by all people, including infants to safe, nutritious, and sufficient food throughout the year b) By 2030, end all forms of malnutrition, including achieving by 2025 the internationally agreed targets on stunting and wasting in children up to five years, and address the nutritional needs of adolescent girls, pregnant, and lactating mothers and older persons
3. Ensure healthy lives and promote well-being for all at all ages	a) By 2030, reduce the global maternal mortality ratio to less than 70 per 1,00,000 live births b) By 2030, end preventable deaths of new-borns and children up to 5 years – with all countries to reduce new-natal mortality to as low as 12 per 1,000 live births and under-5 mortality to as low as 25 per 1,000 live births
4. Ensure inclusive and quality education for all and promote life-long learning	a) By 2030, ensuring all girls and boys complete free, equitable, and quality primary and secondary education, leading to relevant and effective learning outcomes b) By 2030, ensuring all girls and boys' access to quality early childhood development, care, and pre-primary education to make them ready for primary education
5. Achieve gender equality and empower all women and girls	a) End all forms of discrimination against all women and girls everywhere b) Eliminate all forms of violence against all women and girls in the public and private spheres, including trafficking and sexual and other types of exploitation
6. Ensure access to water and sanitation for all	a) By 2030, achieve universal and equitable access to safe and affordable drinking water for all b) By 2030, achieve access to adequate and equitable sanitation and hygiene for all, and end open defecation, paying special attention to the needs of women and girls and those in vulnerable situations
7. Ensure access to affordable, reliable, sustainable, and modern energy for all	a) By 2030, increase substantially the share of renewable energy in the global energy mix b) By 2030, double the global rate of improvement in energy efficiency
8. Promote inclusive and sustainable economic growth, employment, and decent work for all	a) Sustain per capita economic growth in accordance with national circumstances, and at least a 7% GDP growth rate per year in the least developed countries b) Achieve higher levels of productivity through diversification, technological upgradation, and innovation, including a focus on high value-added and labour-intensive sectors
9. Build resilient infrastructure, promote sustainable industrialisation, and foster innovation	a) Promote sustainable industrialisation and by 2030, significantly raise industry's share of employment and GDP, in line with national circumstances, and double its share in the least developed countries b) Increase the access of small scale industrial and other enterprises, in particular in developing countries to financial services, including affordable credit, and their integration into value chains and markets

(Continued)

110 Sustainable Development

Table 3.10 (Continued)

SDGs	Selected Targets
10. Reduce inequality within and among countries	a) By 2030, progressively achieve and sustain income growth of the bottom 40% of the population at a rate higher than the national average b) By 2030, empower and promote the social, economic and political inclusion of all irrespective of age, sex disability, race, ethnicity, origin, religion, economic, or another status
11. Make cities inclusive, safe, resilient, and sustainable	a) By 2030, ensure access to all adequate, safe, and affordable housing, and basic services and upgrade slums b) By 2030, provide access to safe, affordable, and sustainable transport systems for all, improving road safety, by expending public transport, with special attention to the needs of vulnerable sections, women, children, persons with disabilities, and older persons
12. Ensure sustainable production and consumption patterns	a) By 2030, achieve sustainable management and efficient use of natural resources b) By 2030, halve per capita global food waste at the retail and consumer levels, and reduce food losses along production and supply chains, including post-harvest losses
13. Take urgent action to combat climate change and its impacts	a) Integrate climate change measures into national policies, strategies, and planning b) Improve education, awareness raising, and human and institutional capacity on climate change mitigation, adaptation, impact reduction, and early warning
14. Conserve and sustainably use the world's oceans, seas, and marine resources	a) By 2025, prevent and significantly reduce marine pollution of all kinds, in particular, from land-based activities, including marine debris and nutrient pollution b) By 2020, sustainably manage and protect marine coastal ecosystems to avoid significant adverse impacts, including by strengthening their resilience, and taking action for their restoration in order to achieve healthy and productive oceans
15. Sustainably manage forests, combat desertification, halt and reverse land degradation, and halt biodiversity loss	a) By 2020, ensure the conservation, restoration, and sustainable use of terrestrial and inland freshwater ecosystems and their services, in particular, forests, wetlands, mountains, and dry lands, in line with obligations under international agreements b) By 2020, promote the implementation of sustainable management of all types of forests, halt deforestation, restore degraded forests, and substantially increase afforestation and reforestation globally
16. Promote just, peaceful, and inclusive societies	a) Significantly reduce all forms of violence and related death rates everywhere b) End abuse, exploitation, trafficking, and all forms of violence against and torture of children
17. Revitalise the global partnership for sustainable development	a) Developed countries to implement fully their official development assistance (ODA) commitments, including the commitment by many developed countries to achieve the target of 0.7% of ODA/Global Network Initiative (GNI) to developing countries, and 0.15%–0.20% of ODA/GNI to the least developed countries b) Assist developing countries in attaining long-term debt sustainability through coordinated policies aimed at fostering debt financing, debt relief, and debt restructuring, as appropriate, and address the external debt of highly indebted poor countries to reduce the debt distress

Source: UN.org.

not question capitalism that is the root cause of the major environmental problems at local, national, regional, and global levels. No doubt, if the current level and scale of consumption of energy and other natural resources (both per capita and cumulative) continue, we would require three earths in future!

However, on the other hand, it is also true that, first, behind the SDGs firmly stands the top-down initiative by the United Nations that motivates various States to commit and take up eco-friendly policies and action plans to be realised in a fixed time frame. For instance, only the nation-States may impose stringent environment-conserving conditions like green taxes, financial incentives, and disincentives to the exploiters. Since the UN takes a uniform view on major issues, there is less grudge among various nation-States.

Second, various global NGOs and national environmental groups and peoples' movements usually place before the UN the people's views on various environmental problems. This provides the other side of the coin, and hence serious alternatives are duly suggested by them in the interest of the deprived people at large. No doubt, public perception of genuine NGOs and people's new movements is, more or less, positive and trustworthy.

Third, some concrete or even a potential environmental agenda before the national and global stakeholders about the national and global ecological commons is certainly better than no agenda or vacuum – something is better than nothing at all. A global platform from the UN provides an opportunity for all stakeholders to express their cross views or narratives about the equilibrium between ecology and development to fulfil the needs of the present and future generations.

Fourth, a serious negotiation at an international forum is the symbol of cooperation between nations, between nations and NGOs, between NGOs and corporations, between nations and corporations, between scientists and other stakeholders, and so on. There various rounds of discussion, debate, and dialogue often reduce or finally resolve conflicts and thus finally lead to the process of thesis, anti-thesis, and synthesis.

Finally, any international (UN) forum, face-to-face with various stakeholders, undoubtedly provides publicity by the mass media at a massive scale both nationally and globally. This obviously creates critical awareness or 'concientisation' (to use Paulo Freire's term) among the policy-makers, local NGOs/Civil Society Organisations (CSOs), and the people at large for quite some time – and thus the next generation is prepared for facing the environmental challenges locally, nationally, and globally because somehow *experimental knowledge* and *experiential knowledge* supplement and complement each other in the long run of human life as a part and parcel of ecology.

Sustainable Development in Mumbai and Odisha: Myth or Reality

Mumbai is cosmopolitan in character in terms of caste, class, race, ethnicity, region, religion, sex, and language. Various major infrastructure projects have been launched that have serious environmental consequences. Here we may take up, as a case study, five such projects: Mumbai metro in Aarey Milk Colony Forest, Navi Mumbai Airport, Coastal Road, Bullet Train, and Jawaharlal Nehru Port (Dayanand 2020).

(a) **Aarey Milk Colony Forest**: Aarey milk colony is known as the green lungs of Mumbai. There is a forest of 1,280 hectares. In the recent past, a car shed and service centre for Metro 3 was extended to that green forest area where there

112 Sustainable Development

are 86 species of plants, 76 species of birds, a spot to sight rock pythons, etc. Citizens protested and demanded a public debate, but the State government of the day (BJP) did not pay heed to it. Citizens and their organisations argued that out of seven locations four were viable (with minimum ecological damage) as per the Detailed Project Report (DPR) itself. Further, in 2015, citizens' organisations suggested a fresh alternative location that was also supported by the two environmental experts as members of the government committee and gave their dissent note to the government committee's report recommending Aarey forest as the only suitable location! (Dayanand 2020). Article 51A(g) of the Indian constitution mentions the fundamental duties of citizens: 'It shall be the duty of every citizen of India to protect and improve the natural environment including forests, lakes, rivers and wildlife, and to have compassion for living creatures'. But the protesters (including students) were beaten up and arrested by the police and sent to jail in 2019. Further, the High Court of Bombay did not stay the government's action of cutting trees speedily. Even the Supreme Court did not grant a stay and the State government got 2,000 trees speedily chopped off by the machines day and night. Later the Supreme Court asked the State government not to cut more trees! The arrested activists, mostly students, were released from jail after the intervention of the new coalition State government led by Shiv Sena (Uddhav Thakre as Chief Minister [CM]). Shiv Sena's new government of Maharashtra later decided to shift the car shed and service centre of Metro 3 from Aarey to Kanjurmarg (in Mumbai). However, in mid-2022 the new government led by Eknath Shinde (supported by BJP) again decided to re-shift the car shed and service centre of Metro 3 to Aarey! This shows the mafia–politician–corporate nexus.

(b) **Navi Mumbai Airport**: There were two location options for the airport: (i) Rocky and barren area of Niveli (owned by State government) with minimum human occupation; (ii) Panvel/Navi Mumbai area has thousands of acres of wetlands and mangroves, with coastal communities' livelihoods and culture. Further, both locations are at an equal distance from Mumbai. But due to the strong politician–mafia–builder nexus, the costlier and ecologically more damaging option of the airport at Panvel was chosen by the power-that-be in the most non-transparent manner (Dayanand 2020).

(c) **Coastal Road**: It is an example of creating a new problem first and then solving it! If Metro Rail was being planned to be constructed on a large scale in South Mumbai on the plea of reducing pollution, removing traffic jams, and saving fossil fuel, there was no logic in proposing to build a coastal road as it would create more economic, social, and ecological problems due to more number of vehicles. Further, it is more absurd to destroy our ecological heritage and instead build there multi-storey towers with parking spaces. There was a proposal to reclaim 90 hectares of sea land – out of which 20 hectares were meant for the coastal road and the rest 70 hectares of land was to be developed for the city! The High Court of Bombay did not agree with the State government's proposal which was a so-called development project, not an infrastructure project. However, its final adjudication is pending in the Supreme Court; but the Supreme Court has not stayed the destruction of the local ecology (Dayanand 2020).

(d) **Bullet Train**: The Bullet train project from Mumbai to Ahmadabad will involve huge farmlands, forests, wetlands, mangroves, etc. and also a huge cost of

Rs. 1 lakh crore. The public, in general, will not use it normally due to being quite costlier. This Bullet train is supposed to pass below the Thane Creek Flamingo Sanctuary! This Sanctuary is a potential Ramsar heritage site, and India is a signatory to the Ramsar Convention. How can one lakh birds of the sanctuary remain within the limits of the notified sanctuary area? This avoidable, highly costlier, and highly ecologically damaging, without serving a genuine public purpose, the project has been questioned by environmental activists as well as Mumbai residents who desire more local/metro trains for reducing huge crowds and resulting frequent accidents (Dayanand 2020).

(e) **Jawaharlal Nehru Port**: The wetlands of Uran (Mumbai) have been unique due to the mixture of grasslands, water bodies, reeds, marshes, lakes, etc. with more than 100 species of birds. But due to the construction of Jawaharlal Nehru Port there, now one sees only the degradation and 'dust bowl' there. Due to the loss of fisheries there, the local people have lost their livelihood, and instead they suffer from floods. It is found by the ecologists and local citizens that had the landing terminal been built elsewhere on a separate land, the port and ecological sustenance could have existed together (Dayanand 2020). But the central government and State government did not think about sustainable development – rather development at the cost of ecological sustainability!

Thus, due to the unsustainable model of development, lack of Environmental Impact Assessment (EIA) and Social Impact Assessment (SIA), wrong planning, and the undue profit motive of the politicians, technocrats, bureaucrats, contractors, suppliers, etc. in these five projects, ecology of Mumbai has been highly compromised, wounded, and damaged. Here the perceptive observation of E. O. Wilson is quite contextual: 'Destroying rainforest for economic gain is like burning Renaissance painting to cook a meal'. Thus the contradiction is hidden in the very concept of sustainable development – the term 'sustainable' signifies the limits to growth due to the finite natural assets, while the term 'development' in itself does not accept the finite character of natural assets. As rightly pointed out by Saurabh Arora et al. (2019), 'modernising development pathways' extend the unlimited extraction of 'goods' from nature while dumping back the 'bads' (the effluents and wastes generated from the production processes); 'the extension of "extract-dump" modality of human-nature relations through the modernising techno sciences, depends on multiple other forces in society', that are as follows:

(a) Capitalist desire for short-term profit (at the cost of long-term harm)
(b) Scientific/engineering education systems with standardisation and control of nature
(c) So-called efficient innovation policies defined by narrow economic considerations
(d) Rules and regulations to 'manage' pollution and waste – often shifting the location of dumping the wastes
(e) High status given to the individualised engagement with technological artefacts
(f) Social construction of the sense of freedom attached with individualisation

Thus single development pathway usually dominates by cornering most of the natural assets and by shaping specific things as practicable, on the one hand, and, on the other

114 Sustainable Development

hand, it blocks, restricts, or marginalises alternative pathways and thus marginalises the possibilities of realising more sustainable human–nature relations that limit or transcend the extract-dump modality (Arora et al. 2019, 32).

Mining in Niyamgiri (Odisha)

Similarly, corporate investments in Odisha have been more a cultural genocide than 'sustaining' mining and sustainable development (Box 3.2).

Box 3.2: Sustainable Mining or Cultural Genocide in Niyamgiri, Odisha (India)

In East India, specifically in Odisha, Vedanta Resources (a London-based company owned by Anil Agarwal), promoted by the Department for International Development (DFID) of the UK government, and World Bank planned in 2003 for a huge investment in the so-called 'sustaining mining of bauxite mineral' at Niyamgiri range (4,000 feet high mountain) – the native place of the Dongaria Kondhs (primitive tribes). The top peak is considered sacred (due to their supreme deity) by these tribes, and hence traditionally there was a taboo to cut trees/forests thereon – about 80%–90% of the 660-hectare mining lease area on the peak is covered by Sal (Shoera Robusta) forests. Hence these tribal farmers opposed the mining project and they were supported by genuine activists of human rights, political rights, and environmental rights. But the corporate, political representatives, administrators, and police branded the protesters 'anti-development', 'anti-government', and Maoists! On the other hand, the local protesting people and activists label the project-supporters as 'anti-people' due to their 'authoritarian, anti-democratic means' (F. Padel and S. Das 2010). Similar is the situation in other areas of enforced industrialisation and dam-reservoir projects in Odisha like Hirakund, Rengali, Upper Kolab, and Upper Indravati (each of which displaced 40,000–2,50,000 people during 1950s–1990s). It is estimated that three million people were displaced in Odisha alone during 1950s–1990s – 50% of whom were tribals, 25% Dalits, and the remaining 25% others. For such Adivasis (tribals) displacement means 'cultural genocide' due to the destruction of all aspects of their social structure: Economy and identity, political structure, social relationships, religion ('even our gods are destroyed'), material culture, and spatial arrangement of villages. These various aspects are inseparable. 'The sacredness of nature, respect for elders' knowledge, ritual contact with the ancestors, growing their own food on family land, making their own houses and tools, exchanging food with neighbours with an egalitarian spirit: These things are 'swept away by corporate values, which emphasise money and financial power' (Padel and Das 2010, 336). These tribals are highly egalitarian and practise real sustainability in the true sense – e.g. protecting the primary Sal forest at 4,000 feet (Niyamgiri Dongaria Kondh). On the other hand, the mining companies like Vedanta deceptively advertise smiling tribal faces with the slogan 'Vedanta Mining Happiness'! In fact, Britain and other European countries closed coal mines and alumina refineries due to high costs and high levels of pollution, but such extractive industries are promoted fiercely in Odisha and other parts of India, and also in other developing countries.

A case was filed by three environmentalists in the Supreme Court of India for the protection of religious and cultural rights of the Primitive Tribes (Dongria Kondh, Kutia Kondh) vis-a-vis Vedanta Resources' Lanjigarh alumina refinery that gave wrong information. Central Empowered Committee, constituted by the Supreme Court of India, pointed out that, firstly, Niyamgiri hills falls under schedule V of the Indian constitution, prohibiting the transfer of tribal lands to non-tribals; secondly, Vedanta circumvented the law by not disclosing the requirement of forestland. Hence it recommended for revocation of environmental clearance of the project and the banning of the mining operation at Niyamgiri hills. The Supreme Court referred the case to the Ministry of Environment and Forests, Government of India (GOI), that engaged the Wildlife Institute of India to examine the project's impacts. On the other hand, Vedanta continued the construction of the refinery in 2006, resulting in the displacement of more than 100 tribal families; the petitioners showed costs being three times the benefits (3:1), but environmental costs were incalculable, and social impacts could not be converted into money – Niyamgiri hills has incalculable religious and cultural value to the Kondhs (whose sacred deity Niyam Raja resided there). In 2007, the Supreme Court found legal loopholes in the clearance procedures, and also the Vedanta company invited its subsidiary Sterlite to apply for mining; hence, it called for a 'delicate balance between conservation and development'. In 2009 (April), Ministry of Environment and Forests (MOEF) granted environmental clearance but forest clearance was kept on pending. On 16 August 2010, Saxena Committee (appointed by MOEF) reported that granting a mining lease there would deprive two Primitive Tribal groups of their rights and would shake their faith in the laws of the land, as they were not consulted as per Scheduled Tribes and Other Forest Dwellers (Recognition of Forest Rights) Act, 2006 (effective since 1 January 2008). The then-Minister (MOEF) Jairam Ramesh withdrew forest clearance and, later in July 2011, environmental clearance was also withdrawn. Government of Odisha argued for the project and against the romanticising of the environment-friendly tribal way of life and culture, rather to remove poverty, illiteracy, and hunger of such tribals. Finally, the Supreme Court in 2013 (April) ordered for deciding its fate in their local gram sabha. Hence 12 affected villages in Rayagada and Kalahandi districts of Odisha held 'palli sabha' in July 2013 – the environmental referendum unanimously rejected the mining project of Vedanta Resources; thus, they asserted their traditional religious rights and collective 'cultural space' (L. Temper and J. Martinez-Alier 2013). Thus the so-called 'sustaining mining' by Vedanta was stopped forever.

Source: F. Padel and S. Das (2010); L. Temper and Joan Martinez-Alier (2013)

Towards Real Ecological Sustainability (RES)

From the analysis in the earlier sections, it is crystal clear that the concept of sustainable development has also weaknesses along with strengths. However, there are some countries in the world which have pursued the path of 'real ecological sustainability' (RES) and have scored much better in 'environment performance index' (EPI). As per

116 Sustainable Development

Table 3.11 Top Fifteen Most Sustainable Countries in the World (2018)

Rank	Name of the Country	Environment Performance Index (EPI) Score (2018)	Remarks (Areas of Best Performance)
1.	Switzerland	87.42	Mostly hydroelectricity used; 97% of the population connected to sewage treatment – its score in water sanitation is 99.99
2.	France	83.95	30% of the total energy used is from renewable sources – it reduces food waste and promotes access to food
3.	Denmark	81.60	50% of the total energy used is from wind energy; 40% of the population travels by cycle; top air quality (99.16)
4.	Malta	80.90	Out of total energy used, 70% comes from natural gas, 30% from renewable sources – ranks first in water sanitation and water resources in the world with 100 points
5.	Sweden	80.51	100% of energy from renewable sources and recycled sources; low energy-consuming 'passive houses' – use heat energy, from human activities (body heat from daily commuters), electrical appliances, and sunlight – ranks first for heavy metals (100)
6.	UK	79.89	renewable (wind and solar) energy, and recycling of wastes on priority – ranks first with 100 points
7.	Luxembourg	79.12	Addresses ecological issues on priority, its score for water resources is 99.76
8.	Austria	78.97	Addresses ecological issues on priority – its score for water resources is 99.08
9.	Ireland	78.77	Addresses ecological issues on priority, the first country in the world to divest from fossil fuel – in water and sanitation it ranks first in the world with 100 points
10.	Finland	78.64	Addresses ecological issues on priority – its rank in water and sanitation, and heavy metals is first with a score of 100, and its score in air quality is 99
11.	Iceland	78.52	Addresses ecological issues on priority
12.	Spain	78.39	Do
13.	Germany	78.37	Do
14.	Norway	77.49	Do
15.	Belgium	77.38	do

Source: epi.envirocenter.yale.edu.

EPI (2018), based on outcome-oriented indicators the following 15 countries are the top most sustainable countries in the world (Table 3.11)

Thus out of 180 countries evaluated for EPI, all top 15 countries are in Western–Northern Europe, while the most powerful and the largest economy in the world, the USA, ranks 27th with an EPI score of only 71.19, Japan ranks 20th with EPI score of 74.69, China ranks 120th with EPI score of 50.74, and India ranks 177th with the EPI score of just 30.57 – 4th worst sustainable country in the world. Now let us also see the 20 worst sustainable countries in the world (Table 3.12).

Sustainable Development 117

Table 3.12 Twenty Worst Sustainable Countries in the World (2018)

Rank	Country	Environment Performance Index (EPI) Score	Remarks
180.	Burundi	27.43	Poor African country – not addressing ecological issues on priority
179.	Bangladesh	29.56	Poor Asian country – do
178.	Dem. Rep. of Congo	30.41	Poor African country – do
177.	India	30.57	Poor Asian country (the so-called emerging economy) – do
176.	Nepal	31.44	Poor Asian country – not addressing ecological issues on a priority
175.	Madagascar	33.73	Poor African country – do
174.	Haiti	33.74	Poor African country – do
173.	Lesotho	33.78	Poor African country – do
172.	Niger	35.74	Poor African country – do
171.	Central African Republic	36.42	Poor African country – do
170.	Angola	37.44	Poor African country – do
169.	Pakistan	37.50	Poor Asian country – do
168.	Afghanistan	37.74	Poor Asian country – do
167.	Benin	38.17	Poor African country – do
166.	Mauritania	39.24	Poor African country – do
165.	Eritrea	39.25	Poor African country – do
164.	Papua New Guinea	39.35	Poor African country – do
163.	Djibouti	40.04	Poor African country – do
162.	Swaziland	40.32	Poor African country – do
161.	Cameroon	40.81	Poor African country – do

Source: epi.envirocenter.yale.edu.

Thus it emerges from Table 3.12 that 20 worst sustainable countries are from Africa and Asia – 15 African countries (including Burundi as the worst sustainable country in the world with an EPI score of only 27.43, ranking 180th out of 180 countries) and 5 Asian countries, including Bangladesh (179th rank) India (177th rank), Nepal (176th rank), Pakistan (169th rank), and Afghanistan (168th rank). These countries are not giving priority to ecological issues, probably due to financial constraints but also because of a lack of appropriate eco-friendly policies. Interestingly, among these 20 worst sustainable countries, not a single South American country is found. In EPI 2022, India ranked 180th out of 180 countries!

There are mainly two broad approaches to sustainable development: Symptomatic approach and systems approach. In each approach there are two sub-approaches: Fragmented symptomatic approach and comprehensive symptomatic approach, fragmented systems approach and comprehensive systems approach (Chiras 2012). In fact, this reflects the evolution of environmental protection through four stages: Fragmented symptomatic approach, comprehensive symptomatic approach, fragmented systems approach, and comprehensive systems approach – the last being the best stage. On the other hand, as far as the level of action for sustainable development is concerned, it, too, has spread from the local level to the sub-national level to the regional level, and, finally, to the global level by both the formal institutional multilateral agencies of

118   Sustainable Development

United Nations, especially UNEP, as well as international non-governmental organisations like International Committee for Bird Protection (established in 1922), World Wide Fund for Nature (established in 1961), Greenpeace International (established in 1971), and Friends of the Earth (established in 1969). Let us discuss managing ecosystem services in more concrete terms in a developing country.

Ecosystem Services in Mexico

National 'payments for ecosystems' (PES) started in Mexico in the first decade of the twenty-first century. It was based on neo-classic theory, promoting both conservation and development by perceiving ecosystems as factories. Here the basic assumption is that the beneficiaries of cleaner, more water, greenhouse gas sequestration, biodiversity conservation, and other services should make payments to the owners/managers of such ecosystem services. Therefore, some environmental economists and ecologists prepared a PES model wherein transaction costs are low and property rights are well-defined, and hence the markets will then become autonomous and support stable conservation. Thus, the standard definition of PES (adopted in Mexico) has the following features (S. Wunder 2005, cited in Shapiro-Garza 2020):

(a)   A voluntary transaction
(b)   Where a well-defined ecosystem service (or a land use likely to secure that service)
(c)   Is being purchased by an ecosystem service purchaser
(d)   From an ecosystem service provider
 e)   If the ecosystem service provider secures its provision (conditionality)

However, it (the Mexican model), too, assumed that ecosystem services have value and managers of ecosystem services be paid compensation; but, the PES model has been contested by the Compensation for Ecosystem Services (CES) model that has a different understanding of the relationship of peasants and indigenous communities to their rural environment 'to revalue the rural' and contested the role of State and market incorporating and accounting for other values (labour, cultural knowledge, and local practices), nature–society relationships, and considering equity and social justice at multiple scales (Shapiro-Garza 2020). The proponents of the CES model were public or popular intellectuals who worked with and for rural communities and movements, on issues of sovereignty and rights to cultural reproduction and conditions for sustainable land management and livelihoods. They formed an 'epistemic community' and could hold simultaneously multiple positions (as common in Latin America) as professors, directors of NGOs, public administrators, or rural social movement leaders. They have rich field experiences by engaging with the peasants and indigenous communities. CES, as a distinct conceptualisation, was first reported in a report of El Salvador's PRISMA (an environmental think tank). In Mexico, neo-Marxists and critical theorists have questioned the neo-liberal programmes of the World Bank, etc. There is a difference between 'payment' (PES model) and 'compensation (CES model) in the following ways (Shapiro-Garza 2020):

(a)   'Payment' (PES model) is a simple *financial transaction*, while 'compensation' (CES model) is *relational*, involving money, labour, goods, or generation of goodwill

Sustainable Development 119

and strengthening of social relations; thus, the latter addresses both social injustice and environmental degradation.

(b) 'Payment' connotes '*selling*' of something which nature produces, while 'compensation' signifies *recognition* for the work done by the traditional stewards (people) to help nature and benefit others.

(c) In 'compensation' the interaction between the ecosystem service user and service provider is based on *trust*, building something together, while that may not be present in the 'payment' mechanism in reality.

(d) In 'compensation' mechanism there is some sort of *interdependence* of both regarding identifying the problem and helping each other to solve it through '*respectful, interactive dialogue*'; the PES model lacks such a feature, and the PES approach adopts a paternalistic approach of the federal government towards rural communities through disbursing subsidies that promotes passive and 'rent-seeking' attitude among beneficiaries.

(e) CES approach wants to create a 'virtuous cycle' between the rural and urban dwellers – city and rural areas, between agriculture and industries, and between the so-called developed countries ('*producers of waste*') and developing countries ('*producers of oxygen*'); on the other hand, PES approach does not have such a vision at all.

(f) CES approach is *holistic*, encompassing payments' regulations, awareness generation, and presence of alternatives (sowing flowers, ecotourism, etc.), while the PES approach is merely a tool or 'instrument without symphony'; CES gives multifaceted support to the participants who can '*add value*' to the initiatives and overcome purely economic barriers.

Hence a comparison between PES and CES approaches may be seen in Table 3.13.

From Table 3.13, it is crystal clear that in all seven conceptual elements, the CES model is more progressive, holistic, egalitarian, outcome-oriented, and sustainable. However, both models have primarily over-shadowed the inevitable need for State intervention, the significance of accounting for other values that might motivate people for conservation, and the need for equitable distribution of costs and benefits for ensuring both environmental and social outcomes. But such shortcomings were not sorted out by the public intellectuals of Mexico who theorised the alternative CES model. Not only in Mexico but in other developing countries including India, too, one or other pathway becomes dominant (Box 3.3) due to following pressures which eliminate the diverse pathways, as pointed out by various social scientists and environmentalists (Arora et al. 2019):

(a) Incumbent interest behind 'increasing returns to scale' and 'path dependence' due to network effects and interactive learning among users of technological artefacts (e.g. green revolution was dependent on chemical fertilisers, pesticides, HYVs of seeds, massive irrigation by using underground water).

(b) The accumulation of techno-scientific expertise along specific paradigms with a specific methodology of appraisal of viability and performance.

(c) Government practices that replace incalculable uncertainties and ignorance by probabilistic risk.

(d) Development discourses/imaginaries prioritising economic growth, return on capital, speed, efficiency, and standardisation (e.g. antibiotic medicines) over sustainability and distribution.

120 Sustainable Development

Table 3.13 A Comparison Between Payments for Ecosystems (PES) and Compensation for Ecosystem Services (CES) Models (Mexico)

Sl. No.	Conceptual Elements	PES Model	CES Model
(a)	Intention of incentives	to provide financial incentives to rational, selfish actors who would otherwise degrade natural assets	recognition of good stewardship and to cover incremental transaction costs
(b)	Degree of complexity	economic incentives required to induce land, manager's choice to conserve	Holistic suite of approaches necessary for sustainable rural development; must also address structural issues that drive degradation
(c)	Other values	As long as financial payment is greater than the opportunity of conversion to other uses, landowners will conserve	To be effective, compensation should include other types of support, relationships, production of ES of local value, etc.
(d)	Conditionality	Making payments conditional on the measured production of ES is essential for ensuring positive environmental impacts	Conditionality is important, both for ensuring positive social and environmental outcomes and to avoid paternalism
(e)	Targeting priorities	Targeting, based on any criteria other than the ability to produce ES or form markets, dilutes the effectiveness	Targeting based on social criteria ensures equitable distribution of benefits, can help rectify structural injustices, and is necessary if State run
(f)	Role of markets	Markets or market-like policy design more efficient and effective	Markets are a potential source of funding but are difficult to form and are highly volatile
(g)	Role of State	Involvement of State will decrease the efficiency, effectiveness and sustainability of these mechanisms	State must provide regulatory incentives and infrastructure for markets are to form; as public goods and human rights, State must also ensure access to and equitable distribution of ES

Source: Shapiro-Garza (2020).

(e) Neo-liberal notion of seeing people merely as individual consumers (packaged plastic bottle water results in massive plastic waste).
(f) Webs of interdependence between different technologies that have end-of-pipe techno-fixes (catalytic converters in automobiles, carbon capture and storage) over socio-political transformations to sustainability.

Box 3.3: How Dominant Pathway Damages Environment (India)

A concrete example of a dominant pathway in India is opting for the construction of flyovers to decongest the cities. But this pathway has many negative environmental and social consequences:

(a) It pushes for more production of cement, gravels, sand, and iron – hence more mining is done, thus damaging the environment to a large extent.
(b) In many parts of India, especially MP, UP, and Bihar, sand mining has given rise to the sand mafia resulting in loss of revenue, monopoly over the sale of sand from commons and bloodbaths between different gangs.
(c) Flyovers have thus indirectly enhanced air pollution in many cities/towns.
(d) On flyovers common people's modes of transport like cycles, rickshaws, carts are not allowed, and thus such people are further marginalised.

Source: After Arora et al. (2019)

Therefore, diverse sustainable pathways are to be adopted. But there are two fallacies (Arora et al. 2019):

(i) The fallacy of 'technological solution' – actually an alternative development pathway needs social–institutional and politico-economic change and restructuring, but often sustainable development pathway is reduced to a techno solution.
(ii) The fallacy of 'ecological modernisation' – assuming to fit the sustainable development pathway into modernity and its processes like capitalist short-termism, individualisation, standardisation, and control of nature. But actually modernisation needs to be countered to 'transcend the extract-dump modality of human-nature relations'.

Hence, in the recent past, various civil society organisations as well as local farmers have started some successful experiments for agro-ecological sustainability by challenging the green revolution's modernisation pathway. The following examples are notable in this regard (Arora et al. 2019):

a) SRI – 'system of rice intensification' by involving three stakeholders – small farmers, agricultural scientists, and State governments.
b) Sustainable and equitable dry land farming.
c) Zero budget natural farming (ZBNF) movement addressing chronic indebtedness of farmers. Andhra Pradesh Government has scaled up ZBNF; NITI Ayog has also endorsed it.

Thus, real ecological sustainability (RES) goes beyond incremental changes (reform) by adopting a 'paradigm shift' towards associating ecological sustainability (eco-justice) with social distributive justice (in the broad sense of inclusion of all irrespective of caste, class, race, region, language, ethnicity, religion, gender, etc.) in a society with the motto of 'think globally, act locally' – and thus to realise ecological democracy in true sense. In short, multiple sustainability pathways and the CES model show the correct way of real ecological sustainability (RES) that may be adopted by communities, CSOs, and governments (local, State, and national), without any gap in thought, word, and deed. Unfortunately, because of likeness and wrong policies of State administrations in Madhya Pradesh, Uttar Pradesh, Haryana, Rajasthan, Jharkhand, etc., various incidents of killing of even government officers by sand and stone mafia have taken place in the recent past.

Transition to a Sustainable Economy and Society

When Brundtland Commission defined sustainable development (SD) in terms of balancing between the development and the environment, it meant three major objectives: Durability, equitability, and flexibility.

(a) **Durability** – means that any development activity should be steady by not putting various kinds of burdens on society by individuals. Further, durability requires long-term planning for attaining any development task – that is, creating and dumping wastes first, and removing these later is not durable at all. In fact, durability may be ensured through sustainable agriculture wherein nutritious foods are grown with the use of traditional bio-manures (compost, vermiculture, etc.), without using chemical fertilisers, pesticides, and weedicides, and by ensuring integrated biological diversity, that is 'organic agriculture' that ensures four principles of health, ecology, fairness, and care, as International Federation of Organic Agriculture Movements mentions (IFOAM, 2005). SDG 2 emphasises ending hunger, achieving food security, improving nutrition, and promoting sustainable agriculture.

(b) **Equitability** – This objective has two dimensions: 'intra-generational equity' (gender, caste, and class aspects), and inter-generational equity. While the first dimension relates to inclusiveness and fairness to the present generation without any discrimination on the basis of caste, class, gender, ethnicity, religion, language, etc., the second dimension relates to fairness and inclusiveness to future generations – to leave the earth likeable and enjoyable. Thus, the elites, the males, the upper-class social groups, etc. should not deprive the 'marginalised others', but all should use natural assets as per their basic needs only.

(c) **Flexibility** – Nature is fully flexible and resilient in terms of self-generation and self-renewal. However, even this has limits and cannot bear the brunt of unlimited exploitation. Hence there is a need for imposing restraints on individual, social, technological, and institutional activities and practices by adopting new sustainable practices. For instance, in place of thermal power using coal (fossil fuel) and emitting huge amounts of carbon, renewable sources of energy (solar, wind, etc.) may be used more sustainably.

Hence, we need a transition from the culture of consumerism (routed in the capitalist mode of production) to sustainability, as Tim Jackson (2009) rightly observes: 'In consumer societies, people are persuaded to spend money they don't have on things they don't need to create impression that won't last on people they won't care about'. Jan Rotmans defines 'transition' as a social transformation process with three features mentioned below (cited in Ossewaarde 2018):

(a) Change to the structure of part or the whole of society
(b) Large scale technological, economic, and social–cultural influences that reinforce each other
(c) Development at different scale levels (local, regional, and global)

Thus, such changes in different sectors and societies are required in the long run. Otherwise if our present consumption pattern and scale continue, we would require three planets (earths) and, obviously, that is not possible at all. Hence ultimately we have to change our behaviours, habits, and economic activities in consonance with the earth itself. For instance, with the rise of income all over the world, the solid waste from businesses and households doubled during 1950–2010 from 1 kg per capita daily to 2 kg per capita daily. It is a symptom of over-consumption that does not mean satisfaction from needs' fulfilment and thus overall well-being and happiness. On the other hand, many developing countries in the world, including India, have hunger problems substantially. We, therefore, tend to agree with the observation of Franz Fanon (1963) in 'The Wretched of the Earth': 'for a colonised people the most essential value, because the most concrete, is first and foremost the land: The land which will bring them bread and, above all, dignity' (1963, 9). The greatest ecological ethicist of twentieth century, Aldo Leopold, also expressed a deep emotional connection as well as respect for the earth, mother earth as 'Gaia', and talked of 'land ethic' (1949/1986):

> the individual is a member of a community of interdependent part. The land ethic simply enlarges the boundaries of a community to include soils, waters, plants and animals or collectively: the land. In short, a land ethic changes the role of Homo Sapiens from conquering the land community to plain member and citizen of it.
> (1949/1986, 239)

He further adds, 'A thing is right when it tends to preserve the integrity, stability, and beauty of the biotic community. It is wrong when it tends otherwise' (1949/1986, 261). To him, the evolution of land ethic is both an intellectual and emotional process.

Leopold's 'land ethic' means ecosystem ethic. He goes much beyond the concept of sustainable development, to the eco-centric view or deep ecology wherein lies the interdependence of humans, animals, plants, birds, and other entities (rivers, mountains, etc.) as well as the equity, justice, and democracy among all of them.

On the other hand, Tim Jackson (2009) suggests 12 steps for a sustainable economy under three major categories as follows:

1. **Building a Sustainable Macro-economy** – Four policy steps should be taken for building a sustainable macro-economy by shifting from debt-driven materialistic consumption, relentless growth, and expanding material:
    (a) Developing macro-economic capability
    (b) Investing in public assets and infrastructures
    (c) Increasing financial and fiscal prudence
    (d) Reforming macro-economic accounting
2. **Protecting Capabilities for Flourishing** – Within the ecological limits of the earth, creative opportunities for the people should be provided for flourishing and prospering with stability. For this, the following five policy steps need to be taken:
    (a) Sharing the available work and improving the work–life balance
    (b) Tackling systemic inequality
    (c) Measuring capabilities and flourishing
    (d) Strengthening human and social capital
    (e) Reversing the culture of consumerism

124 Sustainable Development

3. **Respecting Ecological Limit** – There is an urgent need to impose environmental limits on various economic activities in different sectors by taking the following three policy steps:
   (a) Imposing clearly defined resource/emission caps
   (b) Implementing fiscal reform for sustainability
   (c) Promoting technology transfer and international eco-protection

The United Nations Environment Programme (UNEP) also recognised ecological decay as early as 2008: 'From 1981 to 2005 the global economy more than doubled, but 60% of the world's ecosystems were either degraded or over-used'. Further emissions during 1990–2008 increased by 40% though Kyoto Protocol had sincerely spoken to reduce it substantially. Due to the prevailing culture of consumerism, most of the upper and middle classes have become the victim of the process of 'Cathexis' (Russ Belk, cited in Jackson 2009) – a process of attachment leading to think of/feel material possessions (like car, electronic gadgets, computer, mobile, and other costly items) as part of the 'extended self'. Conspicuous consumption is about the continual processes of emulation, status competition, and 'self-completion'.

In addition, as Peter Hall and David Soskice (2001) have rightly pointed out, there are mainly two types of capitalism in advanced industrialised countries from the viewpoint of the role of the market vis-a-vis environment:

(a) **Liberal market economies** (the USA, the UK, Canada, and Australia) led the march towards competition and deregulation, especially during the 1980s and 1990s.
(b) **Coordinated market economies** (Japan, Germany, Austria, and Scandinavian countries) depend substantially on strategic interactions between firms – rather than competition – to coordinate economic behaviour.

In coordinated market economies usually the economic policies are more equity-based, less per capita carbon emission, price-maintaining, and needless to say, environment-friendly than those in liberal market economies.

However, a genuine attempt at heritage conservation, native plantations, extensive rainwater harvesting structures, etc. may have a sustainable link between nature and culture, as done in Nizamuddin, New Delhi (Ibrar 2020; see Box 3.4).

Box 3.4: Ecological Restoration at Cultural Heritage Complex, Nizamuddin, New Delhi (India)

Ninety-acre Sunder Nursery was developed at the cultural heritage complex, Nizamuddin, New Delhi (in December 2020). There are 20 heritage monuments and landscaped environs, including a 30-acre biodiversity zone. The project turned the almost barren site into 'an urban oasis' by ensuring 'ecological restoration' through native plantings and an extensive rainwater harvesting system. The Archaeological Survey of India (ASI) and Aga Khan Trust for Culture cooperated with mutual understanding and partnership by creating an urban green area and substantially improving the quality of life for the Nizamuddin Basti community simultaneously. The team applied a multifaceted approach to the World Heritage Site of Humayun's Tomb at Nizamuddin, New Delhi. Hence it was given two Awards – UNESCO Asia-Pacific Award of Excellence for Cultural Heritage Conservation and Special Recognition of Sustainable Development as it 'sets a new bar for heritage conservation and serves as a catalytic model for public-private partnerships in India'. Thus the integrity of the heritage complex was duly restored by sustainably linking nature and culture, and it established a new paradigm for connecting physical fabric to ecology and societal well-being, as UNESCO noted.

Source: Mohammad Ibrar (2020)

Similarly, in Madagascar, small fisher folk sustainably live with the sea and successfully manage local marine areas, and they have restored mangroves, developed alternative livelihoods, and conserved octopus and sea birds (Box 3.5).

Box 3.5: Madagascar's Small Fisher Folk Live with the Sea (Madagascar's)

Madagascar's coastal people, though formally uneducated, have adequate traditional knowledge of marine life – about the ocean's currents, its winds, its weather, meanings of different colours, signs of changes, how marine ecosystem supports each other, and why different species live in diverse habitats. While industrial fishing (by trawlers) is based on exploitation and profit, the small fisher folk respect, love, and live with the sea. Mihari network extends over 5,000 km of coastline and spans over 200 community associations. It builds their capacity in managing fisheries and also leadership and communication, along with new sectors of livelihood like aquaculture and tourism. People also ensure closure of fisheries if found unsustainable. Since 2004–2005, there are 200 Locally Managed Marine Areas (LMMAs) in Madagascar. Local coastal communities in such LMMAs manage mangrove restoration (for storm barriers and fish reservoirs), develop alternative livelihoods, and maintain conventions for sustainable ecosystems. LMMAs also ensure that local communities are at the centre of conservation. LMMAs created octopus reserves in the southern region, and recovery of octopus and return of shrimps, crabs, lobsters, and mackerel through healthy mangroves are major successes. As Whitley Award (2019) winner Vatosoa Rakotondrazafy rightly remarks that the ocean is 'deeply cultural for our fisherfolk communities, who pray to it and conduct rituals based around the seas .... we must work harder towards keeping the oceans plastic-free'.

Source: V. Rokotondrazafy (2020)

Needless to mention here that, first, 85% of global fish stocks are depleted or fully exploited; due to industrial fishing by trawlers marine degradation increased during 1950s–1990s; all large fish species like sharks and tuna declined by 90%; second, fish stocks fell by 95% in South China Sea since 1950s; third, coastal fisheries have shrunk by 50% during 1990–2020; fourth, however, more than 10% marine species are recovering from population decline due to eco-sensitive policies; fifth, giving local fisher folk responsibility for marine reserves is a good strategy; sixth, attaching

126   Sustainable Development

flapping streamers to boats is a simple strategy to save seabirds – this has reduced sea bird losses by 99%, and finally, one more strategy is halting fishing for months – e.g. in 2004, Andavadoaka village in Southwest Madagascar closed fishing for seven months, resulting in producing 700% higher octopus stocks (octopus has 8 limbs and 300 species) (*The Times of India*, 26 December 2020).

In fact, various factors of the environment affect people's well-being: Geography, natural capital, temperature and precipitation, land cover, air pollution, noise pollution, infrastructure, and natural disasters. Undoubtedly, the green natural environment, even in urban areas, helps people to be healthy and happy (see Box 3.6).

Box 3.6: Green, Healthy, and Happy, Pennsylvania (USA)

The natural environment helps the patients to recover fast and have a positive emotional state, physiological activity level, and sustained attention. R. S. Ulrich (1984) studied recovery records of surgical patients in a suburban Pennsylvania (USA) hospital during 1972–1981. Patients in rooms facing natural settings had shorter post-operative hospital stays, received fewer negative comments in nurses' notes, and requested less medication. Later R. S. Ulrich et al. (1991) exposed 120 persons first to view a stressful film and then exposed them to videos of different natural and urban settings. They found that stress recovery was faster and more complete when they were exposed to natural rather than urban settings. Thus natural environment has positive effects on people's well-being.

Source: R. S. Ulrich (1984); R. S. Ulrich et al. (1991)

Further World Happiness Report (2020) correlates sustainable development goals (SDGs) and subjective well-being (SWB) – consisting of six components: Income, social support, generosity, freedom to make life choices, trust in government and business, and healthy life expectancy.

(i)   Three SDG groups – Economic (SDGs 4, 8, and 9), Law (SDG 16), and Health (SDG 3) have strong positive correlations with income per capita.
(ii)   The SDGs representing the environment (SDGs 2, 6, 7, 11, 12, 13, 14, and 15) also have a positive correlation with income per capita, but it is lower at 0.17.
(iii) Social support (a strong determinant of SWB) is very positively related to goals representing social equality (SDGs 1, 5, and 10).
(iv) On the other hand, there is a lower correlation between this social equality group (SDGs 1, 5, and 10) and SWB determinants of values (generosity) and freedom to make life choices.
(v)   Rule of Law has a similar relationship with these three determinants as the social SDGs' group.
(vi) Health determinant has a correlation of close to 1 with Health SDG.
(vii) Environment group is quite significant for health, too, with a positive correlation of 0.63. EPI is a much wider index than environmentally oriented SDGs, and it consists of 24 indicators grouped into two policy objectives and 10 issue categories which are as follows: (a) Biodiversity and habitat, (b) forests, (c) fisheries, (d)

climate and energy, (e) air pollution, (f) water resources, (g) agriculture, (h) heavy metals, (i) water and sanitation, and (j) air quality.

(viii) Though most of the SDGs are positively correlated with well-being, SDG Goal 12 (responsible consumption and production) and Goal 13 (climate action) are negatively correlated with SWB; however, Environment Protection Index (EPI) is positively correlated with SWB.

(ix) SD of a country may have negative consequences (cost) to other countries or the actions of countries may influence the well-being of others.

(x) Nordic countries (Sweden, Finland, Iceland, Denmark and Norway) occupy the top ranks in the happiness ranking because of their welfare State model with extensive social benefit; even the foreign-born inhabitants (migrants) – Sweden has 19% immigrants – have a higher level of happiness. Actually welfare State generosity emancipates them from market dependency in terms of pensions, income maintenance for the ill/disabled, and unemployment benefits – plus labour market is regulated to avoid labourer's exploitation. Further, government quality, both '*democratic quality*' – access to power (e.g. ability to participate in selecting government, freedom of expression, freedom of association, and political stability) and '*delivery quality*' – the exercise of power (e.g. rule of law, control of corruption, regulatory quality, and government effectiveness – promotes citizen life satisfaction and happiness). Further low levels of inequality in Nordic countries contribute significantly to the level of happiness there. Nordic countries have 'virtuous cycle' with real democratic institutions.

As per a UNEP study (2021) the world wastes one-third of food produced – one billion tons of food annually, while one in nine persons lacks food security. Food waste by households and industry leads to 10% of the emissions driving climate change. One apple thrown wastes 125 litres of water used in producing it. Further food decomposing in landfills (like 'hills' in Delhi and other cities) causes greenhouse gas emissions, releasing methane (25 times stronger than carbon dioxide). Food could be reused to produce biogas, etc.; industry can enhance storage capacities: 'A circular food economy can reduce green house gas emissions by 4.3 billion tons of carbon dioxide annually, equalling a billion cars less on the streets' (Times Evoke 2021).

Sustainability Transition and Transformation: A Radical Decarbonising Perspective

At various international fora, ecologists, environmental NGOs as well as environmental movement's activists have, more or less, agreed that ecological sustainability requires structural changes in the mode of economy, practices of society, and public policies. Even IPCC (2014) reported that some gains in lowering carbon emission and energy per unit have been outpaced by a high economic growth rate and high population growth rate; hence, there is a genuine need for decarbonising the economy and society locally, nationally, regionally, and globally. No doubt, as many system ecologists have shown, already humanity is on the threshold of critical planetary boundaries (e.g. biodiversity losses, the nitrogen cycle, and climate change). Hence the radical decarbonising perspective goes beyond the earlier reformative approaches like 'environmental management' in 1970s and 1980s (focussing on command and control regulation and pollution control) or implementing sustainable development (SD) in 1990s (focussing on policy integration and ecological modernisation). It emphasises energy system transformation, transition to a low-carbon economy, a green economy, or a new climate economy. Sweden provides a critical case for ecological sustainability governance in general and for climate governance in particular. It is engaged in greening societal development and developing appropriate responses to climate change. A lower level of per capita emission and carbon intensity makes its performance remarkable compared to other countries. Decarbonisation involves two features – (a) efforts for steering and (b) efforts for enabling. Steering refers to the capacity of public actors to authoritatively make and enforce binding decisions, rules, and regulations upon social actors and sectors of society (law-making, planning regulations, fixing emission norms, taxation, fees, etc.). On the other hand, 'enabling' refers to softer activities (government subsidies, public procurement, public investments, planning innovation policy, research and development support, campaigns, labelling, voluntary agreements, etc.) (R. Hildingsson 2014). Hildingsson (2014) presents a comparative schema of types of environmental states and its capacity to govern decarbonisation (Table 3.14).

According to R. Hildingsson (2014), Sweden is a de facto environmental welfare State, based on administration rationality of governance, but employs other rationalities and forms of governance; it adheres to societal development towards 'ecologically sustainable ends' through decarbonisation, and hence it engages public actors and State institutions in the processes of 'transformative social change'. He finds, 'the Swedish case proves how reflexivity on ecological concerns can be enhanced within political institutions through reforms in State-led forms of environmental governance without necessarily resorting to new modes of (deliberative) ecological governance' (2014, 80). Hence State has to address both the hierarchical forms of steering and softer forms for enabling (e.g. orchestration) the process of change. The Swedish State has actually engaged in gradual (incremental) processes of change and explored ways to support other agents of change and enabled low-carbon transitions – e.g. Swedish transition to district heating (due to cold climate) and investment in renewable energy generation.

In the above typology presented by Hildingsson (2014), we find the fourth model as the best alternative: 'the green State' (post-liberal ecologism) aiming for 'discursive ecological democracy', designing for social change as 'ecological emancipation and reflexibility', by employing strategies of social change like public deliberation and constitutional entrenchment, wherein the civil society is the key change agent and deliberative ecological rationality of governance is opted. However, the radical left ecological politics goes beyond the green State (post-liberal ecologism) by integrating red and green political actors for social equity, sustainability, livelihood enhancing, cultural diversity, and ecological justice for all the people in terms of caste, class, race, ethnicity, religion, and gender (inclusion) to achieve the ethical earth democracy. Thus red-green or social green provides a more radical, comprehensive, and holistic agenda for eco-friendly transformations.

On the other hand, B. Sovacool and Hess (2017) (cited in J. Kohler et al. 2019) have proposed alternative justice theories and conceptualisations for sustainability transition (Table 3.15). The eco-centrism model is the most radical and transformative in this regard, hence to be preferred to others – as it advocates technologies or transitions

Table 3.14 Conceptualisation of the Environmental State and Social Change in Four Traditions

School of Thought (Type of State)	Political Order in Mind	View on Social Change	Strategies for Social Change	Key Change Agents	Main Rationality of Governance
1. Transition management	N/A (network society)	Socio-technical system innovation and 'regime shifts'; long-term restructuration	Goal-oriented modulation, variation selection, transition, experiments; reflexive monitoring	Front runners; rich actors and regime players ('transition arenas')	Managerial (networked governance)
2. Green Liberalism (environmental neo-liberal State)	Liberal democracy	Preferential and norm changes (functional incremental change)	Public deliberation; constitutional entrenchment (e.g. restraint principle, 'polluter pays')	Individual citizen and economic actors	Economic
3. Green welfarism	Social welfare State	Progressive instrumentalism, gradual institutional change from 'within system'	Step-wise policy change and reform; policy integration; institutionalisation; ecological modernisation	Government, public authorities, bureaucracies	Administrative
4. Post-liberal ecological (the green State)	Discursive ecological democracy	Imperative norm and value changes, ecological emancipation and reflexivity	'Ecological discursive designs' (deliberative democracy); legal, institutional, and policy reforms for reflexive ecological modernisation	Civil society (green public discourses); State institutions as 'stewards' and 'facilitators'	Deliberative, ecological

Source: R. Hildingsson (2014).

130 Sustainable Development

Table 3.15 Alternative Justice Theories and Conceptualisations for Sustainability Transition

Sl. No.	Concept	Definition	Application to Transitions
1.	Ubuntu Culture in Africa (I am because we are)	Emphasises act of building community, friendship, and oneness with a larger humanity	Neighbourhood efforts to promote energy efficiency and decisions about food resources within a community
2.	Taoism and Confucianism	Emphasises virtue and suggests that means is more important than the end	Respecting due process in transition decisions, adhering to human rights' protections when implementing infrastructural projects
3.	Hinduism and Dharma	Carries the notion of righteousness and moral duty and is always intended to achieve order, longevity, and collective well-being	Seeking to minimise the extent and distribution of externalities, offering affordable access to technology to help address poverty
4.	Buddhism	Expounds the notion of selflessness and the pursuit of individual salvation (nirvana)	Respecting future generations minimising harm to the environment and society
5.	Indigenous perspectives of Americas	Recognises interdependence of all life and enables good living through responsibility and respect for oneself and the natural world, including other people	Technologies developed cautiously through long-term experience and sovereign cultural protocols avoiding dramatic transformation of ecosystems
6.	Imal-centrism	Values and recognises the rights of all sentient lives	Promoting transition processes/ practices – veganism, vegetarianism, or waste reduction that avoids harm and provides benefits to all sentient animals
7.	Biocentrism	Values and respects the will to live and the basic interest to survive and flourish	Promoting transitions that adhere to a fair share of environmental resources among all living beings
8.	Eco-centrism	Give moral consideration for human and non-human communities and the basic functioning and interdependence of the ecological community as a whole	Advocating technologies or transitions that preserve the integrity, diversity, resilience, and flourishing of the whole ecological community

Source: Modified from B. Sovacool and Hess (2017), cited in J. Kohler et al. (2019).

that preserve the integrity, diversity, resilience, and flourishing of the whole ecological community.

A practical sustainable transition in terms of changes in technology, institution, and individual/community behaviour may be adapted in our everyday life (Table 3.16).

Thus, in three major domains of life, viz., food and agriculture, energy, and urban areas, various practical technological changes, institutional changes, and behavioural changes are suggested. There is, however, a huge gap in water (and other public

Sustainable Development 131

Table 3.16 Plan of Action for Alternative Sustainability Transition

Sl. No.	Domains of Life	Technological Change	Institutional Change	Behavioural Change
1.	Food and agricultural sustainability (production, distribution, consumption, and disposal)	Green manure, zero tillage, agro-forestry, crop cycle, and inter-cropping, use of predators and parasites for pest control	Banning of chemical fertilisers, pesticides, and weedicides, genetically modified (GM) foods, GM seeds, and terminator seeds, etc.; support for drip irrigation, promoting diversity of seeds, especially local indigenous seeds; massive surface (rain) water conservation/harvesting, promoting short-term and drought-resisting crops; promoting animal husbandry and fisheries controlling food wastes; protecting soil erosion	Consciousness raising of farmers not to grow water-intensive crops in water-stressed areas; not to burn paddy stems in fields, rather process it for manure, strengthen water users' associations, vegetarianism to be preferred to non-veg
2.	Energy sustainability	Renewable energy, cooking gas, electric motor, energy efficiency, from -waste-to-energy	Controlling the use of fossil fuel by removing subsidies, and promoting the use of renewable energy (solar, wind, geo-thermal, hydro, tidal waves, biomass) by giving subsidy, removing subsidy on electricity for irrigation, making cycle tracks, popularising animal-driven oil crushers/sugarcane crushers, promoting energy security to all, energy audit, efficiency incentive for electric car, promoting solar water heaters, energy efficiency obligations for utilities	Saving energy, using solar lanterns, avoiding air travel, refrigerators, and air conditioners, popularising walking and cycling to cover short distances
3.	Urban sustainability	Clean public transport (metro rail) rainwater harvesting in large offices, residences, and commercial complexes, green building technology	Use of solar water heaters in all urban houses, climate-smart land use management, good air quality for all, green belts, lakes, gardens, etc. in cities, cycle tracks and walking strips, regulating construction works, Bus Rapid Transit System (BRTS), preventing wastage of water supply, accessibility of toilets at all public places	Popularising walking and cycling, reducing the use of electronic gadgets, popularising reduce, reuse, recycle, recover, and minimum disposal/landfill, using toilet facilities, minimum use of water for a bath (no use of showers)

132 Sustainable Development

utilities too) use per head daily in the developed and developing countries; as per UNDP's HDR 2016, a US citizen uses 575 litres of water daily (in 2014), while 493 litres in Australia, 374 litres in Japan, 366 litres in Mexico, 193 litres in Germany, 187 litres in Brazil, 135 litres in India, 86 litres in China, 46 litres in Bangladesh, 46 litres in Kenya, and only 36 litres in Nigeria! Further this average use has again difference in class, gender, and location terms.

Conclusion

Sustainable Development (SD) has inherent contradictions as while the first term is closer to maintaining and stabilising ecological diversity, the latter term is closer to the economy wherein natural assets are exploited, often in unsustainable ways. History is witness to the decay and death of various civilisations (and cities) like Sumer and Indus Valley due to natural and anthropogenic causes directly or indirectly. Historically as well as contemporarily, there have been various ecozones or agro-climatic zones in India (15 at present) itself – thus different occupations are normally ecologically suitable or unsuitable to these ecozones. On the other hand, in ancient India, '*Aranya*' (forests) and '*Grama*' (settlements) had almost contrasting perceptions in terms of orderliness, discipline, knowingness, occupation, predictability, civilisation, power and status, and location.

The concept of SD got popularity with WCED's report 'Our Common Future' (1987), headed by Gro Harlem Brundtland, hence also called Brundtland Commission Report, that focusses on the need (not greed) and the ability of the future generations (inter-generational equity) to enjoy the bounty of nature as much as the present generation does. We have analysed the Ten Principles of SD by taking almost all the dimensions into account. However, SD is not without criticisms from the viewpoints of distributive justice, and real ecological sustainability: Putting the opposite camps/ideologues into a 'metafix', hijacking the agenda from real radical ecological groups, and not providing a critique of capitalism (by limiting to consumerism only). Here four perspectives on SD have also been analysed for having a critical theoretical appreciation: Bio-environmentalists, institutionalists, market liberals, and social greens. Here also two concepts of holistic sustainability framework (HSF) and real ecological sustainability (RES) have been put forward by going beyond the SD in the interests of the common people, ethics, and ecology. Here not only SDGs and their critique are also duly presented. Finally, radical sustainability transition, especially decarbonisation, is really useful to both developed and developing societies.

However, there is often development-induced displacement. How local people are affected due to big development works causing their dislocation and displacement, and how they are rehabilitated by the government/public administration later will be deliberated upon in the next chapter.

Points for Discussion

(i) What are the positive and negative aspects of SD? Discuss it in local, national, regional, and global contexts.

(ii) What are the major principles of SD? Take empirical data into account in order to explain this.
(iii) What are the parameters of the World Happiness Report? To what extent is it linked to SDGs?
(iv) How is the transition to a sustainable economy and society to be practised in the Indian context?

Chapter 4

Development, Displacement, and Rehabilitation

Introduction

Development is necessarily the 'planned change' in the desired direction, while economic growth, measured in terms of Gross Domestic Product (GDP) and per capita income, is a narrow concept, hiding the class, caste/race and gender disparity, illiteracy, hunger, disease, and infant/child/maternal mortality. Actually development should create and enhance capability, choices, and freedom to earn for well-being. Development should have concern with the inter-dependence of all five basic elements (*Panch Mahabhutas*): Earth, water, fire, space, and air. Further, we may say that development should be socially equitable, ecologically sustainable, politically participative and empowering, culturally acceptable (to local community), and economically income-generating. In short, it should be a holistic progress. In different cultures, the conception of development has been different throughout history. For instance, in developing countries like India, it is not just materialistic but rather moral, cultural, and spiritual, too, unlike the dominance of materialism and consumerism in the West. Bhutan was the first nation to talk of gross national happiness index (GNHI), not GDP because there a person is happy spiritually even if he/she may not be prosperous materialistically. Similarly, in our society, collective progress/well-being in terms of community, region, nation, or even the whole world (*'Vasudhaiva Kutumbakam'* – the whole earth is a family) has been the overall goal, while in the West (especially the USA) individualism has been in the centre of development concerns, and community, nation, or the earth is in the periphery. Further, more in the West, nature is often seen as empty, barren, savage, and detached from social life, hence 'to be filled', 'to be civilised', and 'to be conquered by the humans'! On the other hand, most of the developing societies like India conventionally see nature and culture as interdependent, harmonious, and mutually supportive entities. Finally, in our society time is seen as unbroken and eternal, and hence past, present and future periods of time are perceived as different stages of one whole process of a social continuum. On the other hand, in the West time is often reduced to the present only in terms of 'here and now', as the phenomenologists call it.

Development has different connotations, and hence different theoretical orientations in different contexts:

(i) *Planned change* in a desired direction – e.g. couple protection ratio, gross enrolment ratio
(ii) *Growth* – economic (e.g. GDP – or per capita income, annual growth rate) or technological (teledensity – 100 crore plus mobiles in India)

DOI: 10.4324/9781003336211-4

(iii) *Social change* – unplanned, unstructured, spontaneous as well as planned change in terms of caste, class, gender, family, community, politics, culture, etc.
(iv) *Personality development* – mental, emotional, and spiritual growth – IQ, EQ, and SQ (spiritual quotient)
(v) *Communication* – what is the new development (event) – no news is good news, bad news and views, etc., media's perception of reality
(vi) *More choices, freedom or right, or entitlement* – 'development as freedom' (Amartya Sen)
(vii) *Rural development*: Infrastructure development in rural areas
    (a) *Self-employment*, e.g. Swarnajayanti Gramin Swarojgar Yojana (SGSY) – credit–subsidy (micro finance)
    (b) *Wage employment*, M. Gandhi National Rural Employment Guarantee Scheme (MNREG) – initially Employment Guarantee Scheme (EGS) in Maharashtra
(viii) *Process of transformation of society* – through cooperation – 'pure means for pure end' (M. K. Gandhi) or through struggle – 'end justifies means' (K. Marx).

Thus various views focus on some or other aspects whereas, like poverty, development is also a multi-dimensional concept (economic, social, cultural, political, and ecological) that ultimately leads to human emancipation/liberation from unfreedoms and sufferings.

The Crisis of Development

It may be understood in two ways:

    (a) A crisis in the developing world: Increasing hunger, inequalities, and poverty or a lack of good governance
    (b) A crisis/impasse in development theory

Various social scientists have pointed out various reasons for the impasse in development theory (Shuurman 1993; Slater 1995):

    (a) Growing gap between the rich and poor
    (b) Preoccupation with short-term policies aimed at debt management
    (c) Economic growth devastating the environment
    (d) Delegitimisation of socialism
    (e) The global economy cannot be approached through national policies
    (f) Recognition of differentiation which deconstructs the usefulness of global grand theories or meta-theory
    (g) The advancement and challenges of feminism, post-modernism, and post-colonialism

Hence, let us discuss challenges from feminism, post-modernism, and post-colonialism in detail.

(1) **Feminist Challenge: Gender and Development**

136 Development, Displacement, and Rehabilitation

Various feminist social scientists (E. Boserup 1970; D. Elson 1995; G. Sen and C. Grown 1987; N. Kabeer 1995) challenge the mainstream development theory and provide gender and development perspective (that treats men and women equally in all respects) as follows:

(a) The re-examination of the development process with the awareness that it has impacted men and women differently, with women consistently losing out; for measuring women's empowerment three inter-related dimensions are to be understood – access and future claims to material, human, and social resources, agency (processes of decision-making as well as negotiation, deception and manipulation), and achievements (well-being outcomes), as argues Naila Kabeer (1999).
(b) A gender perspective is important for men and women to benefit equally from the processes of change, and the inequalities between men and women are to be reduced.
(c) Women's subordination is reinforced by the mainstream development policies perpetuating inequalities; during 1976–1985 (UN Decade for the Advancement of women) their position worsened; in economic and political crises – debt, famine, floods, militarisation, and fundamentalism – they suffered more.

But it may be criticised on three counts:

(a) Often it tends to focus on women per se rather than on the relations between men and women.
(b) Problematic representation of 'Third World' women – 'over burdened beasts' (Chandra Mohanty 1991).
(c) The subject has been problematised, technicalised, and marginalised as merely a matter of getting women to integrate more fully into the processes led by men.

(2) Challenge from Post-modernism: A Critique of Enlightenment Thinking

Various post-modernist scholars (Lyotard 1984; Jameson 1984) point out various assumptions of enlightenment thinking:

(i)   The world can be objectively analysed on the basis of *universal principles of truth, justice, and reason.*
(ii) There is a sharp *separation* between objective facts and subjective values
(iii) The *social world* can be captured and analysed *scientifically* just as much as the natural world.
(iv) There is *one path* to civilisation and social development.

Post-modernists challenge such assumptions in the following ways:

(a) They challenge the radical separation of fact and value and argue that all social theories are value-laden. There exists no objective science or truth. There is no single truth but a plurality of perspectives. So abandon the search for universal standards of truth and justice.
(b) A disbelief then in meta-narratives; suspicion of totalising theory and discourse.

(c) All truth claims are simultaneous claims to power.
(d) West is positioned as superior to the rest of the world as it claims to know universal justice.
(e) They also challenge the notion that the West is regarded as a model for other parts of the world to follow.

Hence the post-modernist theory's implications for development are as follows:

(a) Modernist/enlightenment discourse can be seen as a rationalisation for colonialism, development aid with conditionalities ('tied aid'), and Western intervention in the 'uncivilised' world (as 'the whiteman's burden').
(b) They present a critique of modernisation theory and the so-called superiority of the Western model and highlight problems of the modern Western world itself.
(c) Practices in the 'Third World' (developing countries) should be seen in their specific culturally embedded context since there is no way of assessing their truth or falsity apart from people's beliefs (Marglin 1991).
(d) Recognition and celebration of diversity and differentiation is emphasised – tolerance and pluralism, hybridity, fluidity, fragmentation, etc. To quote Lyotard (1984, XXV): 'Post-modern knowledge is not simply a tool of the authorities; it refines our sensitivity to differences and reinforces our ability to tolerate the incommensurable. Its principle is not the expert's homology, but the inventor's parology'.
(e) All development projects are relations of domination based on *power* and/or *knowledge*.

But the post-modernist theory, too, suffers from some fault-lines and infirmities. For instance, D. W. Harvey (1989) points out the following criticism of post-modernism:

(a) Extreme relativism pre-empts some persons to say that female circumcision (female genital mutilation) is not to be removed because it is a cultural tradition, judgement, and evaluation, and small pox in India was claimed as 'sheetla' goddess' curse so it can ignore or even become an apology for all kinds of oppressive practices.
(b) Highly individualised fragmented society fails to address the real issues of social power and inequalities.
(c) The political project gets buried in post-modernism.
(d) Reversals: 'West is bad, Rest is good' romanticisation in the developing world.
(e) Continuation of Western universalism, despite post-modernism.

Further, post-modernism 'deconstructs' the reality but does not 'reconstruct' for a holistic understanding. Finally, it ignores some historical dimensions of the issues.

(3) The Post-colonial Challenge

Many scholars have placed a post-colonial challenge to the mainstream (Western) knowledge and development (Fanon 1967, 1986; Spivak 1987, 1999; Said 1978, 1986; Bhabha 1983; James, C. L. R. 1983; Amin 1988; Thiong'o 1986) in the following ways:

(a) Suspicion of what is generally known as 'totalising discourses' (applies to theoretical systems of the left and the right) – the grand projects and truth claims of enlightenment thinking.
(b) Challenges Eurocentrism and the universality of Western knowledge, and argues for the deconstruction and de-colonisation of the Western intellectual tradition.
(c) The process of de-colonisation (of knowledge) involves, as a starting point, the recovery of the lost historical voices of the oppressed, the marginalised, and the dominated.

Beyond the Development Impasse

The following two categories of theorists go beyond the development impasse:

(A) Reformists: Impasse Theorists (Booth, 1985)

Rather than counter-identify and dismiss development because of the specific way it is articulated in the orthodox discourses of the international system, one may think of the ways in which development can be recast in a quite new kind of project (Slater 1993):
'Reverse discourses' of emancipation talk of six points:

(i)   Responsibility for the environment/sustainable development
(ii) Responsibility for human rights
(iii) Recognition of diversity and differentiation
(iv) Emancipation from oppression, exploitation, and subjection
(v)   Participation and empowerment
(vi) Equity

But the critics of 'Reverse discourses' find faults with these too, as Rist (1997) observes,

> each theory or declaration makes a claim to be original (or novel), to pass itself off as the solution at last discovered to the 'problems of development'. When we look more closely, however, we see that the apparent innovations are merely variations on a single theme which allow various actors to assert their legitimacy within the field of 'development'.
>
> (Rist 1997:5)

(B) Radical Theorists

They take a line, different from reformists, in the following ways:

1. The importance of social movements, alliances, and networks: To Escobar (1997), another meaning of development can be found in the context of resistance movements, and the creation, within civil society, of another kind of democratic innovation.
2. Deconstructing and challenging dichotomies: A historical and political critique of the discourse of development is presented by tracing the relations between imperialism and the construction of knowledge in development.

3. The production of Western knowledge is inseparable from the exercise of Western power.
4. 'Development' is talked about always and everywhere!

As most of us are aware, 'development rarely seems to work' – or at least with the consequences intended or the outcomes predicted. 'What it says it is doing, and what we believe it to be doing, are simply not what is actually happening'. The language of development can be evasive and misleading. Often the crazy slogan and show of development imposed 'from above' results in absurdity and madness (Crush 1995:4).

5. The end of 'Development'? This may be seen in the following ways:

(a) The persistence of the globalisation of capital, leading to the extraction of natural resources
(b) Re-thinking the relationship between different spheres of society: State, market, and civil society

 (i) Influencing government policy for the protection of common property resources (CPRs)
 (ii) The problematic relationship between State, market, and civil society to be changed in favour of civil society

(c) The continuing importance of critique – as Rist (1997) rightly remarks in view of the real experiences of life:

'nothing seems more legitimate than to spotlight what a discourse has been trying to hide, or take a position on the consequences flowing from it'.

(Rist 1997:3)

(d) 'Development' as a global challenge and critique – when development is seen as bringing displacement, climate change, pollution, acid rain, ozone depletion, etc., hence it is damaging the ecology. So it needs to be questioned and replaced by pro-ecology development.

In India, the central government (led by BJP as a major partner of National Democratic Alliance (NDA) since 2014 onwards) has taken a holistic liberal-reformative view of development:

(i) Sabka Sath (participation of all)
(ii) Sabka Vikas (development of all-inclusion)
(iii) Sabka Vishwas (trust of all); and
(iv) Sabka Prayas (efforts of all) (added in 2020–2021 during COVID-19)

However, these dimensions need to be addressed realistically in thought, word, and deed. Further, there should be the first and foremost concern for the ecology – that is, the development should not be carried out at the cost of ecology. India's central government's other motto is, 'perform, reform, and transform' but, in practice, 'reform' dominates over 'transform' – that is, incremental changes are preferred to fundamental structural changes (transformation) in economy and polity.

The Grand Narratives/Perspectives

[i] Development as Modernisation – Westernisation (Talcott Parsons, W. W. Rostow, Lerner, Eisenstadt)

Developing countries are expected here to 'catch up development' following the path of the West – 'west is best'!

(a) Social value system – joint family vs nuclear family, role specificity vs diffused-ness, ascription vs achievement; in such dualisms nuclear family, role specificity and achievement are considered modern values, while joint family, role diffuse-ness and ascription are considered traditional values.
(b) Psychological aspect – motivation, attitudes, and aptitude – individual (not the collective) as the lowest unit of development – individualism (in the USA), or individuality (in Western Europe) is considered a better identity criterion than collective identity.
(c) Economic aspect – capital accumulation – high vs low – free market is more significant in modernity than the State control or regulation for economic development.

W. W. Rostow: He wrote a book entitled *Stages of Economic Growth: A Non-Communist Manifesto* (1960) – every society passes through one of five stages:

(a) Traditional society stage (rigid ascriptive social structure + agricultural economy)
(b) Preconditions for the take-off stage (some impulse from outside, increase in trade, service, and mining)
(c) Take-off stage (growth of trade leads to the rise of investment of 10% of national income; socio-political conditions reshaped)
(d) Drive to maturity stage (consolidation – science and technology is extended to all spheres of the economy)
(e) Age of high mass consumption (consumerism) – per capita high energy consumption (as in the USA)

Key Elements of Modernisation Theory: It has the following key elements:

(a) Evolutionism – change through a fixed set of stages of economic growth
(b) Unilinearity – all societies to pass through the same route and same order
(c) Exposure and diffusion of Western ideas, values, and technology – it prefers nuclear family to joint family, role specificity to role diffuseness, and achievement to ascription
(d) Recapitulation (to follow the path of developed countries) – free market, capital accumulation, individual rationality, and GDP are key to the economy

Development, Displacement, and Rehabilitation   141

Limitations of Modernisation Theory: It has the following limitations:

(a) No actual watertight dichotomy between traditional and modern societies, but it is presumed here to be watertight.
(b) Pre-industrial and non-industrial paths of development are mixed-up/confused here.
(c) Historicity of imperialism/colonialism (development of the West ['metropolis'] is responsible for the underdevelopment of developing countries ['periphery'] as pointed out by the dependency theorists like Samir Amin, Andre Gundre Frank) is ignored.
(d) Often modern communication/transport technologies (films, videos, audio, SMS, social media) facilitate and give impetus to religious rituals or even fanaticism – from visiting religious places more frequently to websites/emails and social media for religious fundamentalism/terrorism.
(e) Blame game of developing countries' backwardness due to fatalist/passive/other-worldly attitude is not correct (in developing countries worshipping a god after purchasing a house, vehicle, land, etc. does not hinder the material development of the purchaser's family).
(f) It encourages a mad rush for over-use and exploitation of natural resources – wrongly sees nature as of 'use or instrumental value' only.
(g) It emphasises the trickle-down approach (top-down) that has been a failure in reality.
(h) Modernisation is wrongly equated with Westernisation.
  Marxist theory has been very critical of modernisation theory. Let us see its arguments below.

[ii] Marxist Theory of Development: Karl Marx's main points of development thought are as follows:

(a) Man makes history, and societal transformation comes through class struggle against the capitalist system.
(b) Existence (environment and socio-economic–political conditions) ultimately determines the human consciousness, not the vice versa.
(c) Mode of production consists of means of production and relations of production and the contradictions between the two lead to change through a class struggle.
(d) Fundamental contradiction exists between labour and capital – the capitalist does not give full wages to labour – as a surplus it is re-invested for more profit.
(e) 'Being' (thesis) contains its own contradictions, 'non-being' (anti-thesis), and the tension between them leads to 'becoming' (synthesis).
(f) Interdependence of humans and nature – 'nature is man's inorganic body' or 'original tool house' – the capacity to produce through tools and words differentiates humans from animals.
(g) Humans transformed external nature with instruments socially organised.
(h) In the capitalist economic system, the land is bought and sold as a private property, while in a pastoral/agricultural society land's use was need-based.

(i) Use, exploitation, and mastery of nature for humans' development.
(j) Distribution (egalitarian and just) matters as much as production.
(k) State to play an interventionist role in the protection of the rights of the deprived as well as to regulate the market in order to prevent monopoly and concentration of wealth in a few hands.
(l) Real development is possible only after the removal of private property – that will be feasible in the socialist mode of production.

Dependency Theory: Neo-Marxists like A. G. Frank, I. Wallerstein and Samir Amin talk of dependency theory or world systems theory in this context. For instance, A. G. Frank (1975) thinks that 'colonial', 'imperial', and 'capitalist' terms refer to a 'system of relations' in which domination, super-subordination, exploitation as well as development and underdevelopment play a key role (Box 4.1). In his view, underdeveloped nations contribute to developed nations in five ways:

(a) Economic 'surplus' and its role in capital accumulation
(b) Inefficiency or wastefulness of spoilation/exploitation and the sacrifice exceeds the contribution
(c) Discontinuities over time in the development of development and/or underdevelopment – the significance of a marginal but qualitatively important upward or downward push at a critical point in time
(d) Organisational or market discontinuities, e.g. monopoly and the possibility that a contribution to a part is greater than to its whole
(e) Contribution to system maintenance

He agrees with the view of Marvin Hariss, Marshal Sahlins, and Paul Baran that it is not so much the total wealth or income of a society, but rather its *surplus* and the *way* it is used that determine the type of development or underdevelopment that takes place. A. G. Frank (1975) further adds that whole civilisations were destroyed and the local people in colonies lost their livelihood when their essential irrigation systems were forced into disuse. Therefore, in his view, the development and underdevelopment are not the summations only of economic quantities, rather the whole social structure and process that determine that accumulation is to be taken into account. Disagreeing with Alfred Marshal's concept 'Natura non facit Saltrum' (nature does not take sudden leaps), A. G. Frank (1975) firmly argues that development is not a continuous ('take off') but rather a discontinuous process and so is underdevelopment that is 'trapped', and hence a jump is necessary. In fact, a marginal increase of investible surplus at the right time may be enough to permit development – and its absence to prohibit development. In his view, the most critical sector for most underdeveloped countries over the longest periods is foreign trade – mostly under foreign control. Such control, not over all the population, but over the bourgeoisie (or even part of it and its instruments of power) may be enough to keep a country underdeveloped indefinitely; unfortunately, the bourgeoisies of the under-developed countries are highly dependent on the metropolitan (centre) power globally. In fact, at the beginning of the mercantile and capitalist expansion, the distinction between conquest, trade, and robbery was not made – all the three were for the good of the metropolitan 'mother' country.

Box 4.1: Consequences of a Single Social System of Capitalism

A. G. Frank (1975) is of the view that the inseparable and inter-related consequences of a single social system of capitalism are as follows:

(a) Capitalism and feudalism – one does not universally follow the other.

(b) Capitalism and mercantilism – their unity is more important than their differences.

(c) Capitalism and colonialism – capitalism inevitably takes some colonial or imperial form, but the form changes with the circumstances.

(d) Capitalism and internal colonialism – essentials of colonial relations inevitably occur within States as well as between States.

(e) Capitalism and exploitation/diffusion – capitalism is a systematisation of an exploitative relation but also a reciprocal relation.

(f) Capitalism and class vs stratification – exploitative system is associated with a single two-class system but is complicated by a multiple stratification system.

(g) Capitalism and development/underdevelopment – development of capitalism produces simultaneous and interrelated development and underdevelopment, and vice versa.

(h) Capitalism and socialism – socialism is the escape from exploitation and underdevelopment made at once necessary and possible by the development of capitalism.

(i) Capitalism and liberation – escape from underdevelopment and subsequent development is no longer possible for them as a part of the capitalist system, and only liberation through socialist revolution offers that possibility – strategy for liberation, development, and peace should be that of unity, not a country strategy that will lead to the 'blood baths on the national level and/or a holocaust on the global one'.

Source: A. G. Frank (1975)

Another Marxist sociologist, John Hariss (2001), on the other hand, has rightly criticised new liberal theorist Robert Putnam's concept of 'social capital', i.e. 'trust, norms, and networks that can improve the efficiency of society for facilitating coordinated actions' – it is not what you know (that counts) but rather whom you know! But, John Hariss rightly argues actually 'capital' means 'assets' or 'resources' that are of value in producing things – not social relationships themselves. The development of capitalism is thus inseparable from the making of the working class that has the means to change the relations of production. Therefore, social capital for one group of people may constitute 'social exclusion' for others. For instance, the caste system may benefit the upper castes but generally excludes the lowest castes. Thus the concept of 'social capital' elevates symptom to cause, ignoring the structural factors like distribution of resources, agrarian production relations, and downplaying the role of political organisation and struggle. Thus 'social capital' is a weapon in the armoury of the 'anti-politics' machine.

144 Development, Displacement, and Rehabilitation

Limitations: With the fall of the Soviet Union in 1990, the practice of Marxist thought got a severe jolt. In practice, there is hardly any socialist society in the world now in the true sense of the term. Marx, too, talked of optimal exploitation of natural resources especially in his later writings. Hence it often overlooks the ecological symbiosis between humans, plants, and animals. However, in his early writings like Economic and Philosophical Manuscripts of 1844, Marx talks of the full emancipation of humans, and their harmony with nature. Further, Marx did not elaborate the gender aspect, though implicitly referred to it.

[iii] World Bank: **Neo-liberal Democratic/Neo-classical Approach**

Here the following notions are considered:

(a) Development equals 'economic growth'; that is, focusing on an increase in GDP and per capita income will lead to more purchasing power.

(b) A high growth rate is essential at any cost – by cutting costs of production, and by removing subsidies that act as distortions/restrictions for market forces.

(c) 'More production, more profit' – size of a cake is more important (than its distribution), maximum profit, and minimum cost (economic) through economy of scale.

(d) The State should act only as a facilitator and need not intervene in a competitive and free market economy.

(e) Free market (trade of goods and services) decides the right allocation of resources.

(f) Taking maximum benefit with minimum costs of production through an optimal exploitation of natural resources.

(g) More quantity of consumption of goods and energy is the indicator of more development.

(h) Development means merely 'incremental change' (not structural) in the economy.

(i) 'Big is beautiful' – big dam, big building, big bridge, big Mall/Bazar, big Highway, big airport/station – size matters for the economy of scale.

It, more or less, follows the modernisation theory, and hence criticism of that applies here too.

[iv] UNDP: **Human Development Approach** (Amartya Sen, Mahbub-ul-Haq, Richard Jolly) – It criticises the neo-liberal/neo-classical approach to development in the following ways:

(a) High GDP is not a guarantee for development (or poverty – and unemployment – reduction) – 8%–9% annual growth rate in India did not lead to a substantive reduction in poverty and unemployment during the first and second decades of the twenty-first century (except 2019–2020).

(b) Higher per capita income does not mean equal income distribution – distributive justice does not accompany faster/higher economic growth.

(c) To build human capabilities – range of choices/entitlements – long life, decent living, education, and participation in the decision-making process, thus ensuring human development.

Development, Displacement, and Rehabilitation   145

(d) Purchasing power alone does not capture one's health, education, and freedom to shape life and enlarge the choice domain.
(e) 'We need a measure of the same level of vulgarity as GNP – just one number – but a measure that is not as blind to social aspects of human lives as GNP is' (Mahbub-ul-Haq).

UNDP's human development report (HDR) came out first with the human development index (HDI) in 1990. HDI consists of longevity (indicating health), school education and adult literacy, and per capita income – three aspects are better than only one aspect of economic growth (per capita income) as follows:

(a) Longevity is measured by life expectancy at birth; thus it reflects health.
(b) Educational attainment is measured by adult literacy (15 years+) (two-thirds weight) – and combined gross (primary, secondary, and tertiary) enrolment ratio (one-third weight).
(c) Real GDP per capita income measures the standard of living.

HDI is measured between 0 and 1, 0.8 and above indicates high HDI, 0.5–0.8 shows medium HDI, and below 0.5 shows low HDI.

In Tables 4.1–4.3, high, lowest, and medium HDI (+ some high) countries are given for proper appreciation of human development.

As per HDR 2020, India ranked 131 out of 189 countries, one rank less than the previous year – with a score of 0.645, while Sri Lanka's score was 0.782, that of China was 0.761, that of Bhutan was 0.654, that of Bangladesh 0.632, that of Nepal 0.602, that of Myanmar 0.583, that of Pakistan 0.557, and that of Afghanistan 0.511. Sri Lanka's HDI rank was 72, China's 85, Bhutan's 129, India's 131, Bangladesh's 133,

Table 4.1 Top 10 High (Above 0.8) Human Development Index (HDI) Ranking Countries (Human Development Report [HDR] 2019)

Rank	Country	HDI	Life Expectancy at Birth (yrs.)	Expected Years of Schooling (yrs.)	Mean Year of Schooling (yrs.)	Gross National Income (GNI) per Capita (PPP$)
1.	Norway	0.954	82.3	18.1	12.6	68,059
2.	Switzerland	0.946	83.6	16.2	13.4	59,375
3.	Ireland	0.942	82.1	18.8	12.5	55,660
4.	Germany	0.939	81.2	17.1	14.1	46,946
4.	Hong Kong	0.939	84.7	16.5	12.10	60,221
6.	Australia	0.938	83.3	22.1	12.7	44,097
6.	Iceland	0.938	82.9	19.2	12.5	47,566
8.	Sweden	0.937	82.7	18.8	12.4	47,955
9.	Singapore	0.935	83.5	16.3	11.5	83,793
10.	The Netherlands	0.933	82.1	18.0	12.2	50,013

Source: HDR 2019.

146   Development, Displacement, and Rehabilitation

Table 4.2 Lowest (Below 0.5) 10 Human Development Index (HDI) Ranking Countries (Human Development Report [HDR] 2019)

Rank	Country	HDI	Life Expectancy at birth (yrs.)	Expected Years of Schooling (yrs.)	Mean Year of Schooling (yrs.)	Gross National Income (GNI) per Capita (PPP$)
180.	Mozambique	0.446	60.2	9.7	3.5	1,154
181.	Sierra Lione	0.438	54.3	10.2	3.6	1,381
182.	Burkina Faso	0.434	61.2	8.9	1.6	1,705
182.	Eritrea	0.434	65.9	5.0	3.9	1,708
184.	Mali	0.427	58.9	7.6	2.4	1,965
185.	Burundi	0.423	61.2	11.3	3.1	660
186.	South Sudan	0.413	57.6	5.0	4.8	1,455
187.	Chad	0.401	54.0	7.5	2.4	1,716
188.	Central African Rep.	0.381	52.8	7.6	4.3	777
189	Niger	0.377	62.0	6.5	2.0	912

Source: HDR 2019.

Table 4.3 Medium (0.5–0.799) and High (Above 0.8) Human Development Index (HDI) Ranks of Selected Countries (Human Development Report [HDR] 2019)

Rank	Country	HDI	Life Expectancy at birth (yrs.)	Expected years of Schooling (yrs.)	Mean Year of Schooling (yrs.)	Gross National Income (GNI) per Capita (PPP$)
85.	China	0.758	76.7	13.9	7.9	16,127
71.	Sri Lanka	0.780	76.8	14.0	11.1	11,611
129.	India	0.647	69.4	12.3	6.5	6,829
134.	Bhutan	0.617	71.5	12.1	3.1	8,609
135.	Bangladesh	0.614	72.3	11.2	6.1	4,057
72.	Cuba	0.778	78.7	14.4	11.8	7,811
147.	Nepal	0.579	70.5	12.2	4.9	2,748
63.	Serbia	0.799	75.8	14.8	11.2	15,218
49.	Russian Federation	0.824	72.4	15.5	12.0	25,036
36.	Slovakia	0.857	77.4	14.5	12.6	30,672
26.	France	0.891	82.5	15.5	11.4	40,511
19.	Japan	0.915	84.5	15.2	12.8	40,799
15.	The USA	0.920	78.9	16.3	13.4	56,140

Source: HDR 2019.

Nepal's 142, Myanmar's 147, Pakistan's 154, and Afghanistan's 169. India's average HD score hides the better position of some developed States like Tamil Nadu, Maharashtra, Kerala, and Delhi and also hides the worse position of backward States like Bihar, UP, MP, Chhattisgarh, Jharkhand, Odisha, West Bengal, Assam, Rajasthan. On the other hand, the top ten countries' ranks are as follows in descending order: (1) Norway, (2) Ireland, (3) Switzerland, (4) Hong Kong, (5) Iceland, (6) Germany, (7) Sweden, (8) Australia, (9) The Netherlands, and (10) Denmark. The HD scores are classified into five categories:

a) Very high HD – 0.898
b) High HD – 0.753
c) Medium HD – 0.631
d) Low HD – 0.513
e) Developing countries as a whole – 0.689

Thus though India aspires to have the third largest economy ($5 trillion) in the near future, the world's knowledge guru, and to be a regional power, in HDI it lags far behind most of the developing countries. Further, India has only 8.6 doctors per 10,000 population, while very high HDI countries have 31.2 doctors, high HDI countries 17, and medium HDI countries 7.9; China has 19.8 and Sri Lanka has 10 doctors per 10,000 population. Similarly very high HDI countries have 52 beds, high HDI countries 31, China 43, Sri Lanka 42, Bhutan 17, Myanmar 10, Bangladesh 8, and India only 5 beds per 10,000 population (HDR 2020)!

From Table 4.1, it is also clear that the three top very high HDI ranking countries in 2019 were Norway (0.954), Switzerland (0.946), and Ireland (0.942) where democracy is functioning strongly. From Table 4.2 it is clear that the three lowest HDI countries were Niger (189th), Central African Republic (188th), and Chad (187th) – thus the position in Africa is the worst. From Table 4.3 it transpires that China ranks 85th, Sri Lanka 71st, India 129th, Bhutan 134th, Bangladesh 135th, and Nepal 147th in the world. However, it is notable that though the gross national income (GNI) of Sri Lanka (11,611 PPP$) is less than China's GNI (11,127 PPP$), Sri Lanka's ranking is higher (71st) than China's (85th) and its HD score is also higher (0.780) than China's (758); though both countries have almost equal score in expected schooling years (14 and 13.9, respectively), mean year of schooling in Sri Lanka (11.1) is higher than China's (7.9); both have an equal life expectancy at birth (76.8 and 76.7 years). Thus democracy in Sri Lanka has a better performance, but in 2022 it faced a systemic crisis due to bad governance, huge debt, and wrong policy decisions (totally banning chemical fertilisers and promoting organic agriculture).

UNDP and Oxford University's Oxford Poverty and Human Development Initiative (OPHI) came out with Multi-dimensional Poverty Index (MPI) first time in 2010 covering over 100 developing countries with the following ten indicators in three broad categories of education, health, and standard of living (complementing income-based poverty measures like $1.90 daily (OPHI 2020):

(I) **Education:** (1) Years of schooling (deprived if no household member completed six years of schooling; (2) child's school attendance (deprived if any school-aged child is not attending school up to 8th standard).

(II) **Health:** (3) Child mortality (deprived if any child in the family died in the last five years; (4) nutrition (deprived if any adult or child is stunted).

(III) **Standard of Living:** (5) Electricity (deprived if a household has no electricity connection; (6) sanitation (deprived if sanitation facility is not improved [as per Millennium Development Goals – MDG-guidelines] or it is improved but shared with other households); (7) drinking water (deprived if a household has no access to safe drinking water, as per MDG guidelines, or safe drinking water is more than 30-minute walk from home; (8) floor (deprived if a household has a dirt, sand, or dung floor); (9) cooking fuel (deprived if a household cooks with dung, wood, or charcoal); (9) assets' ownership (deprived if a household does not own more than one of these – radio, TV, telephone, bike, motorbike, refrigerator, and a car or truck).

Global MPI 2020 report found that:

(a) Sixty-five countries substantially reduced their multi-dimensional poverty levels in absolute terms during 2000–2019; 47 countries were on track to halve their multidimensional poverty, and 18 countries were off track as per sustainable development goal 1.2.2.

(b) Twenty-two per cent of people (1.3 billion people) across 107 developing countries live in multi-dimensional poverty; half of the multi-dimensionally poor people (644 million) are children below 18 years – one in three children is poor compared to one in six adults.

(c) Eighty-four per cent of multi-dimensionally poor people live in Sub-Saharan Africa (558 million) and South Asia (530 million); two-thirds of multi-dimensionally poor people live in middle-income countries.

(d) Seventy-one per cent of the 5.9 billion people covered in global MPI have at least one deprivation – the average number of deprivations is 5.

(e) Four countries halved their MPI value; India halved nationally and among children (during 2005–2006 to 2015–2016), and had the biggest reduction in the number of multi-dimensionally poor people (270 million).

(f) In nearly one-third of the countries studied, either there was no reduction in multi-dimensional poverty for children or the MPI fell more slowly for children than for adults.

(g) The countries with the fastest absolute reduction in MPI were Sierra Leone, Mauritania, and Liberia, followed by Timore-Leste, Guinea, and Rwanda. North Macedonia had the fastest relative poverty reduction, followed by China, Armenia, Kazakhstan, Indonesia, Turkmenistan, and Mongolia – each cutting original MPI value by at least 12% a year.

(h) There is a negative, moderate correlation between the incidence of multi-dimensional poverty and the coverage of three doses of Diptheria, Pertussis, Tetanus (DPT) vaccine.

(i) In Sub-Saharan Africa, 71.9% of people in rural areas (446 million people) are multi-dimensionally poor, and 25.2% (92 million people) are so in urban areas; this is the worst scenario.

(j) Environmental deprivations are most acute in Sub-Saharan Africa – at least 53.9% of the population (547 million people) is multi-dimensionally poor and faces at least one environmental deprivation, and these are also high in South Asia – at least 26.8% of the population (486 million people) is multi-dimensionally poor and lacks access to one of three environmental indicators.

Thus, HDI is a better method for measuring development, but some social and political aspects are left out.

[v] **Social Development Approach (CSD, UNICEF, and WHO Closer to It)**

The Council for Social Development (New Delhi) brought out the first Social Development Report (SDR) in 2006. It pointed out that the human development index (HDI) is not properly indicating all aspects of development. It, therefore, proposed Social Development Index (SDI) with six types of indicators and, later in 2008, two more types of indicators (the last two) were added; hence, SDI has 28 indicators, in total, in the following 8 types:

(A) **Demographic Indicators**

    (i) Contraceptive prevalence rate (CPR)
    (ii) Total fertility rate (TFR)
    (iii) Infant mortality rate (IMR)

(B) **Health Indicators**

    (i) Percentage of institutional delivery
    (ii) Percentage of undernourished children

(C) **Educational Attainment Indicators**

    (i) Literacy rate
    (ii) Pupil–teacher ratio
    (iii) School attendance rate

(D) **Basic Amenities Indicators**

    (i) % of households living in pucca houses
    (ii) % of households having access to safe drinking water
    (iii) % of households having access to toilet facilities
    (iv) % of households having electricity connection

(E) **Economic Deprivation Indicators**

    (i) Gini coefficient ratio for per capita consumption expenditure
    (ii) Unemployment rate

(F) **Social Deprivation Indicators**

    (i) Disparity ratio between SCs (scheduled castes) and general population in literacy rate
    (ii) Disparity ratio between STs (scheduled tribes) and general population in literacy rate
    (iii) Disparity ratio between females and males in literacy rate
    (iv) Ratio between female unemployment rate and the total unemployment rate
    (v) Disparity ratio of per capita expenditure on Muslims to that of total population
    (vi) Child sex ratio

(G) **Social Groups Indicators**

    (i) Demographic indicators (CPR, TFR, IMR)

150 Development, Displacement, and Rehabilitation

(ii) Health indicators (% institutional delivery and the undernourished children)
(iii) Educational attainment indicators (literacy rate and dropout rate of children)
(iv) Basic amenities indicators (% pucca house, safe drinking water, toilet facilities, and electricity connection)
(v) Economic deprivation indicators (people living below the poverty line, and unemployment rate)

(H) **Gender Indicators**

(i)   Health indicators – IMR, % of undernourished children
(ii) Educational indicators – literacy rate, and higher secondary complete and above
(iii) Economic deprivation indicators – unemployment rate, wage rates for males and females

Thus the social development approach is the most comprehensive and realistic. However, an environmental/ecological dimension is to be added here for capturing the Holistic Sustainability Framework (HSF), and Real Social Development (RSD), as discussed in Chapter 3 on Sustainable Development. Sometimes a specific project of development may not lead to displacement as such but may certainly damage the ecology – the habitat of the animals, fish, and birds, hence loss of biodiversity ultimately (Box 4.2):

Box 4.2: Delhi–Doon Express Way's Extension Leading to Cutting of 2,500 Old Trees, Uttarakhand (India)

The National Board for Wildlife (NBWL) gave final clearance for a 19.8-km stretch to Ganeshpur (Uttar Pradesh [UP])–Dehradun Road (NH 72A) – an extension of the Delhi–Dehradun Expressway. It will pass through Rajaji Tiger Reserve and the Shivalik Elephant Reserve. Uttarakhand will forego 2,500 British-era Sal trees and about 10 hectares of reserve forest area (a part of Rajaji Tiger Reserve) on a 3.6-km stretch, while UP will give 47 hectares of forest land (a part of Shivalik Elephant Reserve) – on about 16-km stretch.

Government of India perceives that this extension will reduce the distance between Delhi and Dehradun by 70 km – from 250 to 180 km; thus it will reduce travel time to 2.5 hours. On the other hand, scientists, and NGOs like Green Doon point out its negative consequences:

(a) Destroying the habitats of many wild animals (tigers, elephants, etc.) and birds in eco-sensitive zones.

(b) From Mohund to Asarori on the Dehradun border, at present it takes 28 minutes, and after the extension of the expressway it will take 18 minutes – reduction of only 10 minutes, not much reduction in travel time.

(c) The British era's 2,500 old Sal trees will be cut – Sal trees take a long period to regenerate and form a perfect habitat for birds and animals, and hence cutting off a huge number of Sal trees would be 'a significant loss to the area's biodiversity', as per J. M. Tomar, Principal scientist at Indian Institute of Soil and Water Conservation,

> Dehradun. To this criticism, the Chief Wildlife Warden of Uttarakhand spoke of some mitigation measures like barriers on both sides of the road 'to block and absorb noise and light pollution, bamboo plantation, etc, and monitoring for 2–3 years' to check animal–road killings, and man–wildlife conflicts there! But environmental NGOs hardly trust this statement.
>
> Source: Shivani Azad (2021)

Therefore the main leader of Narmada Bachao Andolan, Medha Patkar (2019) is right of the view that neither there is nor should be a rift between the development and natural continuity; but the nexus between the ruling class and the investors, nationally and internationally, is affecting all our natural resources. She considers that every natural resource or asset is the basis of our life, not merely a means of livelihood. In fact, the holistic perspective of the deprived local people is not limited to the aesthetics of water, nor even restricted to the water conservation, but extends to their need and right to water. Now the question arises of how the environment-induced migration and displacement affect the local people.

Migration and Displacement

Due to various activities of development or environmental change, there is voluntary and/or involuntary migration of the population from the original place of residence to a new destination. The salient features of voluntary and involuntary (forced) migration (first dislocation and then relocation) are notable below (Hugo 2010):

(a) The involuntary migrants do not make preparations.
(b) They maintain a greater commitment to the settlement at the origin.
(c) They are more likely to be in a state of stress.
(d) They are less likely to bring assets.
(e) They are less likely to have connections with their place of origin.

This may be seen vividly in Table 4.4.

Now the question arises whether such migration is uniform or differentiated. In fact, F. Renaud et al. (2007, cited in Hugo 2010) have distinguished between different

Table 4.4 Important Differences Between Voluntary and Involuntary Migrants

Voluntary Migrants	Involuntary Migrants
(a) Make preparations	Do not make preparations
(b) Maintain less (or no) commitment to the origin	Maintain a greater commitment to the settlement at the origin
(c) Not likely to be in a state of stress	Are likely to be in a state of stress
(d) Are more likely to bring assets	Are less likely to bring assets
(e) Are more likely to have connections with origin	Are less likely to have connections with origin

Source: After Hugo (2010).

152   Development, Displacement, and Rehabilitation

Table 4.5 Distinctions Between Environmentally Induced Migrants as per Force or Choice

Type Based on Force or Choice	Characteristics
(a) 'Environmentally motivated' migrants	Who choose to move, and in their choice, environmental factors have a role
(b) 'Environmentally forced' migrants	In situations where environmental change has destroyed or is likely to destroy their livelihood – no choice in moving out, but some choice in the timing of the move
(c) 'Environmental refugees'	Have no choice in moving out and the timing of the move

Source: After F. Renaud (2007, cited in Hugo 2010).

Table 4.6 Principal Causes of Environmentally Induced Migration

Category	Particular Causes
(a) Natural disasters	Floods, earthquakes, volcanic eruptions, landslides, coastal storms, hurricanes, tsunamis
(b) Cumulative (slow-onset changes)	Land degradation, droughts, water deficiency, climate change, sea-level rise
(c) Involuntarily caused and industrial accidents	Nuclear accidents, factory disasters, environmental pollution
(d) Development projects	Construction on rivers, dams, and irrigation canals, mining natural resources, urbanisation
(e) Conflicts and workforce	Biological workforce, intentional destruction of the environment, conflicts due to natural resources

Source: R. Stojanov (2008).

types of environmentally induced migrations according to the degree of force versus choice in the moves. This may be seen in Table 4.5.

Unfortunately the term 'refugee' is 'formed', 'transformed', and 'normalised' in policy discourse by bureaucratic practices (Zetter 2007). Thus these labels are forms of 'stereotyping and social control' (Muggah 2008).

As far as the main causes of migration (dislocation to displacement) are concerned, these relate to natural disasters, cumulative (slow-onset) changes, involuntarily caused industrial accidents, development projects, and conflicts and workforce. This may be seen in Table 4.6.

Further there is a distinction between moving out of people associated with the rapid onset of environmental disasters, and others caused by the slow deterioration of the environment and a reduction in natural resources available to local people. Sudden disasters are more destructive and usually cause major displacement, but these are temporary. On the other hand, in slow and gradual environmental deterioration, the migration response is more complex – labour migration to cities for additional earnings. This may be seen in Table 4.7.

Displacement

Displacement is distinct from migration as the former has three core features:

(a) Involuntariness

Table 4.7 Types of Environmentally Induced Migration and Mobility Response

Environmental Change	Examples	Mobility Response (Displacement)
(a) Sudden, extreme events	Tsunami, flooding, typhoon	Large scale but largely temporary displacement
(b) Gradual deterioration of the environment	(a) Desertification (b) Dam construction	(a) Temporary, circular migration of family members to supplement local income generation (b) Gradual permanent out-migration of households

Source: W. Kalin (2010).

(b) Temporariness in some cases but permanent in large dams, airports, ports etc
(c) Physical dislocation

On the other hand, migration is voluntary, even if it is temporally and spatially varied. Questioning the binary position, and pointing out the differentiation, Gebre (2002), in the study of *villagisation* in Ethiopia, proposed the following new sub-categories:

(i) Voluntary
(ii) Induced voluntary
(iii) Compulsory voluntary
(iv) Involuntary

Therefore, they are labelled differently, mainly by outsiders (experts and common people both), often with disgrace, disrespect, and de-humanising as per Muggah (2008), labelling of migrants and displaced persons has been quite different in various contexts (Table 4.8).

In India, distress migration is prevalent among tribals, Dalits, and other poor sections of society (including upper castes from eastern UP and North Bihar). Due to climate change or development works like big dams, bridges, highways, railways, airports, mining, many tribal migrants are now uprooted. An analysis of 27 tribal-dominated districts in selected States during 2017–2019 reveals the following (Pandit 2021):

Table 4.8 Labelling of Migrants and Displaced Persons

Migrant	Displaced
(a) Economic migrant	Displaced person
(b) Migrant worker	Refugee
(c) Encroacher	IDP
(d) Squatter	De-housed
(e) Seasonal migrant	Evacuee
(f) Economic refugee	Environmental refugee
(g) (g) Labour migrant	Oustee

Source: Muggah (2008).

154   Development, Displacement, and Rehabilitation

(a) Most of the migrating tribals are from tribal-dominated States: Chhattisgarh (with 31% tribal population), Jharkhand (26% tribal population), Odisha (23% tribal population), and Madhya Pradesh (21% tribal population).

(b) About 60% of the Scheduled Tribes (ST) population in Chhattisgarh lives out of the State, 80% of Jharkhand's tribal population lives out of the State, 90% of Odisha's tribal population lives out of State, and 60% of Madhya Pradesh's (MP) tribal population lives out of State.

(c) The destinations of the migrant tribal population are Punjab, Gujarat, Delhi, Kerala, Goa, Maharashtra, and Telangana, where wages are more than the originating native States of the tribals.

(d) Goa has more employment opportunities in fishery and tourism – owners of fishing trawlers prefer cheap labour with stability.

(e) Kerala offers maximum social security to migrant labours (from north India, e.g. West Bengal, Assam, and Nepal), and the language spoken at many large construction sites is often not Malayalam, but Tamil, Hindi, Bengali, Assamese, or Nepali).

(f) Interestingly, migrant labours come to Kerala for work and at the same time many labourers from Kerala go to Goa for work.

(g) Though Telangana is a poor State, its tribal groups do not migrate out for work; it is a destination for migrant workers including tribals from other States – its capital Hyderabad is a hub for jewellery and embroidery works.

(h) Internal remittances from such tribal and other migrants from urban areas have helped in poverty reduction in rural areas: (i) 80% of cash income in some villages came from the migration of people from States like MP, Rajasthan, and Gujarat; (ii) 29% of the income of SCs, STs, and Muslims is earned through migration; and (iii) 33% of the average annual income of the landless and marginal households in Bihar came from remittances of migrants.

(i) There are four trends in tribal migration in India:

   (i)   Family migration triangles – seasonal movement more within Odisha, Chhattisgarh, and Jharkhand, and between Maharashtra, Gujarat, and MP.

   (ii)   Feminisation of migration – Jharkhand's tribal women move to work in the textile industry and as domestic helps in Delhi/NCR.

   (iii) Single-male migration to fisheries – male workers from Odisha migrate to Goa and are preferred for staying on trawlers for long.

   (iv)   Daily wages single biggest draw – more wages in industrialised States attract more migrants from MP, Jharkhand, and Odisha (with low wage rates).

   (v)   More earnings by women as workers improve their status and mobility; on the other hand, the long absence of males in the same communities makes women more vocal in decision-making.

Therefore, the UN Commission on Human Rights (1997) clarified that migration takes place in 'all cases where the decision (to migrate) is taken freely by the individual concerned, for reason of personal convenience and without intervention of an external compelling factor' (cited in Muggah 2008, 17). On the other hand, in displacement choices are restricted, and the individual (or his family) faces more physical and/or psychological risks than opportunities by staying at the original places. No doubt, physical dislocation is associated with the loss of physical home and property, reduced income-earning potential, and productivity.

Three major causes of displacement are (a) natural calamities (floods, famines, landslides, earthquakes, tsunamis, tides, epidemics/pandemics [COVID-19]); (b) conflicts and wars – as per United Nations High Commissioner for Refugees (UNHCR), 15.2 million people became refugees outside their native countries and 26 million were displaced (e.g. in Iraq, Syria, Palestine, Sudan, Chad, Honduras, Nicaragua, Colombia, Ecuador, Azerbaijan); (c) development-induced displacement (DID). After World War II, about 400 million (40 crores) people were displaced due to development works globally and 60 million (6 crores) in India. Every year 15 million people are displaced in the world, mostly in developing nations like China, India, Brazil, Africa, and South America (Peru, Mexico, Bolivia, etc.).

Main Causes of Development-Induced Displacement

Here following causes are notable (Perminski 2013).

(a) Construction of dams, hydropower plants, irrigation projects, artificial reservoirs, and canals. According to a report by the World Commission on Dams, 'construction of large dams has led to the displacement of some 40 to 80 million people worldwide', 26.6% of World Bank-financed projects (active in 1993) were caused by dam-building.
(b) Transportation – roads, highways, and railways – it caused 24.6% of all development-induced displacement in World Bank-financed projects (active in 1993).
(c) Urbanisation, re-urbanisation, and transformation of urban space – over 60% of development-induced displacement worldwide resulted from urbanisation and transportation projects; urban infrastructures caused 8.2% of resettlement worldwide.
(d) Mining and transportation of natural resources – over 60% of the world's natural resources are located on indigenous lands; the large-scale displacement of people is also a consequence of the expansion of mining areas; mining of a goldmine in the Tarkwa region in Ghana displaced up to 30,000 persons.
(e) Deforestation and expansion of agricultural areas – monoculture palm oil plantations displaced, in Borneo Island and Colombia, a huge population.
(f) Creation of national parks and reserves (conservation of nature) – after 1900 AD–2000 AD more than 1,08,000 such parks and reserves were created in the world – many were linked with involuntary resettlement.
(g) Population redistribution schemes – in April 1975 Khmer Rouge relocated Cambodian people from Phnom Penh and other cities to rural areas; in April 1985, in Vietnam (with the unification of North and South regions), the 'return to village' programme led to involuntary resettlement of a huge population.
(h) Other causes – airports, ports, landfill sites, etc. (e.g. New port, metro in Mumbai).
Now the question arises about resettlement and rehabilitation.

Resettlement: Community Relocation

M. D. Lieber (1977) uses the general term 'resettlement' to mean 'a process by which a number of homogeneous people from one locale come to live together in a different locale'. Further John Campbell (2010) used the term 'relocation' to refer to

156   Development, Displacement, and Rehabilitation

the permanent (or long term) movement of a community (or a significant part of it) from one location to another, in which important characteristics of the original community, including its social structures, legal and political systems, cultural characteristics and worldviews, are retained.

(58–59)

Community relocation is one of 'the most radical forms of adaptation to climate change'. Chambers (1969, 5) presents a broad definition of resettlement as 'planned social change that necessarily entails population movement, population selection, and most probably population control'. Thus it has both spatial and control elements. While Chambers (1969) saw resettlement schemes as 'experiments in nation-building in miniature', on the other hand, T. Seuddar (1973) rightly perceived it as 'ecological imperialism' where resettlement schemes represented an extension of the world economic system into new domains. Resettlement failed because, first, it was initiated by the outside social actors (donors, bureaucracy, engineers, planners, political masters) who ignored the local complexity and provided technical solutions – they wanted to civilise and modernise the 'primitive' or 'backward' dislocated people. Hence civil society resisted such elitist intervention 'imposed from above'. Second, unlike 'settlement' and 'colonisation', resettlement is involuntary and also differs from 'assisted migration' (through State planning relocation of individuals ('enterprising') in an uninhabited area. Thus labelling of settlers and resettlers may be seen in Table 4.9.

Further, while displacement may be temporary or permanent (in the case of project-affected), resettlement is seen as a permanent process of relocation or settling again in a new area due to nuclear leakage, natural disasters or conflicts/wars, etc.

As far as adaptation is concerned, it has been defined by Inter-governmental Panel on Climate Change (IPCC) as 'adjustment in natural or human systems in response to actual or expected climatic stimuli or their effects, which moderates harm or exploits beneficial opportunities' (cited in Campbell 2010, 59).

But community relocation or resettlement requires long-term costs to continue for many generations. Further, land has multiple meanings in most of the Pacific Island countries, as elsewhere, and is often difficult to separate from those who 'belong' to it; usually, there land is not owned as a freehold, though it can be exchanged under traditional customary arrangements (see Box 4.3).

Table 4.9 Labelling of Settlers and Resettlers

Settlement	Resettlement
(a) (Spontaneous) settler	Resettler
(b) Coloniser	Re-integratee
(c) Assisted migrant	Relocatee
(d) Pioneer	Project-affected person
(e) Encroacher	Project-dislocated person
(f) Slum dweller	Integratee

Source: Muggah (2008).

Box 4.3: Meaning and Significance of Land (Vanuatu)

(1) Land to a ni-Vanuatu is like a mother to her baby. He defines his identity and maintains his spiritual strength. He allows others the use of his land but always retains the right of ownership.

(2) To Sethy Regenvanu, in Cook Islands Maori, 'enua' means 'land, country, territory, afterbirth'.

(3) 'In Futuna (Wallace) "fanua" means country, land, the people of a place'.

(4) In Tonga, 'fonua' means 'island, territory, estate, the people of the estate, placenta', and 'fonualoto' means 'grave'.

(5) 'In some Polynesian languages, proto-fanua is both the people and the territory that nourishes them, as a placenta nourishes a baby'.

(6) To W. Pond, in 'The Land with All Woods and Water' (1997), 'The People of Nakorasule' (a village in Fiji) cannot live without their physical embodiment in terms of their land, upon which survival of individuals and groups depends. It provides nourishment, shelter, and protection, as well as a source of security and the material basis for identity and belonging. Land in this sense is thus an extension of the self, and conversely the people are an extension of the land.

(7) To Ravuvu, 'Development or Dependence: The Pattern of Change in a Fijian Village', (1988), 'The land has been viewed as possessing a sacred or spiritual quality, expressed in the mental attitudes of Marshallese when they think of the land as the very root of their worldly existence'.

(8) To L. Mason, for the people of Kapingamarangi Atoll 'a land dispute is never forgotten, nor do the opponents forgive each other, nor is the matter ever really settled, even when the litigants are long deceased'.

Source: Cited in Campbell (2010)

Similarly, in India, too, the village people suffer due to forced land acquisition for government, PPP (Private–Public Partnership), or private sectors. For instance, in Pelpa village in Jhajjar district of Haryana (India), for Special Economic Zone (SEZ), under PPP mode, Reliance Industries (with 90% share in the project) started acquiring 25,000 acres of agricultural land (in early 2005–2010) by displacing 22 villages in Jhajjar and 18 villages in Gurgaon. The compensation paid was relatively less and the younger (male) generation took the compensation money from their elders and spent lavishly for motor cars, liquor, guns, and disco music. A farmer's wife perceptively remarked,

> this is our traditional land, cultivated by our ancestors, our only source of permanent economic security. Of what use is money? How long will it last? …. This is not just the forced take-over of our land and ancestral village, it is also the decimation of our culture and roots. …Now with easy money alcoholism is a daily nightmare.

158 Development, Displacement, and Rehabilitation

She further adds, 'Domestic violence is all too common. We do not belong to the city. And our own village seems alien to us now' (cited in Shrivastava and Kothari 2012, 5).

On the other hand, a particular type of development in terms of damming a river or building a barrage on a river has multiple effects on the ecology, economy, culture, politics, and society as a whole. For instance, the building of the Farakka barrage on the river Ganga in the early 1970s brought multi-dimensional effects on the ecology, navigation, agriculture, fisheries, and human health and well-being of Bangladesh in range and depth (Box 4.4).

Box 4.4: Multi-dimensional Effects of Farakka Barrage on Bangladesh

Even before the birth of Bangladesh in 1971, after the partition of India and Pakistan in 1947, there were apprehensions about the effects of the barrage to be made. In fact, in 1944 itself, the Governor of Bengal appointed an inquiry Board, headed by Maharaja of Burdwan and Dr. Meghnad Saha as its member. The Board suggested the creation of the Tennessee Valley Authority (USA) like the authority in Damodar Valley in West Bengal. In 1948, Srischandra Nandy (Maharaja of Kasim Bazar) had noted that after partition major paddy-growing areas went to East Pakistan (later Bangladesh), hence a legacy of 'less food but more to be fed', so his slogan was 'produce or perish' – Farakka barrage was the only hope for efficiency of Calcutta port, navigability of Hooghly-Bhagirathi river, and giving new life to the dead/dying channels in central Bengal. India started constructing the Farakka barrage in 1962.

In a Secretary level meeting held on 16–17 July 1970 in New Delhi, India and Pakistan agreed on the point of delivery of water at Farakka, and a body of representatives of both to be constituted later, but due to turmoil, it could not be done. It was completed at the cost of Rs. 156 crores in 1975 with 75 feet height, 7,000 feet length, and later a 26.5 mile-long feeder canal was added. In March 1972, a Joint Declaration was signed between the PM of India (Shrimati Indira Gandhi) and the PM of Bangladesh (Sheikh Mujibur Rehman); later, on 15 April 1975 allocation of water was agreed upon for mutual accommodation. But with the overthrow of the Bangladesh government in 1975, there was no cooperation and the new government of Bangladesh accused India of water diversion. Maulana Abdul Hamid Khan Bhasani launched 'Farakka Long March' on 19 May 1976 and later Bangladesh government raised the issue in the international fora at the 31st session of the UN General Assembly, and the Islamic Foreign Ministers' Conference in Istanbul but of not any result, as it was a bilateral issue. Later on 5 November 1977, both nations agreed on a formula for sharing dry season water flow. State government of West Bengal criticised the agreement as Calcutta port was not getting adequate water. This agreement expired on 30 May 1982; later it was extended in October 1982, but the minimum guarantee of water to Bangladesh was removed then (1982) due to the inadequate flow of water in Ganga available at Farakka. It continued till 1984 and another Memorandum of Understanding (MOU) for 3 years was signed in 1985; a Joint Committee study was agreed upon by both. After its expiry, India withdrew water unilaterally; Bangladesh criticised India and internationalised the issue. With the new government in Bangladesh in 1996, a 30-year Treaty was signed by the two nations in 1996 for 40% of the dry season period, water flow is shared equally; if flow falls below 50,000 cusecs, two govts. will adjust on an emergency basis – on the principles of fair play, equity, and no harm to either party.

Since 37% of the total area and 33% of the total population of Bangladesh are dependent on the Ganga basin, due to the Farakka barrage it has faced various problems in agriculture, navigation, fisheries, industry, salinity, ecology, etc. in South-West region: 'disruption of fishing and navigation', 'unwanted salt deposits', 'discharge of the river Padma at the point of Hardinge Bridge in Bangladesh fall far below', a decline of groundwater level in Rajshahi, Kustia, Khulna, and Jasior, 'critical problem of salinity intrusion from the Bay of Bengal because of the drastic reduction of fresh water flows in the Gorai river' (Tiwary 2006). Further first yield has also declined, valuable forests have, too, disappeared, and river bank erosion has led to the loss of lands there. Actually the demarcation of common rivers as per borders of new nation-States did not recognise

> the collective ownership of the common rivers … as an integrated unit of resources. Upper riparian States started to use common rivers to the end of their own interest at the cost of the interests of the lower riparian States.
>
> (Kawser and Samad 2016)

Source: Kawser and Samad (2016); R. Tiwary (2006)

Rehabilitation

While resettlement refers to the process of relocation or settling again in a new area due to natural disasters, nuclear plant leakage (Japan), or wars/conflicts, rehabilitation means restoration to the former state of living. Rehabilitation of the displaced people due to development works is necessary because displacement brings multi-dimensional poverty and suffering due to loss of lands, homes, livelihoods, foods, and access to common natural resources, as well as social isolation, increased morbidity, and mortality due to stress in them. Displacement is a forced/involuntary migration. Hence, the rehabilitation of such people needs to be holistic. In the case of the Sardar Sarovar Project on the Narmada River (MP, Gujarat) World Bank gave the following four guidelines to the State of Gujarat:

(a) There should be an improvement in the standard of living of the displaced people.
(b) They should be relocated to the units as per their preferences/choices.
(c) They should be fully integrated with the host communities.
(d) They should be provided compensation, and social and physical rehabilitation, including community services and facilities.

However, these guidelines are not sufficient; rather following points should also be incorporated in a fair and just rehabilitation policy, keeping in mind *preventive, restorative, supportive,* and *palliative* components of rehabilitation – full restoration is the absolute rehabilitation:

(i) Displaced persons/families should not be treated as objects – a 'top-down' approach to imposing such a policy will be anti-people; therefore, a comprehensive policy should be prepared with a genuine consultation and consent of such local people as an *agency* and to be pursued at all stages – from formulation to implementation to monitoring and evaluation.

(ii) It should be broad-based both in range and depth-multidimensional (economically, livelihood creating, culturally, acceptable, politically participative, socially egalitarian, and ecologically sustainable).

(iii) Local progressive practices and indigenous knowledge systems should be restored optimally.

(iv) There should be proper mechanisms to ensure equality, fairness, and justice among the rehabilitated persons/families as well as between the host communities and the rehabilitated persons/families.

(v) Rehabilitation is a dynamic social process, and hence in addition to one-time compensation, there should be lands for lands, house sites, core civic amenities, infrastructure facilities, income-generating schemes, and adequate vocational training for up-skilling, multi-skilling, re-shilling, etc. – in fact, restoration as absolute rehabilitation.

(vi) Rehabilitation perspective should be human rights and need-based, not simply welfare-based.

(vii) There should also be a fair provision of rehabilitation for the 'secondary displacement' – persons who reside outside the submergence but have been dependent on that area for their livelihoods (artisans, small traders, landless labourers, etc.).

(viii) There should be adequate provision for having the cultural identity of the project-affected people in terms of belief systems, gender identity, community, or ecological identity, especially of the tribals displaced.

(ix) They are not 'anti-development', but rather against an imposed unfair and unjust 'development' policy or project. Hence, empathy, not sympathy, with them is required. Hence, both 'environment impact assessment' (EIA) and 'social impact assessment' (SIA) need to be carried out at the beginning of the project itself.

Large Dams: Rational or Irrational?

The oldest dam in the world is a 2000 year old Kallanai dam on the Kaveri river in Tiruchirappalli district in Tamil Nadu State (India); it was constructed by Chola King Karikal. It irrigates ten lakh acres of land. Its height is 5.4 metres, 329 metres long, and 29 metres wide; it protects crops, people, and animals from floods. In 1835, Lord Macaulay had reported to the British government about the advanced state of engineering like this in India, and hence Indian education system was destroyed deliberately. It is still in operational form. However, then India's population was far less and there was abundant land. But now the construction of large dams is not desirable in India from the viewpoint of a large number of small landowners, especially the tribals in remote areas, who are the ultimate losers and displaced persons, whereas the upper-class industrialists, rich farmers (interested in cash crops) as well as urban water users and electricity consumers usually corner most of the benefits of the large dams.

Here our focus is on the displacement of human beings due to the construction of large dams in India, in general. Our objective is to analyse the issues of the rationale of dam-building, the processes, and procedures of land acquisition from the original small landowners, especially the tribals residing in the remote areas, either inside or adjacent to the forests, the ills of the so-called compensation, relocation, resettlement, and rehabilitation. Therefore, the following questions have been attempted here: Is it necessary to construct large dams? Why should small dams not be constructed which are time-tested, eco-friendly, cheaper, and do not bring negative consequences of displacement? Why should the State have the last say in land acquisition? Can one be compensated for one's basic source of livelihood? If yes, how and to what extent? If no, then why is there a land acquisition process at a large scale? Is there any uniform policy for relocation, resettlement, and rehabilitation of the land oustees in India? Development for whom and at whose cost? We will analyse these aspects one by one in order to reach a better understanding of the problem of displacement due to large dams.

The first Prime Minister of India, Jawaharlal Nehru, talked of dams and industries as the 'temples of modern India'. But the experience has actually shown otherwise. A large dam is defined by the Government of India as one with a command area of more than 10,000 ha, while a medium dam is one with a command area of more than 2,000 ha but less than 10,000 ha, and a minor dam has a command area of less than 2,000 ha (Dogra 1986). Whether a large dam is to be built or not remains a very contentious issue. Because large dams artificially change the natural course of rivers and streams, on the one hand, and, on the other hand, instead of local people, outsiders like those residing in urban areas or pursuing industrial–commercial operations or potential energy consumers elsewhere are more interested in the building of dams in the name of modernisation and development. In fact, the scholars working on this issue are highly divided and could be put into three categories (Sharma 2016):

(a) Development optimists
(b) Conditionalists
(c) Negationists

For the **development optimists** like Pearce, and World Bank economists, essentially the neo-classical economists, liberal policy-makers, government engineers, and working administrators, first, the primary concern lies in the mottos like 'big is better', 'large scale is cost effective', 'development should be quantifiable, scalable and visible', dam-building is multipurpose encompassing energy generation, flood control, assured irrigation of crops, and water supply in urban-industrial complexes. Thus dams are projected as necessary for the essential development of a society.

Second, the new reservoir would give a congenial habitat to water birds, fish, and other animal species as well as plants.

Third, the degraded catchment area gets restored and there is a micro-climatic change in terms of lowering the levels of humidity and temperature.

Fourth, there arises an opportunity to develop tourist resorts, and hence a new trade of 'ecotourism' could be easily realised.

Fifth, the price of land shoots up, and hence the original landowners get the benefit.

Finally, due to the big reservoir, groundwater is constantly recharged.

The **conditionalists** like B. Blackwelder, Philip Williams, B. Bramble, and B. M. Rich put several conditional recommendations as pre-requisites for the construction of large dams (Sharma 2016):

a) An adequate environmental impact assessment (EIA) be prepared and be available to the people concerned.
b) Only if it benefits the large population, not merely the urban elite.
c) It should have labour-intensive economic activities.
d) It should help the production of food crops for local people, not the cash crops for exports.
e) It should not affect people's health and safety.
f) It should not affect national parks, heritage sites, areas of scientific and educational importance, rainforests, endangered species, etc.
g) It should not silt up within 100 years.
h) It should not salinise agricultural lands.
i) It should not displace sustainable long-term resource enhancement.
j) It should not displace indigenous people and the compensation should not be inadequate.
k) It should not be built in seismic zones and landslide areas.
l) It should not cause significant damage to the ocean fisheries, etc.
m) It should not harm the environment of a neighbouring country without its consent.

The **negationists** like E. Goldsmith and N. Hildyard (1986), Smitu Kothari and P. Parajuli (1993), S. Sanghvi, and others are of the view that large dams are not desirable because they bring various environmental, social, economic, cultural, psychological, and other insoluble problems. They also argue that the so-called development optimists compromise with all the equity issues by not going beyond the short-sighted petty goals of 'here and now'. Further, they are also of the view that if the above-mentioned conditions posed by the conditionalists are complied with, large dams cannot be built at all. For instance, in the case of most of the large dams in India, like the Sardar Sarovar Project (Madhya Pradesh, Gujarat and Maharashtra), or Silent Valley dam (Kerala), Environmental Impact Assessment (EIA) was not carried out at all. In fact, the Government of India's Ministry of Environment and Forests had given conditional approval to the Sardar Sarovar Project, but those conditions were not fulfilled before the actual work started. Further in the case of the Suvarnarekha Multipurpose Project (in Jharkhand, West Bengal, and Odisha) restoration of mining and quarrying sites was not done, though it was recommended.

In fact, the Central Board of Irrigation and Power (Government of India) has reported that the annual rate of siltation from a reservoir is usually two or three times more than the assumed rate at the time of project planning. For instance, in Maithon dam (Jharkhand) the assumed rate of siltation was 9.05 ha m per 1,000 sq km, but actual siltation is 12.39 ha m per 1,000 sq km, and thus there was an increase of 37%. Similarly, in Panchet dam, the assumed rate of siltation was only 6.67 ha m per 1,000 sq km, but the actual siltation was 10.48 ha m per 1000 sq km, and thus an increase of 57% was really worrisome (Singh and Banerji 2002). Further, many crop varieties and methods of cultivation are things of the past due to the submergence of lands.

In India, as per data from the Central Board of Irrigation and Power (Government of India), there were 429 large dams (including those under construction) and in the world about 45,000 large dams were built till 2000. In 1967 an earthquake (in Maharashtra) of the magnitude of 6.5 on the Richter scale killed 117 persons. It was induced by the Koyana reservoir (Maharashtra) whose dam height is 103 metres. In a

Table 4.10 Causes of Dam Failures in the World (1900–1975)

Type of Dam	Cause of Failure	% Share
1. Concrete dam	Over-topping	29%
	Faulty foundation	53%
	Seepage	0%
	Other causes	18%
2. Fill dam	Over-topping	35%
	Faulty foundation	21%
	Seepage	38%
	Other causes	6%
3. All types	Over-topping	34%
	Faulty foundation	30%
	Seepage	28%
	Other causes	8%

Source: World Bank (1990) cited in S. Singh and P. Banerji (2002).

study, it was found that 17 cases out of the 75 cases of Reservoir Induced Seismicity (RIS) reported in the world were from India (Singh and Banerji 2002). In addition, there has been waterlogging in the command areas of big dam projects. For instance, in Bihar, the Gandak project and Kosi project caused waterlogging in 0.40 million ha and 0.12 million ha in the command area, respectively (Singh and Banerji 2002). Not only this, the failure of a dam may have a catastrophic effect. Often the failure is due to the faulty foundation, use of materials of sub-standard quality, over-topping due to excess water, sabotage, or earthquake. The World Bank in its study of dam failures during 1900–1975 (in excess of 15 m height) found the following major causes (Singh and Banerji 2002) (Table 4.10).

In case of any dam bursts, it would cause a heavy loss of humans, animals, and property in a very large area. Further, all the investment in such work would be lost in one stroke.

Some social scientists, therefore, have rightly raised the issue of equity at five levels regarding the construction of the large dams (Singh and Banerji 2002):

(a) Some people are compelled to pay the cost, while others corner the benefit (intra-generation class benefit analysis).
(b) The costs of the dam are distributed among the adversely affected people (class benefit analysis among project-affected people).
(c) The benefits of the dam are distributed among the beneficiaries of the project (class benefit analysis among beneficiaries).
(d) The costs and benefits are distributed among various generations (inter-generation equity).
(e) The costs and benefits are distributed among species (inter-species equity).

However, in India as elsewhere, the issue of equity is often not considered, and hence a situation of inequity and injustice prevails among the oustees. Thus the construction of large dams results in more harm than benefit. Globally half of the rivers were

164   Development, Displacement, and Rehabilitation

dammed at the rate of one per hour, and at an unprecedented scale of over 45,000 dams, more than four storeys high as World Commission on Dams (2000) noted. Its Chair Kder Asmal perceptively remarked in the Preface of the report 'Dams and Development': 'Dams remove water from the Ganges, Amazon, Danube, Nile or Columbia to sustain cities on their banks. For parting or imparting the waters, dams are our oldest tool. Yet are they our *only* tool, or our *best* option?' (2000, ii). He further adds,

> real development must be people-centred while respecting the role of the state as mediating, and often representing, their interests ... we do not endorse globalisation as led from above by a few men. We do endorse globalisation as led from below by all, a new approach to global water policy and development.
>
> (2000, iii)

World Commission on Dams (WCD) (2000) identified five key decision points for the energy and water sector: (a) Needs assessment-decentralised consultation process; (b) selecting alternatives-identifying preferred development plan from many options; (c) project preparation (detailed planning and design); (d) project implementation – procurement and construction; and (e) project operation adapting to changing contexts – based on a participatory review of project performance and impact. World Commission on Dams (WCD 2000) gave five messages as follows:

(a) Dams have made a significant contribution to human development.
(b) An 'unacceptable', 'unnecessary', and 'high' price has been paid to secure benefits, especially in social and environmental terms, 'by people displaced, by communities downstream, by taxpayers, and by the natural environment'.
(c) Lack of equity in the sharing of benefits.
(d) By ensuring the participation of all stakeholders, conditions for a positive resolution of conflicts and competing interests are created.
(e) Negotiating outcomes will improve the development effectiveness of water and energy projects by removing non-beneficial ones at an early stage, and by offering only those options agreeable to key stakeholders to meet their needs.

Though as a member of WCD, Medha Patker, the main leader of Save Narmada Movement (Central India), signed on the report due to 'its many positive aspects', she attached her comprehensive comments as follows:

(i)   'Basic systemic changes needed to achieve equitable and sustainable development' and to challenge 'the forces that lead to the marginalisation of a majority through the imposition of unjust technologies like large dams'.
(ii) 'Frequent failure' of large dams and 'poor performance' to be recognised – just rehabilitation of the displaced with land for land and alternative livelihoods not provided; large dams have hindered 'human development'.
(iii) An 'inclusive, transparent process of decision-making' with 'equal status to all stake holders', 'equal significance to social, environmental, technical, and financial aspects' of planning not ensured; developers like the World Bank or State push large dams without complying with risks, rights, social, environmental

parameters, etc.; 'principle of subsidiary' (working from ridge to river, from origin to sea) not recognised.

(iv) Priority be given to more 'equitable, sustainable, and effective options to satisfy basic human needs and livelihoods for all' before the additional luxuries of the few.

(v) The role of the State is diminishing, national laws and institutions are marginalised, and human rights are crushed due to the increasing role of private capital and free trade.

(vi) The sovereignty of both people and nation-State not to be compromised at the behest of corporations and international finance; to critique the privatisation of water and power and corporate domination over national resource-based communities, resulting in the marginalisation of local people.

(vii) Ideologies, strategies, and vision of people's movements should be given due place.

(viii)To reject the assumptions of a development model that has failed, and to caution against the gulf between a statement of good intent and a change in practice by the vested interests.

Lack of Transparency in Land Acquisition Act 1894 and LARR Act 2013

After discussing the irrationality of the construction of large dams, let us discuss the second dimension of the problem, namely the complex process and procedure of land acquisition for the so-called development projects in general and dam-building in particular. For entire India (except Jammu and Kashmir), a law of the land, namely the Land Acquisition Act, was enacted during British rule in 1894. Acquisition differs from confiscation because of three in-built conditions: The landowner is to be given a reasonable hearing under the basic principle of 'audi alteram partem' (let the other party be heard); there should be a specific 'public purpose', and the landowner should be paid compensation.

In Section 3 of the Land Acquisition Act 1894, the term 'public purpose' had been defined to include provisions for:

(i) Village sites including their planned development

(ii) Town or rural planning

(iii) Any scheme or policy of the central or State government

(iv) A corporation owned or controlled by the State

(v) Residential purpose to the poor, the landless, or persons residing in areas affected by natural calamities or persons displaced/affected due to any scheme of government or any local authority or corporation owned or controlled by the State

(vi) Carrying out education, housing, health, or slum clearance scheme sponsored by the government or its authority or local clearance scheme sponsored by government or its authority, or local authority or society (registered under Societies Registration Act 1860) or other law or cooperative society

(vii) Any other scheme of development sponsored by government or local authority

(viii)Locating a public office (but did not include acquisition of land for companies)

166   Development, Displacement, and Rehabilitation

Thus, the term 'public purpose' was very vast though under Section 3 (f) it did not include land for companies as public purpose yet indirectly the companies used to get land under the rubric of industrialisation policy of the State when a company was to provide houses to its members (Keshav Pal vs State of Bihar AIR 1985). Under Section 5A or 40, the purpose of a company was to be inquired only after issuing of notification under Section 4 as a court ruling held so (Shyamanand Prasad vs State of Bihar, 1993 SCC). Land Acquisition (Bihar Amendment) Act 1960 further provided that public purpose included provision for (i) sanitary improvements including reclamation as well as (ii) laying out of village sites or townships or extension or planned development or improvement of existing village sites or townships.

Regarding 'public purpose' following problems arose during the actual task of land acquisition: First, land required for State or 'Government purpose' was unfortunately considered synonymous with 'public purpose'. Often the interests of the State clash with those of the local people at large due to difference in their perception and conception of 'development'.

Second, since this law was enacted in 1894, that is much before the enforcement of the Indian Constitution on 26th January 1950, its provisions did not consider the dissent of the people as a violation of fundamental rights (especially Article 21) of the constitution.

Third, State has the right of 'eminent domain', that is, it shall assert its overriding power over any piece of land situated in the State on account of public exigency and for the public good, as upheld by various High Courts and the Supreme Court of India. But, unfortunately, this right has been often misused in the name of the State's discretion to choose any land, exclude any land, take all lands of one or some families, and take even the lands of places of worship, or other commons.

Fourth, though under Section 4(1) of the Act, 'public purpose' was to be specified otherwise various matters mentioned in Section 4(2) could not be carried out (Munshi Singh vs Union of India 1973, SCC), yet this was not done in practice either due to lack of knowledge, or pressure of work or political consideration, or irresponsible action or hidden agenda. Unfortunately some of the court rulings also help such reckless acts of the State and its functionaries by stating that it may be practically difficult to specify the particular purpose for which each item of land is needed (Pt Lila Ram vs Union of India, 1975, SCC).

Fifth, it has also been interpreted by the courts that if the government contributes to acquisition of land, it is presumed to be a 'public purpose', even if it is for constructing residential houses of a company – and in such a case it is not necessary to go through the procedure laid down in part VII of the Act – land acquisition for companies (Girija Dube vs State of Bihar, 1989, PLJR).

Sixth, 'the emergency provision' (Section 17) did not require issuing of a proper notice for being heard.

Finally, even if notices were issued, or published in newspapers, due to illiteracy or lack of access most of the farmers were not well aware and often notices showed even the fertile lands as waterlogged, barren, or wasteland.

Inadequacy of Compensation (Under LA Act of 1894)

Under Section 11 of the said LA Act of 1894, on any objection raised after the serving of notice regarding measurement, value, and interest of the person in the said land, the Collector (under the Act) was to give an award with the prior approval of the appropriate government (State or Union government). Further, such an award was to be made within a period of two years from the date of publication of the declaration; otherwise, the proceeding would lapse on its own. After such an award, under Section 16 of the Act of 1894, the Collector would take possession of the land, which would then vest in the government 'absolutely'. However, in urgent cases, at the direction of the appropriate government, the Collector, even without making the award, could take such possession (under Section 17) after 15 days from the publication of the notice. Land Acquisition (Bihar Amendment) Act 1960 widened the scope of Section 17 further: Either after 15 days of publication of the notice for persons interested under Section 9(1) or at any time after the publication of preliminary notice under Sec. 4 with the consent, in writing, of the persons interested, could take possession of any waste or arable land needed for a public purpose or a company. Such a vast power was violative of one's right to equity and right to life, on the one hand, and was also a symbol of arbitrariness on the part of the Collector (on behalf of the State), on the other.

However, under Section 18 of the Act of 1894, if someone did not accept the award, he might request the Collector on specific grounds within six weeks of the award who would refer the matter to a court for determining the measurement of the land or amount of compensation, or the persons to be paid or apportionment of compensation among persons interested. However, due to legal illiteracy, this provision was often not used by the small landholders. Further, under Section 23 of the Act of 1894, the referred court should take the following points into consideration:

(a) Market value of the land on publication of the preliminary notification under Section 4(1)
(b) Damage of standing crops/trees at the time of taking of possession
(c) Damage of severing such land from one's other land at the time of possession
(d) Damage, at the time of possession, by affecting one's other property (moveable or immovable) or earnings
(e) If, due to acquisition, one is compelled to change residence or business place
(f) The damage resulting from diminution of profits of the land between time of publication of declaration and taking possession of land

In addition to the market value, interest of 12% per year from the date of publication of preliminary notification to the date of award or of taking possession whichever was earlier (as per amendment in 1984) and in case of compulsory nature of acquisition a sum of 30 per centum on such market value was to be paid.

Actually under the old law of 1894, the term 'market value' had not been defined. Hence for every case the court used its discretion in awarding compensation, hence the element of subjectivity existed there. However, several judgments of various High Courts and the Supreme Court of India have pointed out the following guidelines for consideration:

(a) Location of land, its importance, prospect of being developed, purpose for which it was used, and purpose for acquisition (Shambhu Nath vs State of Bihar, 1989, PLJR)
(b) Potentialities of the land (State of Bihar vs Parasuram Prasad, 1976, BLJR)

(c) Sale deeds of a pre-notification period of adjoining lands (State of Bihar vs Thakur K. P. Singh 1968, BLJR)

(d) The price which a willing vendor might reasonably expect from a willing purchaser (Sushila Devi vs State of Bihar, 1963, BLJR)

(e) Income derived by taking the gross income and deducting therefrom his expenses and the net income was to be multiplied by 20 years purchase (Ramchandrariah vs Land Acquisition Officer, Sagar, 1973, BBCJ)

(f) Land includes things attached to earth (like trees), so its value was also to be calculated and an allowance of 15% be given (Chaturbhuj Pandey vs Collector, Rajgarh, 1969, BLJR)

(g) Its condition, suitability for building, proximity to residential, commercial, and industrial areas, and educational, cultural, or medical institutions, existing amenities like water, electricity, and drainage and the possibility of future extension (Collector Rajgarh vs Hari Singh Thakur 1979, SCC)

(h) Average price in different sale deeds of relevant time (Harbans Singh vs State of Bihar, 1974, PLJR)

(i) Continuous rising inflation and rising land values (State of Bihar vs S. K. Thakur 1981, AIR)

However, many lower revenue courts, due to the pressure of a large number of cases, insensitivity to the landowners, as well as the prevailing malpractices hardly decided such cases judiciously. Further, since in most of the cases, the land records were not up-to-date, and hence the landowners suffered because their total lands and category of valuation ('maliyat') were not properly shown. Their oral statement and actual possession did not weigh much against the government 'land records'. Moreover, some local cronies, in connivance with local government staff, often succeeded in wrong measurement and evaluation. In addition, payment of compensation was often delayed due to red tapism and some sort of cuts (commission) were also practised in some cases. Finally, the compensation paid was far less than the actual sale rate and even the rate written in registered sale deeds was far less than the actual rate paid by the purchasers in order to avoid paying higher stamp duty.

RFCTLARR Act, 2013

The archaic law, Land Acquisition Act 1894, was finally replaced by a more comprehensive law, RFCTLARR Act 2013 (Right to Fair Compensation and Transparency in Land Acquisition, Rehabilitation and Resettlement Act 2013), wherein for acquiring a piece of land, 70% farmers' consent is necessary for public sector works and 80% farmers' consent is required for private sector works, and they are to be paid adequate compensation at market rates. Its salient features are notable below:

(1) *Transparency of processes and fairness of outcomes* (standardised compensation)

(2) *Humane and participative* (consented) land acquisition system – addressing long-standing asymmetries of power between acquiring bodies and affected people

(3) *Time bound, prompt, and effective* land acquisition with a timeline for compensation, R&R

Development, Displacement, and Rehabilitation   169

(4) *Pro-people progressive law* – safeguarding the interests of the landowners – to prevent land conflicts
(5) *Rehabilitation and resettlement being an integral part* of the land acquisition process
(6) In all cases of land acquired for a public purpose, *Social Impact Assessment* (SIA) being an integral part of the land acquisition process – keystone for transparency
(7) *Genuine concern for the landless* other than landowners – *whose livelihoods are dependent on lands and they lose livelihoods* due to such land acquisition – both are 'affected families'
(8) Ensuring *sustainable livelihood of all the affected families*
(9) *Option for the lease* to be explored before the acquisition
(10) *Dedicated authority for the disposal of disputes and monitoring* (at State and central levels)
(11) *Provision of rehabilitation and resettlement in case of private purchase of land –* 100 acres or more in rural areas and 50 acres or more in urban acres
(12) Formation of a *committee under the chairmanship of the Chief Secretary* if the land acquisition is for 100 acres or more land
(13) *Limitation on the acquisition of multi-crop land for safeguarding food security*
(14) *Consent* of 70% of farmers required in case of acquisition for a *public purpose* and 80% of farmers consent required in case of acquisition for a *private purpose*
(15) *Provision of 25% of shares* as part of compensation in case the requiring body offers shares to landowners
(16) *Provision of penalties for companies and government* in cases of false information, malafide action, and contravention of the provisions of the Act

We may compare the old (Act of 1894) and new (Act of 2013) laws of land acquisition in Table 4.11.

Thus in the new Act (2013), there are new and detailed provisions for SIA, safeguard for multi-crop land for food security, R&R Award, additional compensation in case of multiple displacements, special provision for SCs/STs, State and National Monitoring Committees for R&R, etc. that were not existing in the old law (Act of 1984). Further in the case of land acquired for public purpose works (plus PPP) 70% of farmers' consent is needed, while in the case of private (companies) works 80% of farmers' consent is needed. They will be paid at market rates multiplied by two in rural areas and by one in urban areas plus 100% market value as solatium; in case of acquisition for urbanisation 20% of the developed land will be offered to the affected landowners. If the land is not used in ten years, it will be transferred to State's landbank. On every transfer of land, without development, 20% of the appreciated land value shall be shared with the original landowners. Further, 25 infrastructural amenities are to be provided in the resettlement area including schools, playgrounds, health centres, roads, electricity, safe drinking water, panchayat ghar, anganbadis, places of worship, burial and cremation grounds, fair price shops, post office, seed-cum-fertiliser storage facilities, etc.

Further, in the interest of the landholders, in the following four situations, Sec. 24 of the new Act (2013) provides for its retrospective effect:

(a) Where award u/s 11 of LA Act, 1894, has not been declared.

170 Development, Displacement, and Rehabilitation

Table 4.11 Comparison of Old and New Laws of Land Acquisition in India

Provisions	Sections in New Land Acquisition, Rehabilitation and Resettlement Act (2013)	Sections in Old LA Act (1894)
(a) Social impact assessment (SIA) and SIA-related studies	Sec. 4(1) to Sec. 9	No provision
(b) Special provision for safeguarding food security (limitation on multi-crop lands)	Sec. 10	No provision
(c) Publication of Preliminary Notification	Sec. 11 to Sec. 18 (with SIA report)	As in Sec. 4 and Sec. 5(a), except notification with SIA report
(d) Publication of declaration and Summary of R&R Scheme	Sec. 19 to Sec. 22	As in Sec. 6(1), except R&R scheme
(e) Land Acquisition Award	Sec. 23 to Sec. 30	As in Sec. 11
(f) R&R Award	Sec. 31 to Sec. 37	No provision
(g) Power to take possession of land	Sec. 38	As in Sec. 16
(h) Additional compensation in case of multiple displacement	Sec. 39	No provision
(i) Emergency provisions (restricted for Defence, national security, and natural calamities)	Sec. 40	As in Sec. 17
(j) Special Provision for SC/STs	Sec. 41 and Sec. 42	No provision
(k) Process and Procedure of R&R	Sec. 43 to Sec. 47	No provision
(l) State and National Monitoring Committees for R&R	Sec. 48 to Sec. 50	No provision
(m) Establishing State Authority for Land Acquisition Resettlement and Rehabilitation	Sec. 51 to Sec. 74	Sec. 15 and Sec. 30 – but not as comprehensive as in the new Act
(n) 80% farmers' consent in private cos. and 70% farmers' consent in public purpose + PPP, and in scheduled areas consent of Gram Sabhas/Panchayats, Autonomous Dist. Councils are mandatory	Sec. 2	No provision
(o) Definition of Affected families widened: (i) Loser of land/other immoveable property (ii) Loser of livelihood (iii) Tribals and other traditional forest dwellers losing traditional rights under ST and OT FD Act 2006 (iv) Land allotted by government to a family	Sec. 3	Limited to land loser's family only

Source: Authors.

(b) Where award u/s 11 of LA Act, 1894, had been made five years ago or more before the commencement of new Act, 2013, but physical precession of the land not taken or compensation not paid.

(c) If an award has been made but compensation in case of the majority of landholdings has not been deposited in the account of beneficiaries, all notified land losers will be entitled to new compensation calculation.

(d) If any land purchased on or after 5 September 2011 being contrary to the provisions of Sec. 46(1) or within three years from the date of commencement of the Act (1 January 2014), 40% of the compensation paid for such land acquired shall be shared with the original landowners.

Critical Appreciation

It is notable that u/s 14 of the Act of 2013 the Social Impact Assessment (SIA) report will lapse after 12 months; however, the government may extend it if there are genuine reasons. The Commissioner of Rehabilitation and Resettlement (under Section 17 of the Act) shall approve the scheme of rehabilitation and resettlement. But u/s 18 of the Act Commissioner will publish his/her report in the locality for Panchayats/municipal bodies for their suggestions, if any. Under Section 21, Collector under the Act shall issue a public notice to invite objections from the interested persons before finalising the award of compensation and also the benefits under the rehabilitation and resettlement scheme. Collector has to make an award within 12 months from the date when the final declaration u/s 19 was issued. Further, u/s 2(1) of Act 2013, the 'public purpose' has been defined widely as follows:

(i)   For strategic purposes (armed forces, including central paramilitary forces) or works vital for the national security of Defence, or State police, or safety of people
(ii) For infrastructure projects
(iii) Projects involving agro-processing, supply of inputs to agriculture, warehousing, cold storage facilities, marketing infrastructure for agriculture and allied activities, e.g. dairy, fisheries, and meat processing, owned by appropriate government, or farmers' cooperative, or by an institution set up under a statute
(iv) Projects for industrial corridors, mining activities, national investment, and manufacturing zones
(v)   Projects for water harvesting and water conservation structures, sanitation
(vi) Project for government administered, government-aided educational and research schemes or institutions
(vii) Project for sports, health care, tourism, and transportation of space programme
(viii)Any infrastructure facility notified by the central government
(ix) Project for project-affected families
(x)   Project for housing for such income groups, as may be specified by the appropriate government from time to time
(xi) Project for planned development or improvement of village sites or any site in urban areas or provision of land for residential purposes for the weaker sections in rural and urban areas
(xii) Project for residential purposes to the poor or the landless to persons residing in areas affected by natural calamities, or to persons displaced or affected by reason of the implementation of any scheme undertaken by the government, any local authority or a corporation owned or controlled by the State

Thus, first, the definition of 'public purpose' has been widened, and hence the social and environmental impacts of land acquisition would increase undoubtedly.

172   Development, Displacement, and Rehabilitation

Second, the timeline in the new Act of 2013 has also been increased to 56–59 months – e.g. SIA study – 6 months, appraisal by expert group – 2 months, preliminary notification – 12 months, publication of declaration – 12 months, award of Collector for land compensation, and award of Collector for R&R compensation – 24 months, and only then possession of the land would be taken over. On the other hand, in the old law of 1894, it was less – 38 months, including preliminary notification – 2 months, publication of declaration – 12 months, award by Collector – 24 months, and only then possession of the land was taken over. In the new Act of 2013 total period of land acquisition has been increased because of several new provisions as follows:

(a)   Social Impact Assessment (SIA) – two public hearings and six public consultations
(b)   Environmental Impact Study (EIA) – after completion of SIA only
(c)   Expert committee of officials, two non-official social scientists, including one technical expert for ensuring transparency
(d)   Provision of public hearing for fair compensation and transparency – two public hearings and six public consultations are mandatory; then a hearing by the forest dept.; for diversion of forest land seven steps are to be taken
(e)   Provision of rehabilitation and resettlement – earlier in 75%–85% of cases, no rehabilitation was done and among these affected families, the maximum were tribals
(f)   Provision of compensation to be given to those dependent on the said land for their livelihood (tenant, sharecropper, labour, etc.)
(g)   Provision of avoiding irrigated multi-crop lands' acquisition
(h)   Provision of the consent of the landholders – it is now a dialogic process
(i)   Additional compensation for multiple displacements

It is our experience that even under the old Act of 1894, Collector used to take more time except under emergency provisions (Sec. 17), and there the land acquisition authority (Collector) had hardly any responsibility for the delay or for not following the due process and procedure (due to discretion), but now Collector and other authorities have to strictly follow it (mandatory). As per sources in Dept. of Land Resources, Ministry of Rural Development, Government of India (GOI), under the new Act of 2013, the actual mean time of land acquisitions is about two years, not five years (the maximum time given in law).

Third, the new Act of 2013 is inclusive and participative, while the old (British) Act of 1894 was State-centric ('eminent domain') – now a balance between the *power of the State* for development, and the *right to property and livelihood of the individuals* (hence fair compensation and rehabilitation). However, the gender aspect (widows, unmarried daughters, women-headed families) has not been duly considered in the new Act (as earlier).

Fourth, within five years of the acquisition, the said land is to be used for the purpose it was acquired, otherwise the State government to take it back to the land bank.

Fifth, in 2014, an Ordinance was promulgated to add Section 10A, due to the pressure of the industrialists' lobby, wherein the State government was empowered not to carry out SIA as well as not to limit the acquisition of the multi-crop irrigated land in projects of 'public purpose' and afterwards such an amendment bill was passed by the Lok Sabha. However, Rajya Sabha referred it to the Joint Select Committee.

Consequently many State governments (Gujarat, Telengana, Andhra Pradesh, etc.) have amended it by adding Section 10A as per the Ordinance 2014 and the President of India, at the advice of the central government, gave assent to it. Hence many critics are of the view that such an amendment is the dilution of the original provisions of transparency, participation, and inclusion of the people (landholders, etc.); at the most for Defence projects shorter time for acquisition could be provided, and farmers' consent percentage could be reduced to 50% – but it should not be done away with outright.

Finally, the bodies of industrials, their developers, etc. point out, on the contrary, that there are now more litigations and more delay, on the one hand, and emerging of cheat, crony property dealers/mafia/middlemen who manipulate by notionally mentioning higher rate in registered sale deeds in order to get more award based on the market value, on the other.

All said and done, the new RFCTLARR Act 2013 is a 'paradigm shift' in favour of the losers of lands and the losers of livelihoods (based on those lands), but the amendment by inserting a new Section of 10A is the dilution of that law against the interests of the people. Hence the Supreme Court is the only hope in this regard as pointed out by legal activists as it can declare such an amendment null and void.

Pangs of Displacement

Now we will discuss about the pangs of displacement and inadequate rehabilitation. The displaced persons suffer on various counts, especially social, economic, political, psychological, cultural, and spiritual deprivation:

First, most of the large dams are built on agricultural lands, and hence the displaced families lose their permanent sources of livelihood and employment forever. People also lose various indigenous varieties of seeds and techniques of cultivation – thus loss of livelihood and biodiversity.

Second, loss of land is ultimately a loss of social security in case of natural calamities or other contingent situations. Further, a displaced person being landless cannot get a loan from banks or traditional moneylenders as land's highly valued function of collateral is important.

Third, due to displacement, people also lose common property (natural) resources like forests, rivers, fields, grazing grounds, ponds, lakes, hence their supplementary source of livelihood (for tribals' primary livelihood) is lost, and they cannot engage in fish-catching or cow, camel, sheep and goat-rearing, or rope-making. This is not compensated in any way.

Fourth, displacement destroys their formal and informal educational institutions which were also used for other community purposes like public meetings, stays for marriage parties, festivals, feasts, but these losses are not compensated and simply money compensation is not sufficient.

Fifth, due to micro-climatic changes in the humidity and temperature at the new sites, they are unable to adjust and are unable to get their preferred choices of foods. So this results in malnutrition and morbidity among them. Moreover, they are not provided adequate hospital/health facilities in new places.

Sixth, at new sites, the previous neighbourhood setting is not maintained, nor are previous sizes of houses and homestead lands given. Therefore, there often arise inter-personal and group conflicts between the newcomers and hosts or between the newcomers themselves. Moreover, the collective identity lost is not replaceable.

Seventh, due to displacement people lose their age-old religious places like temples, mosques, gurudwaras, churches, and 'Sarnas' (religious places of Jharkhand tribes). They are not given compensation for these. Again the graves, tombs, burial/cremation grounds, etc. are also lost, and many festivals and rituals revolving around ancestors and agriculture or ponds are also lost.

Eighth, in most of the cases, only cash is given, and in some cases house sites are also given, in very few cases built houses as well as agricultural lands are given and only in rare cases all these cash, house sites, houses, and agricultural lands are duly provided. Wherever cash for a house is paid, it is a too meagre amount to construct a house. For instance, in Majalgaon, Warna, and Hirakud rates were Rs. 105, Rs. 500, and Rs. 3000 per house, respectively (Singh and Banerji 2002). Even if agricultural land is provided, it is usually barren, unirrigated, waste, degraded, or desert land. In Suvarnarekha Multipurpose Project (two dams constructed at Icha and Chandel in West Singhbhum, Jharkhand) revised rehabilitation package provided that every oustee of 18 years of age or more owning less than five acres of land was to get two acres of land or Rs. 10,000/- for buying land and in addition Rs. 20,000/- was given for house construction. In the case of Maithon dam (Jharkhand) even for paddy lands only Rs. 600/- per acre was paid in 1990s – that was inadequate.

Ninth, for rehabilitation purposes a family is the unit, but 'family' was defined (in the LA Act 1894) in a way suitable to the project management. In some cases, all the persons living under one roof or kitchen are considered as only one family (e.g. in upper Kolab), and thus many adults in a joint family are at a disadvantage. In other cases, only the male head of the family, in whose name the land is recorded, is considered (Tehri), and thus it is against the females. In the third category, only married males (Lok Tak) or adult males (Almatti, Srisailam) are considered, and hence it is anti-women. In the fourth category, all adult sons and only unmarried adult daughters are considered, and hence married daughters, widows, and divorcees suffer. However, in the new Act of 2013, 'affected family' has been defined comprehensively, yet it is against women in a real sense.

Tenth, there is a lack of correct information and transparency, leading to various kinds of rumours and malpractices (like middlemanship, bribery). Therefore, the oustees are unable to participate at various stages from land acquisition to rehabilitation and resettlement, and hence a sense of alienation prevails there. In addition, usually most of the displaced persons in multipurpose projects are the poor and marginalised tribals (up to 100% in some cases). This may be seen in Table 4.12.

Thus, first, it transpires that most of the displaced victims are tribals who are further marginalised, e.g. in Karjan Project (Gujarat) all the displaced persons (11,600) were tribals. Second, in the Icha dam, Chandil dam, Koel Karo project, and Upper Indravati, the share of tribals in the total displaced persons ranged from 80% to 89.2%. Third, in the Maheshwar dam, Bodhghat, Mahi Bajaj Sagar, and Ichampalli projects, the share of tribals in the total displaced persons ranged between 60% and 80%. Fourth, in Tultuli, Polavaram, Pong, Maithon, and Panchat, and Sardar Sarovar dam, the share of the tribals in total displaced persons ranged between 50% and 60%. Fifth, in Masan Jor Reservoir, Bhakra dam, and Daman Ganga, the share of the tribals in total displaced persons ranged between 30% and 50%. Finally, in only one project, Ukai Reservoir, the

Development, Displacement, and Rehabilitation 175

Table 4.12 Displaced Population in Different Development Projects in India

Sl. No.	Project	State	Displaced Population (Number)	% Share of Tribals in Displaced Population
1.	Karjan	Gujarat	11,600	100.0
2.	Sardar Sarovar Dam	Gujarat/MP	2,00,000	57.6
3.	Maheshwar	MP	20,000	60.0
4.	Bodhghat	MP	12,700	73.91
5.	Icha	Bihar (now Jharkhand)	30,800	80.0
6.	Chandil	Bihar (now Jharkhand)	37,600	87.92
7.	Koel Karo	Bihar (now Jharkhand)	66,000	88.0
8.	Mahi Bajaj Sagar	Rajasthan	38,400	76.28
9.	Polavaram	Andhra Pradesh	1,50,000	52.90
10.	Maithon and Panchet	Bihar (now Jharkhand)	93,874	56.46
11.	Upper Indravati	Odisha	18,500	89.20
12.	Pong	Himachal Pradesh	80,000	56.25
13.	Ichampalli	Andhra Pradesh	38,100	76.28
14.	Tultuli	Maharashtra	13,600	51.61
15.	Daman Ganga	Gujarat	8,700	48.70
16.	Bhakra	Himachal Pradesh	36,000	34.76
17.	Masan Jor Reservoir	Bihar (now Jharkhand)	3,700	31.0
18.	Ukai Reservoir	Gujarat	52,000	18.92

Source: Shodhganga (2020).

share of tribals in total displaced persons was less than 20% (18.92%). Various scholars' estimates show that from 1947 to the early twenty-first century about 4–6 crores (40–60 million) of people in India were displaced by the so-called development projects, and 75%–85% of them were not rehabilitated at all. Hence people radically resisted such projects in Bastar, Nandigram, Singur, Jagatsinghpur, Posco, Narmada (MP), Niyamgiri (Odisha), and so on. In China, during 1950–2010 more than 80 million (8 crores) people were displaced due to development works like hydropower, highways, urban development, and renewal projects; about 1.3 million people were displaced by the Three Gorges Dam project itself (world's largest power station with an installed capacity of 22,500 mw); its construction cost was $10.2 billion, and the work started in 1993 and was completed in 2009 (Guoqing 2014). Their resettlement policies aimed at minimising land requirements and displacement. Compensation was paid to all collective owners and users, and compensation and resettlement costs were included in the project. There the compensation standard was Annual Output Value (AOV), hence 16 times the AOV or equivalent of 32 years of net income; further compensation was paid for individual assets and subsidies were provided for the relocation and sustainable livelihood training; further, post-relocation support of 600 yuan per head was also given for 20 years. In addition, community infrastructure was provided (Guoqing 2014). According to Guoqing, the following key factors led to their successful resettlement:

176 Development, Displacement, and Rehabilitation

(a) Good resettlement legislation and policies
(b) Comprehensive Resettlement Plan prepared in consultation with the participation of affected communities
(c) Effective implementation of Resettlement Plan
(d) Independent monitoring and supervision mechanism
(e) Effective participation of affected people in the entire resettlement process
(f) Transparent consultation and grievance mechanism
(g) Adequate mitigation measures for impacts on social and cultural livelihoods, especially for the affected indigenous people and minorities
(h) Sustained capacity building through research and training, and education for affected persons to reconstruct their livelihood in a more sustained way

On the other hand, an oral testimony resettlement project was carried out in a comprehensive manner about several projects in eight countries and in 15 languages by Panos, London: Kariba dam (Zambia and Zimbabwe), Tarbela dam (Pakistan), Lesotho Highlands Water Project (Lesotho), Coal mining project in Jharkhand (India), San Conservation and Development project (Namibia, Kenya and Botswana), and mining projects (Madagascar and Jamaica). Its findings are as follows (Siobhan Warrington 2014):

(a) Feeling of powerlessness – losing control over many aspects of their life, traumatic experiences.
(b) Failure in resettlement to grasp the significance of soil, land, vegetation, and rivers as essential components of villagers' sense of security and identity – people value land economically as well as in terms of heritage or identity; land-based institutions (helping in time of need) fail to be effective at relocation sites.
(c) People's perceptions influence their choices and actions.
(d) Deficit of trust with State/State agencies responsible for land acquisition and resettlement.
(e) Monetary compensation does not give security for the future.
(f) Change of status 'from being masters of their destiny to slaves working for others'.
(g) Crisis of identity, particularly among women in tribal communities – earlier they worked as equals in fields but post-resettlement, the jobs mostly went to males, hence subjugation of the women; further internal family dynamics changed as the job was given to only one son (hence there emerged sibling rivalry).

For the real sustainable livelihood of the victims of the development-induced displacement, global best practices in different components of resettlement are worth consideration by the policy-makers and development practitioners in India and other developing countries (Table 4.13).

A Case Study of Displacement and Rehabilitation in Jharkhand (India)

Now let us discuss some concrete cases of displacement due to large dams in Jharkhand. Table 4.14 gives the profile of the area submerged and population displaced, and the tribals affected.

Development, Displacement, and Rehabilitation   177

Table 4.13 Best Global Practices for Sustainable Livelihoods of the Displaced People

Sl. No.	Best Practices	Country	Provisions
1.	Beyond one-time compensation	China	(a) Annual allowance 600 Yuan ($100) per head for 20 years (b) Economic measures for correcting past under-payments in the last five decades (c) Post-Resettlement Development Fund created with regular contributions from the power companies
2.	Land leasing	Japan	(a) Upfront payment to the landowners (b) Regular rent payments to landowners for the leased land for the life of the project
3.	Taxation	Norway	(a) All electricity companies need to pay 28% tax on their profits with the shares divided equally among the central and country budgets (b) 4.75% of taxes goes directly to the local municipality (c) Tax on the use of natural resources (based on average power generated over previous seven years) distributed to municipalities
4.	Sharing of project benefits	Norway	Electricity companies to provide 10% of the electricity produced to the local municipality
5.	Irrigated land options	Japan	Development of dry lands into cultivable lands by introducing irrigation at government cost for distribution to resettlers
6.	Sharing of project revenue	Colombia	(a) 3.8% to regions' watershed agencies for investment in water saving and local irrigation (b) 1.5% to municipalities bordering the reservoir (c) 1.5% to upstream municipalities
7.	Sharing of royalties	Brazil	Public hydropower plants: (a) 45% of royalties given to the overall budgets of the affected States (b) 45% of royalties directly given to municipalities within those States
		Papua New Guinea	(a) Mining companies: payment of royalty by a mine leaseholder at the rate of 2% of the sale of minerals (b) 20% of the royalty to be distributed between the landowners of the project area and the rest to be spent for community sustainable development plan in the area
8.	Equity sharing	Canada, Australia	Equity stake that entitles the communities to a share of project benefits for the long-term proportionately with their land share

Source: Renu Modi (2010) and other sources (cited in Nair 2014).

From the above table, it transpires that these dams took a very long time to complete, and hence a multifold increase in costs, a large area (including forests) submerged, and a large population was displaced and out of them the maximum oustees were the poor tribals! Usually, on average, two to three persons per ha are displaced in large dams in India and the average submerged area per dam is about 10,000 ha. This itself speaks volumes of the misery of the land oustees. We keenly observed the following tendencies

178 Development, Displacement, and Rehabilitation

Table 4.14 A Profile of Area Submerged and Displaced Population due to Large Dams in Jharkhand (India)

Name of Dam (1)	Year of Completion (2)	Area Submerged (ha) (3)	Displaced Population (No.) (4)	Tribal Population Displaced (No. and %) (5)
1) Telaiya	1953	7,500	13,455	NA
2) Konar	1955	2,800	5,747	1,224(21%)
3) Maithon	1957	10,700	28,030	15,837 (57%)
4) Masanjore	1955	6,950	3,700	1,147(31%)
5) Panchet Hill	1959	15,300	41,461	22,000 (56%)
6) Chandan	1972	1,080	NA	NA
7) Tenughat	1981	6,372	NA	NA
8) Chandil	1995	17,409	48,500	46,075 (95%)
9) Icha	1995	16,769*	30,800	26,640 (80%)

Source: Based on S. Singh and P. Banerji (2002) and Agarwal and Narain (1982).

and problems (as 'oral history') among the displaced persons of Chandil and Icha dams (West Singhbhum, Jharkhand) under the Suvarnarekha Multipurpose Project in 1990s:

a) The houses constructed were small and of poor quality – these houses did not have provision for cattle keeping, storage of articles, front spaces (homestead), and community spaces.

b) The cash compensation often put them under the clutches of local thugs, money-lenders, and traders.

c) Availability of a large sum suddenly led them to drink alcohol heavily and consequently increase domestic violence (males used to beat their wives and children unnecessarily). In some cases, they deserted their wives and remarried other women, leading to family tension and conflicts.

d) Many tribals changed their lifestyles by purchasing costly ornaments, clothes, transistors, goggles, etc. as symbols of consumerism; but these vanished soon.

e) The dowry rates went up alarmingly.

f) Some non-oustees (about 600) managed illegally to get jobs; some oustees cornered more jobs than prescribed in the package (husband and wife, two brothers, two sisters, father, and son/daughters) so the district administration provided separate government jobs to more than 100 deserving oustees after due verification (in lower grade) in different offices at the district level.

g) The old pattern of social relationships changed; old neighbourhood settings changed and old connections with weekly markets (haats) disappeared.

h) There were no adequate facilities for schools, health centres, pucca roads, drinking water, electricity, sanitation, drainage, etc.

Our keen empirical observation of Chandil and Icha dams (in West Singhbhum in Jharkhand) found not only differences but contradictions in the perception of the local people, District administration, State government, and the project authority of Suvarnarekha Multipurpose Project (SMP). This is shown in Table 4.15.

Table 4.15 Different Perceptions on Dam-Building, Displacement, and Rehabilitation in Suvarnarekha Multipurpose Project (SMP) (Jharkhand), India

Stake Holders	Large Dam-Building	Displacement Problems	Rehabilitation Package	Sensitivity	Transparency
1. Project authority	Essential for irrigation, electricity, water supply, and flood control (Odisha and West Bengal)	Inevitable – no gain without pain	Cash of 10,000/- for land or two acres of agri-land, cash of 20,000/- for house, and jobs	Very low	Least
2. State government (Bihar, Odisha, West Bengal)	Essential as above	Inevitable	Same	Low	Less
3. District administration	Partly essential, but with eco-sustainability	should be minimum displacement	Insufficient – scope for better	Medium	Medium
4. Local community	Not required	Avoidable – local people suffer, others enjoy	Negligible, market value of land not paid	Highly sensitive	Full transparency demanded

Source: Authors.

180 Development, Displacement, and Rehabilitation

Needless to mention here that the Suvarnarekha Multipurpose Project (SMP), a joint venture of the then-united Bihar (now Jharkhand), Odisha, and West Bengal governments, was planned in 1973 at a cost estimate of Rs. 129 crores, but it ultimately crossed Rs. 2,500 crores over two decades (completed in 1995). Local people, especially the tribals, protested strongly against it, but during the emergency period (1975–1976) the work was carried out forcibly. But, in 1978, the protest took the shape of a full-fledged movement; four protesting persons were killed in the police firing and many others were badly beaten up, tortured, and arrested. Further, several activists of the anti-Icha dam were also killed, injured, and arrested by the police; the mass leaders like Nirmal Mahto, Ganga Ram, and Sidiu Tiyu were also killed. In July 1991, the protesters sat on fast unto death and only then project authorities gave assurance to fulfil their demands. But later they were falsely implicated in theft and arson cases! However, their long protests ultimately pressured the government authorities to revise the rehabilitation package, by adding 'land for land', increasing the amount for house construction, and also providing government jobs to the oustees (Agarwal and Narin 1999).

Another project, Koel Karo hydel project, was started in Ranchi and Gumla districts in united Bihar (now Jharkhand) in the late 1970s, but on a petition from a tribal leader, the Supreme Court of India stayed the project work. The Planning Commission of India had cleared this project in 1973 itself for promoting lift irrigation, electricity generation, and industrialisation with an estimated cost of Rs. 175 crores, but that escalated to Rs. 17,000 crores in 1994. Two large dams on the south Koel river at Bansi and across north Karo river at Lohajini were proposed to be built. It was to submerge 10,522 ha of cultivable private land, 5,666 ha of so-called barren land, and 364 ha of government land. It was to displace about 4,500 families in Ranchi and Gumla districts (about 25,000 persons) and to submerge 150 'Sarnas' (tribals' places of worship), and more than 300 'sasandiris' (burials of the ancestors), thus of quite a significance for nature-worshipping (animistic) tribals. Hence, tribals protested by arguing that land, water, and forests belonged to them. They were happy with their traditional irrigation system of 'aahar' and 'pyne', and manki-munda-Parha Raja system of customary rights of jurisprudence. There was, however, a communication gap about the submergence, displacement, and rehabilitation. The Koel Karo Jan Sangathan, backed by Jharkhand Mukti Morcha, put forth 18 demands which were not considered by the then-State government of Bihar. So the local people imposed 'Janata curfew' and ploughed the roads to prevent the entry of the project officials. Hence, the National Hydel Power Corporation (NHPC) withdrew from it and ultimately the then-government of Bihar had no option but to close it in November 1997, though by that time it had spent about 18 crores of rupees (Agarwal and Narain 1999). This is how a huge amount is wasted in the name of development against the aspirations and needs of the local people.

Conclusion

Thus from the above discussion, the following points of conclusion emerge that have many policy implications for India:

First, small dams, especially earthen ones, as in China earlier (87,000 dams in total) are more economically useful, people-friendly, egalitarian as well as ecologically sustainable because there is no question of mass displacement, soil erosion, siltation, deforestation, or desertification. It clearly implies that local people actually need water conservation in small/medium kachcha check dams, reservoirs, lakes, and rivers, not large concrete dams. Hence the so-called development model conceptualised and 'planned from above' and market-driven should give way to an alternative model of developing need-based ecologically sustainable agriculture, with adequate water and soil conservation. That is to be 'planned from below' where the local community should be fully involved at all stages of work. The 'waterman' Rajendra Singh has shown the utility of a building and reviving small and kachcha check dams, lakes, and revival of seasonal rivers for rainwater conservation in dry Rajasthan State. We need to learn and adapt that model in order to have real sustainability.

Second, displacement should be avoided in the first place, only in the rarest of rare cases minimum displacement should be permitted and the land oustees and land losers and losers of livelihood must be fully rehabilitated with equal quantity and quality of land, houses with necessary amenities, suitable government jobs for oustees as well as adequate infrastructure development at the new sites of relocation, and resettlement be provided as a genuine human right and need. Fortunately, RFCTLARR Act, 2013, takes care of these, and hence their compliance must be ensured by central, State, and local authorities.

Third, multilateral agencies like the World Bank, Asian Development Bank should not finance large dams which are, 'in the ultimate analysis', anti-people. Actually, the Sardar Sarovar Project on the Narmada river was initially financed by the World Bank, but when its own Morses Committee submitted an adverse report, the World Bank withdrew from it. But this lesson should be learnt by all financial institutions for not financing large dams.

Fourth, the hide-and-seek tendency in major development projects, especially large dams, leads to all kinds of rumours, misinformation, and disinformation. This lack of transparency gives birth to distrust of the people, on the one hand, and malpractices by the staff and the middlemen, on the other. Therefore, for any development project, the community concerned should be treated on equal, if not better, footing with other stakeholders, sharing all relevant information including the experiences of other schemes/projects of similar type and magnitude. This will ensure that the assumed, potential and real, unequal and exclusive power in decision-making will finally give way to the egalitarian, fair, and inclusive process of decision-making at various stages of formulation, planning, implementation, monitoring, and evaluation of development schemes/projects.

Finally, in the present age of liberalisation, privatisation, and globalisation (LPG), the market has cornered the larger space of civil society, and the State has, too, encroached upon the arena of civil society. The State, and its machinery, as the local affected people claim, is withdrawing from its social responsibilities to its citizens' quality education, health, employment, housing, drinking water supply, and so on. Consequently the cruel hands of the free market are ruling the roost, inclined to the traders, industrialists, suppliers, contractors, and rich consumer class. The local community has full rights over common property (natural) resources like lands, forests, rivers, minerals, and other water bodies in the real sense. In a parliamentary democracy the so-called 'representatives' of the people have unfortunately made and got implemented various laws and projects against the interests of the people at large, whom they claim to represent. This hiatus between the people and their representatives needs to be removed fully, only then 'real social development' (RSD) could be planned and practised. This will be real self-reliance and empowerment of the community, as Tarun Bharat Sangh, under the leadership of Rajendra Singh ('waterman'), has done in parts of rural Rajasthan by building earthen dams and recharging and reviving dried rivers. Such 'real social development' needs to be culturally acceptable, economically livelihood creating, ecologically sustainable, technologically feasible, and socially equitable. And this alternative paradigm of development is possible only in real deliberative ecological democracy.

When there is development-induced displacement or environmental devastation, local people protest against these and launch movements there in an organised way. This aspect will be deliberated upon in the next chapter.

Points for Discussion

(1) Discuss and compare major concepts of development, especially in the Indian context.
(2) Discuss major types of displacement, especially in the context of developing countries.
(3) What are the major causes of displacement nationally and globally; give some examples.
(4) What are the consequences of displacement in India, especially regarding large dams.
(5) What is the role of resettlement and rehabilitation for displaced persons? Give some examples of best practices of rehabilitation and resettlement in the world.

Chapter 5

Ecological Movements in India

Introduction

Since the inception of mankind, humans have been dependent on the natural environment in different ways for their 'critical life issues' in terms of earning a livelihood from forest produce, making a socio-cultural space, as well as a space for spirituality (Sharma 2016). However, over several millennia, people learnt agriculture (about 12,000 years ago) on some portions of land and, in the course of time, with the increase in population as well as the inception of the concepts of private property, there resulted various kinds of environmental degradation in terms of denudation and desertification of land, pollution of air and water, consequent to activities like the rapacious mining of minerals and felling of forest trees, etc., in the whole world. The industrial revolution in Europe and North America in the eighteenth century compounded the problem in range, depth, frequency, and intensity. Its consequences are visible in both developed and developing countries, especially due to the colonial empires the agricultural system suffered a lot because of compulsive cash crops grown (e.g. indigo in Bengal and Bihar) during the colonial period, and constantly 'unequal ecological exchange', on the one hand, and the high rate of carbon emission due to conspicuous consumption of a huge amount of energy per capita in the developed countries, on the other hand. This has led to global warming and, finally, to adverse changes in climatic conditions. However, there are also class divisions in developing countries like India, especially three ecological classes: *omnivores, ecosystem people,* and *ecological refugees* (Gadgil and Guha 1995). Even after achieving Independence on 15 August 1947, various colonial policies of the State have continued wherein most of the common property (natural) resources have been put under the control and ownership of the State, e.g. forests, rivers, lakes, minerals. Therefore, socio-ecological inequities led to social conflicts that fuelled ecological movements in different parts of India, especially since the early 1970s. These have varied in terms of the proximate causes, kind of participation, type of leadership, ideological disposition, and strategy/tactics of action.

Needless to say that usually traditional social movements have been peasant and labour movements, but new social movements have emerged with environmental issues, peace, gender, and identity or these are new expressions of old ideologies. Harsh Sethi (1993) makes five categories of environmental struggles in India:

(a) For forest conservation
(b) Against inappropriate use of chemicals (fertilisers, pesticides, weedicides, etc.) in agriculture, resulting in land degradation and water pollution

DOI: 10.4324/9781003336211-5

184   Ecological Movements in India

(c)   Against dams construction causing huge displacement of the people (mostly tribes)
(d)   Against waste and pollution caused by industrialisation
(e)   For judicious utilisation of marine resources

Furthermore, Gadgil and Guha (1994) analysed ecological movements in India in three contexts: material, political, and ideological. Material context refers to the struggle over natural resources between the industrialists, rich farmers, urban consumers, on the one hand, and the poor small peasants, labourers, tribal and fisherfolk, on the other. Second, political basis means that the activists mobilise the victims of environmental degradation because of adopting a specific development model and ultimately question it. Third, ideological orientation is Marxist or Gandhian.

However, Janki Andharia and Chandan Sengupta broadly classify India's ecological/ environmental movements into eight categories on the basis of issues in focus with tensions, paradoxes, and contradictions (Table 5.1).

Before discussing three major ecological movements in India, namely Chipko, Narmada Bachao Andolan, and Silent Valley movement, we shall discuss in brief other such movements too.

Plachimada Movement against Coca-Cola (2000), Kerala

In March 2000, the multinational Coca-Cola company started its bottling plant on 34.64 acres of land at Plachimada, in the Palakkad district of Kerala to produce 5.61 lakh litres (5,61,000) of beverage daily. It required 3.8 litres of fresh groundwater to prepare 1 litre of Coca-Cola, i.e. about 2 million litres of groundwater was extracted daily on an average, but exceeding this more often. After some months, local people found that there was a change in the colour of the water and wells were contaminated, hence not suitable for drinking, cooking, and irrigation. Hence, local people, mainly the tribals under the leadership of C. K. Janu, organised protest rallies and mobilised all concerned, especially on 4 August 2002. Later a Public Interest Litigation (PIL) case was filed in the Kerala High Court that decided in local villagers' favour by observing that water is a common property resource (CPR), hence a private company has no right to extract groundwater. In August 2005, Kerala State Pollution Control Board ordered the company to stop the production after the confirmation of the adverse findings by the University of Exeter (UK). In 2011, the Kerala government enacted for establishing a tribunal for compensation and relief for the environmental degradation caused by the Coca-Cola company there. There was lead and cadmium at very high levels in the sludge of the plant. Earlier in January 2004, the International Water Conference was organised there and environmentalists like Maude Barlow and Vandana Shiva attended it and the Plachimada declaration stated: 'water is not a private property, not a commodity', rather a common natural resource and a fundamental right of the local people. On 3 April 2003, Perumatty panchayat revoked the licence of the plant. Due to continuous protest, the company stated in the Supreme Court that it intended to close the plant. A high-powered committee determined the damage done to the community amounting to Rs 216.26 crores ($28 million) in terms of agricultural loss, health problem, cost of providing water, wage loss, opportunity cost, and cost of pollution of water resources. Thus, this closed company was used as a COVID-19 treatment centre in 2021.

Ecological Movements in India   185

Table 5.1 Categories of Environmental Movements by Issues and Examples in India

Sl. No.	Categories	Issues	Examples
1.	Forest and land-based	(i) Right to access forest resources (ii) Non-commercial use of natural resources (iii) Prevention of land degradation (iv) Social justice/human rights	Chipko, Appiko, tribal movements (e.g. Jharkhand – Sal tree, giving fodder, nut, and fuel wood, versus teak (timber only)) and Bastar Belt
2.	Marine resource and fisheries (aquaculture)	(i) Ban on trawling, to prevent commercialisation of shrimp/prawn culture (ii) Protection of marine resources (iii) Implementation of coastal zone regulations	National Fishermen's Forum (Kerala), Chilka Bachao Andolan, Odisha
3.	Industrial pollution	(i) Stricter pollution control measures, compensation (ii) Prevention of reckless expansion of industries without considering design, location factors, and livelihood of local people	Zehrili Gas Morcha, Bhopal; Ganga Mukti Andolan, Bihar; movement against Harihar polyfibre factory in Karnataka; movement against pollution of Sone river by Gwalior Rayon factory (Vidushak Karkhana Group) Shahdol, MP; movement against poisoning of Cheliyar river in Kerala (by Kerala Shastra Sahitya Parishad)
4.	Development Projects (a) Dams and irrigation projects	(i) Protection of tropical forests (ii) Ecological balance (iii) Destructive development (iv) Rehabilitation and resettlement of the displaced people	Silent Valley movement (KSSP); Narmada Bachao Andolan; movements against Tehri dam (Tehri Bandh Virodhi Samiti); Kosi-Gandhak Bodhghat and Bedthi, Bhopalpatnam Ichampalli in the West; Tungabhadra, Malaprabha, Ghataprabha Schemes in the South; Koyna project-affected committee
	(b) Power projects	(i) Ecological balance (ii) Rehabilitation and resettlement, high costs	Jan Andolan, Dabhol against Enron, Koel-Karo Jan Sangathan, Bihar (now Jharkhand)
	(c) Mining	(i) Depletion of natural resources (ii) Land degradation (iii) Ecological imbalance	Anti-mine movement, Doon Valley, Anti-bauxite mine movement (Balco project), Odisha
	(d) Industrial plants/railways/airport projects	(i) Re-alignment (ii) Rehabilitation and resettlement of the displaced people (iii) Ecological balance	Protests/demands of Konkan Railway Re-alignment Action Committee, citizens' group against Dupont Nylon 6.6, Goa Amravati Bachao Abhiyan against a large chemical complex
	(e) Military bases	(i) Ecological balance (ii) Rehabilitation, resettlement and safety	Anti-missiles test range in Baliapal and at Netarhat (Bihar) (now Jharkhand)
5.	Wildlife Sanctuaries, National Parks	(i) Displacement, rehabilitation and resettlement (ii) Loss of livelihood	Ekjoot in Bhimashankar (Maharashtra), Shramik Mukti Andolan in Sanjay Gandhi National Park, Bombay

(Continued)

186 Ecological Movements in India

Table 5.1 (Continued)

Sl. No.	Categories	Issues	Examples
6.	Tourism	(i) Displacement, cultural changes, social ills	Himachal Bachao Andolan, Bailancho Saad, Goa
7.	Advocacy groups individual campaigns, citizen's action groups	(i) Policy inputs, stricter measures for protected areas (ii) Clear policy on national park/wildlife sanctuaries, lobbying research, training/ documentation on wildlife, conservation education, community-based environmental management (iii) Intellectual support to grassroots movements on environmental issues	Society for clean cities, Bombay Natural History Society, Centre for Science and Environment, Delhi, Bombay Environmental Action Group, Save Bombay Committee, Save Pune Citizens' Committee
8.	Appropriate technology/ organic farming	(i) International debates (ii) Sustainable development, eco-friendly model of development (iii) Low-cost, environment-friendly housing, and technology	Ralegan Siddhi (Anna Hazare's village) SOPPECOM, Laurie Baker's Housing experiments, People's Science Institute, Dehradun

Source: Janki Andharia and Chandan Sengupta (1998).

Anti-nuclear (Power) Movement

India and the USA agreed in 2008 for civilian nuclear trade, hence 30 nuclear reactor plants are to be established in India by 2050, of which Jaitpur (Maharashtra), Gaurakhpur (Haryana), and Kudankulam are most significant ones. Local people see such plants against their livelihoods, with land acquisition causing loss of agricultural lands, radiation, and displacement. During Rajiv Gandhi's regime, in 1988, India and the Soviet Union had agreed for a nuclear plant at Kudankulam, but farmers, fishermen, intellectuals, and social activists protested against it. When people came to know that Pechiparai Reservoir water would be used for the Kudankulam nuclear plant, many farmers got mobilised. Then the Social Equity Movement organised massive rallies in Tirunelveli, Kanyakumari, and Tuticorin districts of Tamil Nadu. Due to the disintegration of the USSR, the plant was closed. But, later, the then Prime Minister of India, H. D. Deve Gowda, and the Russian President Boris Yeltsin agreed to restart it as Russia decided to supply two water-cooled and water-moderated reactors. Hence, again there was a protest, especially because the flawed design of Chernobyl unit 4 was to be installed here. People's movement against nuclear energy came up and, in 1999, Tamil Nadu Fish Workers Union called for a massive strike against their livelihood loss. In 2003 there was again a massive rally, and, consequently, the National Alliance of People's Movement (led by Medha Patkar of Narmada Bachao Andolan (NBA) fame) joined it. In December 2004, a tsunami caused huge floods there, and on 19 March 2006, tremors of the earthquake were also felt there. Again, on 12 August 2011, tremors were felt in seven districts of Tamil Nadu. In 2014, at a national conference at Idinthakarai, the Indo-US nuclear deal was declared to be fatal. The activists supported protests at other nuclear plants in India too (Kumar 2015).

Similarly, the nuclear plant at Gorakhpur in Fatehabad district, Haryana, too, was opposed by the local people, especially local farmers who are to lose 1,500 acres of fertile agricultural land, and secondly, the irrigation water from the Bhakra-Nangal dam is to be diverted for cooling down of this plant. Further, the plant will damage the flora and fauna of the region – the local Bishnoi community referred to the decline of Chinkaras. Finally, as per Atomic Energy Regulatory Board, a nuclear plant is not to be set up within a radius of 5 km of a place with more than 10,000 population and the Gorakhpur village has 13,000 population (2011 census), hence health issue too. Various organisations like Haryana Parmanu Sanyantra Virodhi Morcha, Haryana Janhit Congress, Indian National Lok Dal, CPI (ML) have been protesting against this plant. Various cases have also been filed in the National Green Tribunal and the Punjab and Haryana High Court (M. G. Devasahayam, cited in Kumar 2015).

Ganga Mukti Andolan, Bhagalpur (Bihar)

Although India achieved freedom from the British colonial rule on 15 August 1947 and Zamindari (intermediary land tenure system) was abolished in the 1950s in Bihar as elsewhere, in its Bhagalpur district two 'waterlords' continued to establish their authority over the water of the Ganga river by collecting taxes from fishermen and women for fishing in the Ganga river – Mahesh Ghosh collected taxes on the Sultanganj–Barari stretch of Ganga, and Musharraf Husain Pramanik of Murshidabad (West Bengal) did so on the Barari–Pirpainti stretch of Ganga. Though an earlier case was filed by the aggrieved party in Patna High Court in 1961 itself, the High Court decided that such water tax on the flowing Ganga river is not imposed on the immovable property like land, and it has been continuing since the Mughal period, hence Zamindari abolition law is not applicable here. Later, in February and March 1982, hundreds of fishermen and women protested and demonstrated against such an unjust taxation practice. Further, on 4 April 1982, various fishermen and women from different areas held a meeting and decided to eliminate such a tax collection system as well as to remove caste system, liquor prevalence, and ensure equal participation of women. Consequently, Anil Prakash of Chhatra Yuva Sangharsh Vahini emerged as its leader who mobilised the fishermen, women, and other local youth. On 22 March 1983, forty boats of activists/fishermen moved from Kahalgaon to Bhagalpur in the Ganga water with a slogan: 'Jaal Baans Aujar Hamara, Ganga par Adhikar Hamara' (fishing net, bamboo, and instruments are ours; we have right over Ganga water). Various folk singers and artists, students, and labourers of nearby Jhuggis also joined this movement. An illiterate folk artist (Dileep Turi) composed a song for this movement – 'Ganga ki god base Bhaiya ho dukh bhelo apar' (residing in the lap of the Ganga river, has a lot of sorrow). In the adjacent Ganga *diara* (riverine), there used to be disputes for sowing and reaping of crops, hence the local peasants were also organised there. This movement used the symbol of Eklavya to motivate the aggrieved people. It had twofold strategies: those having problems should lead; and without women's cooperation every change is incomplete ('Naari ke Sahbagh bina har badlav adhura hai').

It also opposed the installation of the NTPC plant in a highly dense population area and on highly fertile land at Kahalgaon, and the fly ash from the plant will ruin the Ganga and the local environment. A pamphlet was also circulated among the local

188 Ecological Movements in India

people and administration with a slogan 'save Bhagalpur from becoming a new Bhopal' (Bhagalpur ko Bhopal Hone Se Bachayiye). On 4 September 1984, a massive meeting of the fishermen and boatsmen was organised. Since 1985 the movement also raised the 'diara' problems and thus it expanded a lot among the local people. On the other hand, the landlords, contractors and their cronies filed many false cases against the movement activists. Hence, they gave a new slogan: 'Zulm ka Chakka aur tabahi kitane din/ye lathi, bandook, sipahi kitane din' (for how long will be the wheel of oppression, sticks, guns, and police). A boat demonstration started on 8 April 1987 at Kursela and reached the Patna College river bank on 21 April 1987 after covering places like Bateshwar Sthan, Kahalgaon, Bhagalpur, Sultanganj, Munger, Simaria, Khusrupur, Fatuha, and Gaighat. Further, the fishermen and women stopped paying taxes to the waterlords who, therefore, deployed their goons to assault them. On 21 May 1987, three boats were looted at Lailakh by their goons; again on 6 July three boats were looted at Lailakh; on 26 September 1987, three boats were looted at Tintenga; on 30 September 1987, 11 boats were looted at Karari Tintenga, and on 29 September a boat was also looted at Oriyap and the boatman Baijnath Mahto was kidnapped. In addition, Ganga Mukti Andolan's (GMA) leaders like Prasadi Sahni and Durga Sahni were arrested.

On 19 December 1989, hundreds of fishermen and women peacefully took the possession of the waterlord's local 'Kachehri' (office) and various leaders and activists were implicated in police cases, but due to a strong united front of GMA, the children and women stood in the front to resist the local police to arrest their leaders, rather demanded them too! On another occasion, some criminals looted their fishing nets, so they went to their (criminals) houses and, ultimately, the nets were recovered. Again on Yogendra Sahni's boat contractors fired, hence women followed them on boats and caught them near Pakadtalla; but they begged pardon.

Finally, in 1990, Lalu Prasad, the then Chief Minister of Bihar, accepted the demands of the fisherfolk and declared their right to free fishing in all rivers of Bihar; earlier two Chief Ministers of Bihar had verbally assured this. However, later due to the declaration of Dolphin Sanctuary by the Government of India on the Patna–Bhagalpur stretch of Ganga, the fisherfolk cannot fish in the flowing water of Ganga there. In fact, Tehri dam was built on the initial stretch of Ganga (in Uttarakhand) and Farakka dam was constructed on the last leg in West Bengal, hence breeding of various types of fish declined substantially there (Yogendra 2022).

In this chapter, now we shall discuss the context, genesis, formation, and consolidation of three major ecological movements in India in detail. These are as follows:

(a) Chipko Andolan (Hug the Tree Movement), Garhwal, Uttarakhand (in the former State of Uttar Pradesh), North India (1973–mid-1980s)
(b) Silent Valley Movement, Palakkad, Kerala, South India (1976–1983)
(c) Narmada Bachao Andolan (Save Narmada Movement), Madhya Pradesh (MP), Central India (1987–September 2017).

Chipko Movement, Garhwal Uttarakhand, North India

Context

The philosophy of Mahatma Gandhi has been an inspiration for the Chipko movement since its very beginning in 1973. Gandhiji had once aptly remarked: 'The earth has enough to sustain everyone's need but it has got too little to satisfy everyone's greed' (cited in Bandyopadhyay 1992). The Chipko movement was started by the Sarvodaya ('welfare of all') activists (followers of Gandhi's disciple Vinoba Bhave) of Dasauli Gram Swarajya Mandal (DGSM) in the Mandal village in the Chamoli district of Garhwal in the Himalayan region of the State of Uttarakhand in North India. The State of Uttarakhand is a new state carved out of the former State of Uttar Pradesh on 9 November 2000 by an Act of Parliament: 'The Uttar Pradesh Reorganisation Act, 2000 (Act 29 of 2000)', notified on 25 August 2000. The historical legacy of Chipko strategy of tree-saving goes back to 1763 when in Rajasthan (Western India) 300 persons, belonging to the Bishnoi sect of Hindus, sacrificed their lives for saving (by hugging) their Khejri trees which were being felled under the orders of the then King of Jodhpur. But, actually, there is no link between this historical event and the Chipko movement except in the similarity of action. The analogy obscures Chipko's origins in specific conditions of the hill region of the former State of Uttar Pradesh, now Uttarakhand (Guha 1989). During the colonial period under British rule, the creation of the Indian Forest Department in 1864, the British government's contract of long-term felling rights with the Tehri State (Garhwal region) in 1865, and enactment of an anti-people law, 'The Indian Forest Act, 1878 (Act VII of 1878)', actually, in the name of scientific management of forests, ended the system of local community's use, control, and collective self-management of forests as common property resources (CPRs). This Act gave the British Raj a monopoly over all kinds of forest produce, and thus the livelihood of the peasants, tribals, and other forest-dwellers was severely threatened. The basic goal of colonial rule was the commercial exploitation of timber for ship-building industries of the British Royal Navy and for sleepers for the expansion of the Indian Railways, and this led to the massive felling of trees, especially in the Himalayas (North and Northeast Indian forests) and Western Ghats (South Indian forests) (Gadgil and Guha 1992). For instance, during 1869–1885, 65 million railway sleepers were exported from the deodar (*Cedrus deodara*) forests of the Yamuna catchment (in the Garhwal Himalayas) alone (Guha 1989, cited in Bandyopadhyay 1992). On the other hand, the colonial Raj restricted '*shikar*' (hunting) by the traditionally hunting people though it permitted freer hunting for sports by the British and Indian elites. Further, the forests and hills were made more accessible to outsiders (tourists, army, contractors, and entrepreneurs) through the construction of roads, bridges, and even rails to 'the hill stations' (tourist spots).

Finally, the introduction of the Zamindari (feudal-intermediary landlordism) system of land tenure created private property in land. And, therefore, most of the village commons were either brought under the State ownership or were privatised for the sake of collecting more revenue. Consequently, the local people from the Kumaon region of the Himalayas resisted the reservation of the Kumaon forests (in 1911–1917), and the protest culminated in the paralysing of the administration in 1921, first by a strike against *utar* (statutory labour) and later by a sustained campaign in which the Himalayas' pine forests 'were swept by incendiary fires almost from end to end' (Guha 1998, 275). This encroachment, under the dictates of European capitalism in general and British capitalism in particular, in the 'space of civil society' by an all-powerful leviathan State (the British Raj) had severe consequences:

> a *political* watershed, in that it represented an enormous expansion of the powers of the state, and that a corresponding diminution of the rights of village

communities; a *social* watershed, in that by curbing local access it radically altered traditional patterns of resource use; and an *ecological* watershed, in that the emergence of timber as an important commodity was to fundamentally alter forest ecology.

(Gadgil and Guha 1994, 104)

Thus, the new British forest laws and colonial administration brought about a triple misery: political, social, and ecological.

These factors led to several protests through *satyagraha* (peaceful resistance) by the local people. Many protests took place in the Tehri Garhwal region (of United Provinces, now Uttarakhand) in 1904, 1906, and 1930 against the so-called scientific forest management, and for an alternative conception of rights and focus on village autonomy. On 30 May 1930, many protestors were killed by the royal army of Tehri (Guha 1985; Bandyopadhyay 1992).

Genesis

India, after achieving independence in 1947, adopted almost the same path of use of forest resources for railways and other industrial development as essentially did the colonial British Raj. But later the natural resource conflicts became more intensified due to the following factors:

(a) A rush for *modernisation* of India through industrialisation, i.e. non-agricultural factories and big dams often described as 'modern temples' of India, to use Jawaharlal Nehru's terms, meant more intensive use of natural resources as raw materials. As Jawaharlal Nehru said in 1956: 'We are not going to spend the next hundred years in arriving gradually, step by step, at that stage of development which the developed countries have reached today. Our pace and tempo of progress has to be much faster' (cited in Akula 1995). So industrialisation led to a massive deforestation in the 1960s which meant deterioration of the conditions of the peasants and tribals largely dependent on the forest produce. Consequently, during 1951–1973, India lost four million hectares of forests, thus leaving 37–45% forest cover against the minimum requirement of 66% (Bhatt 1983).

(b) Industrialisation also required more energy, hence mining in forests and offshore-drilling (for coal and oil, respectively), and construction of dams for electricity intensified.

(c) In the name of 'scientific management' of forests, massive monoculture afforestation projects were started in many parts of the country, especially during the early 1960s.

(d) There was an associated process of industrialisation and urbanisation, i.e. many suburban areas, army camps, contractors' wood depots, tourist spots, were developed as hill stations. Along with these developments, there also emerged an illicit liquor trade. This ruined many families as whatever the male workers earned as wages in the felling of trees and other activities during the day, they spent on liquor by the evening. These illicit liquor shops had a linkage with outside forest contractors. Hence, the local people, especially women, of Garhwal raised their voice for prohibition in the Ghansyali village as early as in 1965. In November 1965, many women demonstrated and picketed liquor shops in Tehri, Uttarkashi, Chamoli, Garhwal, and Pithoragarh hill districts of the then State of Uttar Pradesh, now Uttarakhand (Shiva 1988).

(e) In 1970, a natural disaster in the form of floods in the Alaknanda river destroyed 6 motor bridges, 16 footbridges, a road length of 30 km, 604 houses, and 200 hectares of standing crops and affected people's lives and property, especially terraced agricultural fields in Garhwal; however, the then State Government of Uttar Pradesh did not come to their rescue in terms of compensation for victims of disasters, and as development assistance to the Garhwal Himalayan region in Uttar Pradesh (Bhatt 1983; Rangan 1996; Akula 1995).

In 1972, there were several protests against the commercial exploitation of forests in the Garhwal region of the then State of Uttar Pradesh (now Uttarakhand): in Purola on 11 December 1972, in Uttarkashi on 12 December 1972, and in Gopeshwar (the headquarters of Chamoli district) on 15 December 1972; then the local folk poet Ghanshyam Raturi 'Shailani' composed the following poem which became a means of generating awareness amongst the people and uniting them (Shiva 1989):

> Embrace our trees
> Save them from being felled
> The property of our hills
> Save it from being looted.

Shailani's poem was an expression of a conflict between 'insider' local people's survival/livelihood and 'outsider' commercial profit, i.e. need versus greed.

Formation: Early Phase or 'Private Face' of Chipko Movement

Three major issues were at the root of the Chipko movement:

First was the inter-regional gap between backward hills and the developed plains. For instance, 96% of Chamoli district's population lived in villages; 42% were not gainfully employed; 60% of the total female population was working while only 55% of men were working; against 97% of working women engaged in cultivation and only 72% of working men were engaged in cultivation. Further, nearly every family owned on an average less than half hectare of land; it did not suffice for more than six months' subsistence, hence their dependence on forests (Jain 1991). Consequently, in search of gainful employment in the armed forces, etc., there was a massive migration of the local population from the hills to the plains.

Second was the intensification of the perception of the 'insider–outsider', i.e., the hiatus between 'us' (local peasants), and 'them' (non-local contractors/traders/indus trialists/State). As one old man said,

> For sure, the government is only working in its own selfish interests, and it has no aim of benefitting the people. All the same it is up to us to benefit from the new developments, and if we want to take advantage of the new schemes we must prepare ourselves to come forward and *push the outsiders out*.
>
> (Cited in Jain 1991, 168, emphasis added)

192   Ecological Movements in India

Third was the difference in the vision of development: small-scale, local, indigenous knowledge-based development versus large-scale, national/international, modern (western) scientific development. How was it formed and mobilised? In the 1960s, in the Garhwal region, the local people having Gandhian orientation had formed co-operatives, based on forest resources, viz. Dasauli Gram Swarajya Sangh (later Dasauli Gram Swarajya Mandal), Purola Gram Swarajya Sangh, Kathyur Swarajya Sangh, Taluka Gram Swarajya Sangh. However, the interests of these cooperatives, it is some-times alleged by some scholars, clashed with those of local women because while the former wanted more forest resources as raw materials for their cooperatives, the latter wanted protection of the forests (Shiva 1989). Other scholars have criticised Shiva's naive view as a simplistic dichotomisation between 'feminine' principle of 'conserva-tion' and the 'masculine' principle of 'destruction': 'Shiva's analysis seems to be an effort to impose a decadent Western model of gender conflict on a Gandhian move-ment characterised by unique gender collaboration. This makes Shiva's work sensa-tional but largely unrealistic" (Bandyopadhyay 1992, 267–268). Other researchers also assert that the Chipko movement was officially begun in early 1973 by some Sarvodaya workers (all males) at the Mandal village in Chamoli district, and only later was it joined by women (Jain 1993).

Earlier DGSM had started an unskilled and semi-skilled workers' cooperative in 1960 to create more employment by processing forest resources for farm implements. The DGSM thought of purchasing timber, sold in a public auction by the Forest Department, for its small workshop that aimed to make farm tools for the local use of the peasantry. But it was out-manoeuvred by the rich contractors. For instance, in 1971, this cooperative started a small processing plant in Gopeshwar (district head-quarters of Chamoli) to manufacture turpentine and resin from pine sap. But the Forest Department did not supply enough pine sap even when the price paid was higher than that paid by a partly State-owned producer in the plains; hence it remained closed for eight months in 1971–1972. So on 22 October 1971, the villagers demonstrated against the forest policy of the State government (Jain 1991). Later, the Forest Department allotted 300 ash trees to Simon Company, Allahabad (a sports goods manufacturer), in an auction, though it had rejected the DGSM's annual request for ten ash trees for its farm tools' workshop. In March 1973, when the agents of the company came to fell 300 ash trees (for making tennis rackets) in the Mandal village of Chamoli dis-trict, the villagers protested against it and gathered in the forest by beating drums and singing songs. They were mobilised by the same Gandhian cooperative, DGSM. They declared that they would rather die than allow the felling of the trees, and to stop the trees from being felled they hugged all those trees. The labourers, hired by the con-tractor, then left the site. Thus was born the Chipko (hug the tree) movement. Later the manufacturer got an alternative contract from the Forest Department at Rampur Phata forest in Kedar Ghati (a valley in the region). But, there too, the local people, who had been mobilised by the villagers who had resisted and prevented the felling of ash trees in the Mandal village of the Chamoli district, did not allow the felling of the trees, and the Chipko activists kept a vigil from June to December 1973. Then the Forest Department offered one ash tree to DGSM if it allowed Simon Company (of Allahabad) to fell those trees, but DGSM rejected the offer, and when offered ten trees, they firmly rejected that offer too. Ultimately, the Forest Department cancelled the Company's permit and the trees were allocated to DGSM instead. It also withdrew the ban on pine sap supplies. The third incident took place at Reni forest where an auction of 2,500 trees was announced by the Forest Department. The pioneer of the movement, Pandit Chandi Prasad Bhatt of DGSM (who later got the Magsaysay Award in 1982 and Padma Shri from the Government of India in 1986), consistently mobilised the people and reminded the villagers of the floods and landslides in Alaknanda river in 1970 caused by the massive deforestation, and, therefore, he also suggested that they should hug the trees as a tactic. The women took his message in letter and spirit, as they had been the worst sufferers during the disasters earlier. So Gaura Devi of the Lata village organised the women and prevented the contractor's agents from felling the trees. She compared the forests to her mother's home (*maika*). Consequently, the government set up an Experts Committee to look into the issue, and the contractor for the time being withdrew from the scene. After two years, the Experts Committee reported that the Reni forest (in Chamoli district, the Garhwal region of the then State of Uttar Pradesh, now Uttarakhand) was an ecologically sensitive area, hence no trees should be cut in this region. Thereafter, the government placed a ten-year ban on that area of about 1,150 sq km (Guha 1989; Shiva 1989; Jain 1991). Hence, Guha (1989) calls the earlier phase of Chipko the '*private face*', i.e. it was essentially a peasant movement.

Consolidation: Later Phase or 'Public Profile'

The Chipko movement consolidated in range and depth from the mid-1970s to the early 1980s when it achieved a '*public profile*', as Guha (1989) calls it, as one of the most celebrated ecological movements in the world. This phase began in 1977 in a Chipko meeting in the Adwani village in the Henwal Valley of Tehri Garhwal district (of Garhwal region) with the slogan 'What do the forests bear? Soil, water and pure air'. Firstly, the Virendra Kumar Committee (appointed by the then State Government of Uttar Pradesh to investigate the causes of floods in Alaknanda in 1970) reported widespread deforestation as a major cause of disaster. Secondly, the writings of Eckholm (1975, 1976) regarding linkage between Himalayan deforestation and the floods in Bangladesh (though Hamilton and King (1983) did not support this view) also had some impact. Sarla Devi, a European disciple of Gandhiji and an activist-leader of the Chipko movement, who had insights in both local and global issues, too, had drawn the attention of the Planning Commission of India to the soil and water conservation properties of the forests. These were the 'cognitive roots of sustainability consciousness in Chipko' because it became 'a global campaign focussing on sustainability of forests, on the one hand, and sustainability of the agro-pastoral economy of Garhwal Himalaya, on the other hand', and 'no longer a hill-people's movement against forest felling. It has evolved into a philosophy' (Bandyopadhyay 1992, 270).

In the later phase, the collective action became radicalised and broad-based. For instance, a woman activist of the Adwani village, Bachni Devi, led a protest campaign against the village headman (Gram Pradhan), her own husband, who had obtained a local contract to cut the forest trees. In the protest campaign when women held lighted lanterns in their hands in the broad daylight, the foresters ridiculed them by asking the question: 'do you know what forests bear? They produce profit, resin and timber', and then the women replied with a song in chorus: 'What do the forests bear? / Soil, water and pure air / Soil, water and pure air / sustain the earth and all she bears' (cited by Shiva 1988, 1977). Afterwards, protests against the quarrying of limestone in the Doon Valley, and the Tehri Dam project, got support from the formal scientific community. This capacity of 'thinking globally and acting locally' was the basic strength of the Chipko movement, in which Pandit Sundarlal Bahuguna's contribution, too, is appreciable (Bandyopadhyay 1992). The Chipko movement thus spread in range and depth. It, later, expanded to the Uttara Kannada district in Karnataka as Appiko (Kannada word for Chipko – 'hugging') movement in 1983, Rakshasutra movement in other parts of Uttarakhand (in 1995–1996) as well as the anti-Tehri Dam movement in the then State of Uttar Pradesh, now Uttarakhand. In depth, too, it was strengthened.

The Chipko movement had three major strands (R. Guha):

The first strand was the 'appropriate technology' group of Chandi Prasad Bhatt of DGSM, which saw non-destructive but sustainable use of forest resources for local purposes through cooperatives.

The second strand was the eco-centric 'Crusading Gandhian' group (S. L. Bahuguna, A. Nandy, Vandana Shiva) that sought a total ban on tree felling, and relied heavily on the Hindu religious-spiritual tradition to reject modern (western) lifestyle of consumerism and values of materialism.

The third strand was the Marxist-oriented Uttarakhand Sangarsh Vahini (USV) that focused on a need for redistribution of productive resources, and it also included wider issues of alcoholism and mining in the region (Gadgil and Guha 1992; Omvedt 1984).

However, despite these ideological differences, all of them joined together to fight against tree felling in the Himalayas. The climax of the Chipko movement reached in 1981 when the activist S. L. Bahuguna went on an indefinite hunger strike for a total ban on tree felling in the forests situated on the hills above 1,000 m. It had the support of thousands of activists, allies, and supporters; activists were reciting from religious scriptures as well as reciting poems and singing songs suitable for the occasion. The then Prime Minister of India, Shrimati Indira Gandhi, met with S. L. Bahuguna, and ordered a 15-year ban on the commercial felling of trees in Garhwal Himalayan forests in the former state of Uttar Pradesh (Shiva 1991; Akula 1995).

Martin J. Haigh (1988, 1999) has rightly observed:

> The Chipko movement is the Third World's most celebrated indigenous non-government, non-urban, environmental protection group. It is a rare example of a popular organization that campaigns against forest destruction and works to manage and replant forests for rural use .... The Chipko movement looks to the philosophy of a Third World leader, Mahatma Gandhi.

Various world views/perceptions of different strands of the Chipko movement as well as that of the Forest Department (Uttarakhand) may be seen in Table 5.2.

From the above, it is crystal clear that the views of the Forest Department and those of the Chipko movement (with differences in orientation and focus) were, by and large, poles apart because the Forest Department (and their contractors, agents, buyers, etc.) was the prime exploiter of forests for profit (selling valuable wood and other forest produce to outsiders but preventing local people from collecting these for their daily needs).

Table 5.2 A Comparison of Worldviews of the Chipko Movement and the Forest Department, India

Sl. No.	Theme	*Crusading Gandhians (CG) of Chipko*	*Appropriate Technology (AT) of Chipko*	*Marxist (USV) of Chipko*	*Forest Department*
1.	Worldview	Deep ecology (spiritual/ sacred bond between humans and nature)	Stable environment – sufficient bond with nature	Stable environment plus social development by removing inequity and questioning capitalism	Forests for development of nation through more infrastructures
2.	Change	Transformation – towards pre-capitalist, non-industrial indigenous system	Reforms – for eco-democracy	Transforming the Capitalism as root cause – for socialism	Status quo – Forest Department as 'necessary protector' and 'scientific manager'
3.	Forest's commercial exploitation	Total ban on commercial forest exploitation	Forests be reserved for local hill people	Forest produces be allowed for local people	Forests for nation's full development
4.	Need versus greed	People's needs be fulfilled and people's control be over forests	Basic needs of people be fulfilled at once	Redistribution of productive resources	Villagers as enemies of forests – national govt's control over forests required
5.	Perception of Forest Department	Incompetent, corrupt, hostile	Incompetent, corrupt, hostile	Incompetent, corrupt, hostile	Politically neutral and 'necessary protector' of forests
6.	Examples by actions	No practical actions on the ground, only campaign	Eco-development camps, fuel-efficient stoves, biogas plants, massive afforestation, micro-hydropower plants	Radical action for prohibition, and for banning mining/ deforestation	Government nurseries – but less survival of plants (than AT group's)

Sources: Based on Torsten Koval (1984) cited in Martin J. Haigh (1988), and M. Gadgil and R. Guha (1992).

196 Ecological Movements in India

Silent Valley Movement, Palakkad, Kerala (South India)

The historical context of the Silent Valley movement in the Palakkad district of Kerala, South India, is the same as that of the Chipko movement. Therefore, the context will not be repeated here.

Genesis

Silent Valley is a narrow valley of Kunthi or Kunthipuzha river in the State of Kerala, South India, at an elevation between 2,400 m and 1,000 m. The Silent Valley forests are locally known as Sairandhrivanam, and are considered one of the last representative tracts of virgin tropical evergreen forests in India. It has 8,950 ha of rain forests with rare flora and fauna including the lion-tailed macaque, a threatened primate. In 1972, Steven Green, a scientist from New York Zoological Society, studied the primates, especially the lion-tailed macaque in the Silent Valley in the Palakkad district in Kerala and expressed his concern about the threats to the primates from the proposed hydroelectricity project there. Similarly, Rom Whitaker wrote to Bombay Natural History Society (BNHS) about the necessity for conserving Silent Valley. The local people, especially the journalists, intellectuals, and teachers, who considered the proposed project to be anti-ecological, in letter and spirit, protested against its implementation. In 1973, the then State Government of Kerala actively considered to dam the narrow gorge at the lower end of the valley, to fill a reservoir upstream and generate hydroelectricity. It would have submerged an area of 830 ha, including 500 ha of prime tropical forest. The project would have generated 240 MW (four units of 60 MW each) of electricity to facilitate industrialisation in the region, irrigate some 10,000 ha of land in the relatively underdeveloped districts of Palghat and Malappuram, and would have given employment to 3,000 persons during the construction. The project was approved by the Planning Commission of India for Rs 25 crores; however, due to lack of funds, it was delayed till 1976 (Swaminathan 1983).

Formation

The (local Marxisant) Kerala Sastra Sahitya Parishad (Kerala Science and Literature Society – KSSP), a 'science to people' movement by school and college teachers, resisted implementation of the proposed hydroelectricity project, and, therefore, drew the attention of the central government and the ecological task force to the aspect of the consequent heavy loss of rare flora and fauna. A task force of the National Committee on Environmental Planning and Co-ordination (NCEPC), under the chairmanship of Dr M. S. Swaminathan, the then Secretary, in the Agriculture Department, Government of India, and several non-governmental conservation organisations including Bombay Natural History Society, Indian Science Congress, and KSSP queried the cost estimates of the project and were not happy with the submergence of the forest (Mahesh .com 1998). Hence they urged the Government of Kerala to drop the project but all went unheeded. Later, in October 1976 (during the emergency in India), the National Committee on Environmental Planning and Co-ordination set up a task force, chaired by Zafar Futehally, to study the impact on ecology, and hence the work on the project was suspended. The task force recommended that the project be dropped with the rider so that if the dropping of the project was not possible, some safeguards should be ensured. The State Government of Kerala wanted to continue the project with suggested safeguards and argued that only 10% of the ecosystem would be damaged. But the environmentalists and NGOs argued three points:

(a) The entire lower valley will get submerged by the dam proposed, hence it will destroy the biodiversity of Silent Valley
(b) 10% loss is projected by the State Government but it will be far more
(c) The construction workers will stay in the area for many years and destroy the forest and biodiversity by wood cutting, poaching, cattle grazing, and encroaching.

In 1977, Sathish Chandran Nair, after visiting Silent Valley, launched an awareness campaign in academics. V. S. Vijayan of the Kerala Forest Research Institute studied it and recommended not to start the project but his report was suppressed. The Government of Kerala, however, decided to go ahead. S. Prabhakaran Nair visited the villages to create awareness amongst the people about the adverse ecological consequences of the project. Professor John Jacob trained young nature lovers, hence various Nature Clubs emerged.

Consolidation

KSSP mobilised the people on a wider scale in order to challenge a wider problem, as the project was not only supported by the State Government but all the political parties in Kerala. KSSP broadened the base of its struggle and involved

> an entire gamut of experts – botanists, zoologists, economists – in arguing that not only would the scheme have adverse environmental impact on a rare ecosystem rich in biological and genetic diversity, but also that the required power generation could as easily take place by setting up thermal power units in other locations and improving the efficiency of the transmission system.
>
> (Sethi 1993, 132)

On the other hand, international conservation organisations, like the World Wide Fund for Nature (WWF) and the International Union for Conservation of Nature (IUCN), also became involved in the campaign 'Save Silent Valley' and requested the Kerala government to conserve the forests and save the lion-tailed macaque, a rare breed of monkey residing in Silent Valley. Thus, the movement got an international dimension. However, on the contrary, the Kerala Assembly passed a unanimous resolution for the speedy implementation of the proposed hydroelectric project. But the implementation was delayed due to scientific controversy, lobbying through the press, parliamentary and expert committees, massive campaigns, *several court orders*, and the requirement of the approval of the central government. Salim Ali, Madhav Gadgil, C. V. Radhakrishnan, M. S. Swaminathan, Subramaniam Swamy, Piloo Modi, Krishna Kant, and others wrote against this 'short-sighted' project. The Bombay Natural History Society and the Geological Survey of India argued for declaring Silent Valley as a Natural Bio-reserve. However, the then Prime Minister Morarji Desai rejected these appeals. In 1979, the project work started again. In January 1980, a Public Interest

198　Ecological Movements in India

Litigation (PIL) case was filed before the Kerala High Court, but the Kerala High Court refused to grant a stay and allowed the project to continue. The activists then met the Governor of Kerala who stayed the work till submission of the report by the Experts Committee which had earlier been constituted by the Prime Minister Morarji Desai's successor in office, namely Prime Minister Charan Singh. In order to placate the conservationists, the Government of Kerala created a national park in December 1980, which excluded the proposed project site from the park area. Meanwhile, in the early 1980s, Shrimati Indira Gandhi became the new Prime Minister of India again, and being relatively sympathetic to the environmental conservation issues, in January 1981, declared that Silent Valley would be protected and, therefore, referred the matter to a new scientific committee, headed by M. G. K. Menon, which in 1983 opined in favour of its conservation. Finally, the project was shelved in November 1983 in deference to the weight of public protest/campaign and the sentiments of Shrimati Indira Gandhi, the then Prime Minister of India. So on 15 November 1984, the notification dated 26 December 1980 whereby a decision had been taken for the formation of the Silent Valley National Park constituted the areas in the Silent Valley Reserve Forests by excluding the submergible area falling on either side of the Kunthipuzha River falling below the maximum water-level contour of 4,979 m above mean sea level (MSL), and the proposed dam site, colony site, and other ancillary sites required for the Silent Valley Hydroelectric Project were amended to include the areas of the proposed project (Sethi1993; Gupta 1988; Agarwal and Narain 1985; Mahesh.com 14.7.98). In 1985, the new Prime Minister, Rajiv Gandhi, inaugurated the Silent Valley National Park extending to an area of 8951.65 ha.

Thus, two major features of this movement are notable: first, this successful ecological movement was unique because the proposed dam in this uninhabited area did not involve any displacement of the local people, rather ecological issues of saving a rare animal (simian) species (lion-tailed macaque) and rare plant species (last surviving natural evergreen tropical rainforests) were the basis of this ecological movement; hence, it also struggled for a new paradigm of 'development without destruction' (Sethi 1993; Gadgil and Guha 1994; Agarwal and Narain 1985). Second, this movement's success depended on the support received from various levels (local, national, and international) and various strata of society like leftist NGOs, wildlife conservationists (scientists and political activists), press, government experts, and the central government.

Narmada Bachao Andolan (Save Narmada Movement), Madhya Pradesh (Central India)

The historical context of this movement, too, is the same as that of the Chipko movement.

Genesis

Narmada river is the largest westward-flowing river in India; it originates from a holy tank (Narmada Kund) located at Amarkantak in Anuppur district, Shahdol Zone, in the midst of Hindu temples in eastern Madhya Pradesh, and flows into the Arabian Sea covering a distance of 1,300 km through forests and plains. About 20 million (2 crores) people live in the Narmada basin depending on the river for their survival in daily life.

Secondly, it is considered by many Hindus to be more sacred than the Ganges, since it is believed to evolve from the body of Lord Shiva (Kalpvriksh 1985). Thirdly, more than 80% of the people live there in the villages and there is a sizable population of tribal peasantry, namely Bhils, Gonds, Baigas, and others. Tribals have a fairly equitable distribution of land while non-tribals have a marked differentiation – e.g. in some districts of lower reaches of Narmada, 70% of the land is owned by 20% of the farmers (Kothari and Bhartari 1984). Sardar Sarovar Project (SSP) was the dream and vision of Sardar Vallabh Bhai Patel (the then Deputy PM) and its foundation stone was laid out by Pandit Jawaharlal Nehru, the then Prime Minister of India, on 5 April 1961.

Though the Narmada basin is resource rich, it has remained underdeveloped with lower agricultural yields; lack of medical, educational, and banking facilities; lower energy consumption (only 50% of the national average); high illiteracy; lower life expectancy; and so on. This is mainly because of political reasons. For instance, since 1946 the three States of Madhya Pradesh, Gujarat, and Maharashtra had been quarrelling over the sharing of water, the area to be irrigated in each State, and the level (height) of the major dam Sardar Sarovar. Later the Narmada Water Disputes Tribunal was constituted in 1969 and it submitted its report on 7 December 1979 and was notified by the Government of India on 12 December 1979. After studying the plans for about a decade, the Tribunal decided on a system of water distribution as provided in Table 5.3.

Thus, it awarded Madhya Pradesh 65%, Gujarat 32%, 1% to Maharashtra, and 2% to Rajasthan. As per the claims of the governments of Gujarat and Madhya Pradesh, Sardar Sarovar Project and Narmada Sagar Project would irrigate 1.9 million ha and 0.14 million ha of land and would generate 1,450 MW and 1,000 MW of power, respectively. Most of the villages in Madhya Pradesh, whose lands were submerged, have been of the view that while they are losing their main source of livelihood and socio-cultural–spiritual space forever, most of the fruits will be cornered by the outsiders (in terms of irrigation of lands, drinking water supply, electricity, etc. especially in Gujarat, Maharashtra, and marginally in Rajasthan). Hence, the Narmada Valley Development Project wanted to 'develop' and transform the valley into 30 major, 135 medium, and 3,000 minor dams! Hence, Claude Alvares correctly called such a huge Narmada Valley Project the 'world's greatest planned disaster' (cited in Gadgil and Guha 1992, 112). Its two major dams were built earlier but the third major dam, Sardar Sarovar Project (SSP), is the largest (1.2 km long, 138.68 m high, depth of 163 m, and two power generation units installed in it with a combined generating power of 1,450 MW with Gujarat getting around 16%, Madhya Pradesh around 27%, and Maharashtra around 57% of the electricity produced). Its foundation stone was laid by the then Prime Minister of India, Jawaharlal Nehru, on 5 April 1961, but it took 56 years to be built and finally Prime Minister Narendra Modi dedicated the dam to the nation on 17 September 2017. Construction of the dam began in 1979 but the work was stopped in 1983 and restarted in 1987. Actually, the Planning Commission (Government of India) sanctioned it only in 1987 at an estimated cost of Rs 6,700 crores (67,000 million rupees) but unofficially it was estimated to cost between Rs 1,30,000 million to Rs 2,00,000 million (Kothari 1995, 123). The World Bank had agreed in 1985 to lend $450 million for the dam and the canal components of SSP but, in 1993, under the pressure of international environmental NGOs, it gave strong signals to the Government of India regarding its withdrawal. So, in order to avoid humiliation, the Government of India announced in March 1993 to terminate its contract with the World

200 Ecological Movements in India

Table 5.3 Narmada Water Distribution to Four States in India

Sl. No.	States	Flow MAF (Million Acre Feet)
(a)	Gujarat	9.00 MAF (32%)
(b)	Madhya Pradesh	18.25 MAF (65%)
(c)	Maharashtra	0.25 MAF (1%)
(d)	Rajasthan	0.50 MAF (2%)
	Total	28.00 MAF

Source: Narmada Water Disputes Tribunal.

Bank. Thus, only $280 million had been disbursed by then (Kothari 1995, 123–124). Thereafter, a private company (S. Kumar group) continued with the construction works. Work on the Sardar Sarovar Project was started after the Supreme Court gave the green signal to raise the height of the SS Dam in October 2000, with a rider that the dam-oustees will be satisfactorily rehabilitated, and the process be repeated every 5 m increase in height. Supreme court also observed:

> The project has the potential to feed as many as 20 million people, provide domestic and industrial water for about 30 million, employ about one million, and provide valuable peak electric power in an area with high unmet power demand (farm pumps often get only a few hours of power per day).

But, NBA activists reiterate that the Hon'ble Supreme Court did not look into the significant aspects like Environmental Impact Assessment (EIA) and Social Impact Assessment (SIA), rather SSP was actually started even without EIA!

As the dam was to displace about 2,00,000 people in 243 villages of whom 60% are tribals (as per the Narmada Bachao Andolan), the people vehemently protested against its construction. The number of displaced families has been assessed differently by different agencies. When the Narmada Water Disputes Tribunal gave its award – the award was notified by the Government of India in the month of December 1979 – only 6,147 families were estimated to be displaced, but by the early 1990s, the Khosla report of a five-member group assessed 40,245 families affected by the Sardar Sarovar Project (Sahoo et al. 2014, 889). This was, firstly, because of the following factors:

(a) Wrong assessment of the social impact of the dam
(b) Substantially raising of dam height later
(c) The state-level agencies (Narmada Development Authority/Nigam/Department) always underestimated the number of families likely to be displaced, while the agencies of the Government of India were relatively more scientific and reliable, and, the Narmada Bachao Andolan activists, who were well aware of the ground realities in the States of Madhya Pradesh, Gujarat, Maharashtra, and Rajasthan were the most reliable.

Secondly, an environmental impact assessment (EIA), e.g. the submergence of 13385.45 ha of land, soil erosion, siltation in the dam (both leading to degradation of water quality, and thus reducing the life span of the dam), was not carried out prior to the start of the Sardar Sarovar Project.

Thirdly, there were no clear policies or guidelines laid down for fixing of accountability of the various agencies involved in the construction of the dam and canals, and in the rehabilitation of the project-affected people (PAP); in addition, there was no proper and comprehensive coordination, evaluation, and monitoring of such multiple agencies involved therein.

Out of total 243 villages affected, 193 villages fall within Madhya Pradesh, while the command area, within which the major beneficiaries live, lies in Gujarat (Kothari and Bhartari 1984). Further, to add insult to injury, an additional 4,200 ha of forests, which provided the source of the livelihood to thousands of tribals, were cleared for the resettlement and rehabilitation of the dam-displaced people (Kothari 1995). According to the Narmada Bachao Andolan, when the reservoir is filled to its optimum capacity, the level of water in the submergence area of the dam in Barwani and Dhar districts of Madhya Pradesh will cause the displacement of around 40,000 families in 192 villages. According to the World Bank, the project was started with virtually no assessment of resettlement and rehabilitation, and even environmental impact assessment. On the flip side, around 8,00,000 ha of land in Gujarat and about 2,46,000 ha of land in Rajasthan are supposedly to be irrigated by water from the dam.

Formation: Earlier Phase

An organisation, befitting its purpose, Narmada Bachao Andolan (NBA) was formed in 1987 to launch a movement against SSP. But even a decade before that, in 1977, the villagers of the Nimad region of Madhya Pradesh had protested as the potential threat of eviction loomed large on them. The main issue of the protest in the first phase then was the demand for adequate *compensation* for the lands to be submerged in the SSP. This was a moderate kind of mobilisation for a functional minor change (more quantitative rather than qualitative) within the framework of the project. This was the pre-NBA phase of '*moderate protest*'. With the formation of the NBA in 1987, under the leadership of Medha Patkar, the second phase started where the main issue became proper '*rehabilitation*' of potential oustees. NBA, at first, targeted the funding by the World Bank to SSP. This was a radical change as till then, in India, there was no national rehabilitation policy at all for the displaced people from development projects. This may be called the 'identity' phase of moderate mobilisation.

Consolidation: Later Phase

The Narmada Bachao Andolan (NBA) got momentum in 1989–1990 in its third phase in two ways. First, its base was broadened, and second, it adopted and applied more radical strategies of protest. The process of '*movement dynamics*' involved four major events (Roy and Sen 1992, Gadgil and Guha 1994).

First, in 1989, activists from the neighbouring villages of Barwani town uprooted the stone-markers from the submergence area of the dam, took them away, and flung them outside the Madhya Pradesh Legislative Assembly in Bhopal.

Second, the Harshud rally, which was the most important event and where more than 60,000 volunteers had gathered, was held on 29 September 1989. This town had significance because it, too, was to be submerged in the dam, if the height was not reduced. Further, famous Gandhian social activists, Baba Amte, S. L. Bahuguna (of the Chipko movement), Medha Patkar of NBA, Dr B. D. Sharma (a social activist of Bharat Jan Andolan, and ex-Commissioner for the Scheduled Castes and Scheduled Tribes, Government of India), and many other important activists participated in this rally under the banner of NAPM (National Alliance of People's Movements). There the NBA raised two fundamental issues: (a) total opposition to the dam (no big dams), because dams displaced a large number of people and there was no land available in Madhya Pradesh, Maharashtra, and Gujarat for implementing the declared 'land for land' policy of rehabilitation; this disillusioned the movement activists; and (b) instead it then thought for an *alternative development paradigm*, as activists took a collective oath to oppose the kind and pattern of 'destructive development' exemplified in Sardar Sarovar Project. It is very important to note here that the name of Harshud town was lost forever in the deep waters of the Narmada. Unfortunately when the local dwellers were not vacating their houses and sources of livelihood (shops, offices, and other establishments), they were divided by the regime of the day and asked to break their houses first, and take the rehabilitation package thereafter! Undoubtedly, the local people had spent most of their lives in making all necessary arrangements for building 'houses' – and afterwards those houses were made 'homes' – but were then forced to break and destroy those. To quote the famous Urdu poet Dr Bashir Badra: '*Log Toot Jaate Hain Ek Ghar Banane Mein / Tum Taras Nahin Khate Bastian Jalaane Mein*' (People break in building just one house, but you don't mind in setting a whole settlement on fire)! In this case, of the dominant development model that holds out the chimerical promise of material wealth, all settlements (e.g. in the town of Harshud) were submerged by raising the height of the Sardar Sarovar dam, eventually causing the displacement of about 2,00,000 people.

Third, in May 1990, villagers from the Narmada basin gathered in New Delhi and sat on a '*dharna*' (sit-in protest) at Gol Methi Chowk (close to the then Prime Minister, V. P. Singh's residence) for several days, singing, dancing, and delivering speeches. With the assurance of the then Prime Minister to review the SSP, they went back.

Finally, on December 25 1990, the NBA organised a *Sangharsh Yatra* (struggle march from Rajghat (in Delhi) to the Madhya Pradesh – Gujarat border village of Ferkuva (originally they planned to reach Kevada colony, the site of the Sardar Sarovar dam, but were stopped from entering Gujarat by the Gujarat police). However, about 100 activists in four sub-groups, including Baba Amte, with their hands tied to symbolise the use of non-violent methods of resistance, entered Gujarat. On the Madhya Pradesh side of the border, Medha Patkar and other activists sat on a hunger strike for three weeks but the Gujarat government did not relent. This was what may be called the 'collective self-empowerment' phase. However, the SSP was completed and dedicated to 'the nation' in September 2017 – after 56 years! On 27 August 2020, for the first time the dam level was at 130.53 m, and hence Riverbed Powerhouse has resumed generation of hydropower in a phased manner. According to Sardar Sarovar Narmada Nigam Ltd, Narmada dam has an inflow of 62,723 cusec of water, and about 34,856 cusec of water was released after generation of power at the RBPH (Riverbed Powerhouse) while 5,178 cusec water was released from the Canal Head

Power House (CHPH). The energy is transmitted to Gujarat (16%), Maharashtra (27%), and MP (57%).

Analysis of Major Ecological Movements

The three major ecological movements, viz. Chipko, Silent Valley, and Save Narmada (NBA), are analysed here in detail from the multidimensional critical disempowerment perspective (Sharma: 2016). In other words, these movements are analysed in terms of seven dimensions: context, causes, composition (participants), goals, mobilisation process, control, and outcome. We have proposed six hypotheses:

(a) The ecological movements do not emerge spontaneously.
(b) Ecological movements in the developing world mainly focus on 'critical life issues' of livelihood, culture, spirituality, and morality.
(c) Usually the subaltern groups (from different communities) are in the forefront of ecological movements in the developing world.
(d) The ultimate goal of the protesting people is to change the existing State hegemony and dominance, and to create an egalitarian and autonomous space in civil society by pursuing the ethics of diversity of nature and culture, and an alternative indigenous paradigm of development.
(e) The State is usually not neutral regarding ecological issues raised by the protesting people.
(f) Political parties may or may not be averse to ecological movements.

Context: Roots of Ecological Movements in the Developing World

Major ecological movements in the developing world originated in ex-colonial countries, viz. India, the Philippines, Malaysia, and Brazil. To be precise, India was under the colonial rule of Britain, Sarawak (Malaysia) was under that of Brunei and (later) Britain, the Philippines was under the rule of Spain and (later) USA, and Brazil was under the rule of Portugal. This historical fact brought a massive change in the economic, political, social, cultural and ecological spheres. However, the fact is that even the nature of colonial rule in these countries varied in terms of economic exploitation, political structure, and socio-cultural hegemony. For instance, while the Brunei and Portuguese Empires were non-industrial economies implying less rigorous exploitation of natural resources of their colonies (Sarawak and Brazil, respectively), the British and American Empires were industrial economies implying more rapid and rigorous exploitation of natural resources of their colonies. The British Empire even illegally brought the rubber plantation from Brazil to Malaysia (later) for colonial exploitation. The case of the Spanish Empire was different: It was primarily more interested in the colonial control (from the sixteenth century to the late nineteenth century) by religion (Christianity) as well as gold; however, it, too, created private property in land and Spanish *friars* emerged as 'absentee landlords' (Hurst 1990). Secondly, while Brazil achieved independence from colonial rule in the early nineteenth century itself, others achieved the same quite late in the mid-twentieth century. However, the rubber-producing state of Acre (Brazil), where the rubber tappers' movement started, was annexed from Bolivia (which had achieved independence from the Spanish Empire in the nineteenth century) in 1903 (Hall 1996). This fact also has a bearing on the pattern of the use of natural resources in these developing countries.

During the pre-colonial period, these four countries had, more or less, decentralised community management of common property (natural) resources like rivers, ponds, lakes, pastures, and forests. There was no concept of private land ownership in Sarawak (Malaysia) and the Philippines. These systems had evolved the well-formulated norms and practices for use of their commons for meeting vital needs of everyday life, and those found violating the norms, could be punished in some way or the other. However, these industrial empires began to use and exploit their colonies' natural resources for four main purposes (Guha 1989; Bandyopadhyay and Shiva 1987; Shiva 1989; Gadgil and Guha 1994; Arnold and Guha 1995):

(a) As a primary source of revenue
(b) As sleepers for expansion of railways for the colonial purpose of better transportation of goods from remote areas
(c) Use of timbers as raw materials for military–industrial purposes like Navy ship-building industries
(d) Use of agricultural products like indigo (farmers were forced to grow indigo, at first, in Bengal, and then in Champaran, and other districts in Bihar, India), or plantations like rubber (Malaysia) or tea/coffee (India), in the industries established by the Empires.

To achieve these purposes, they took three measures: First, they usurped the communal control over, and customary access to, natural resources and the State became the ultimate owner of the forests and other natural resources through the application of the doctrine of '*eminent domain*'. Land Acquisition Act, 1894, in India is the fittest example in this regard; under this law, lands of farmers were acquired for the so-called 'public purpose' without paying adequate compensation at the market rate. Even for the nominal compensation ownership, documents were required but most of the illiterate farmers did not have papers.

Second, they introduced the so-called 'scientific management' of forests, hence the 'modern' management of forests was declared rational and useful while the traditional system of local and community conservation was condemned as irrational and backward. New techniques, for instance in India, were 'designed to reorder both nature and customary use in its own image', and in this process 'both legislation and silvicultural technique were designed to facilitate social control' (Guha 1989). As a Secretary to Governor, United Provinces (India) noted in 1922 that if villagers looked after a forest well, 'a passing forest official will say – here is a promising bit of forest, government ought to reserve it. If, on the other hand, they ruin their civil forest, they feel free from such reservation' (cited in Guha 1989).

Third, they commercialised the forests through the creation of private property; hence industrialists, traders, ranchers, and contractors took over the forests for production for profit, and therefore, more revenue to the State. It had a wider connection with the growing demands of natural resources in the global market (Kothari and Parajuli 1993). In the post-colonial period, these countries unfortunately adopted the same path of modernisation through industrialisation which, in the ultimate analysis, means a system of unrestrained use of natural resources as raw materials in industries

(Chipko), exports of logs to earn foreign exchange (Sarawak), forest land's sale for resettlement or cattle ranching in forest areas (Amazonia) as well as additional energy generation (for both production and consumption) through big dam construction (anti-Chico dam in the Philippines, Save Narmada, and Silent Valley movements in India). Therefore, local people protested in these regions, not spontaneously, but as a lesson learnt from earlier peasant's protest movements and also in a collectively organised way (Shiva 1989; Guha 1989; Hurst 1990; Hecht and Cockburn 1990; Jojola 1984; Colchester 1993; Sethi 1993; Gadgil and Guha 1994, Eccleston 1996). These movements have their historical-contextual roots in primarily local peasant movements that were concerned with local people's customary rights to use natural resources for their vital needs of everyday life. Thus, the first hypothesis proposed here that 'the ecological movements do not emerge spontaneously' is proved. Rather these movements clearly show that ecological movement is a social process of organised resistance over a sustained period of time.

Causes: Why people protest?

Chipko

People launch a movement when they are disempowered in one way or the other. The Chipko movement was launched because of two types of causes: First 'general causes': an exploitative contract system under the Forest Department (outsider contractors were also bringing outsider labour) as well as massive floods in the Alaknanda river that had caused a heavy loss of lives and property in 1970 which was linked to the deforestation. Second, a 'proximate cause' arose when the Forest Department took the unjust step of denying the sale of ten ash trees (at a reasonable price) to the local cooperative (DGSM) for manufacturing local agricultural implements (this was a kind of 'social injustice'), but later it auctioned 300 ash trees to an outsider industrial company (hence 'ecological injustice', deforestation also causing flood). Thus, social injustice and ecological injustice made a 'total injustice' or 'absolute deprivation' – a stage of 'critical disempowerment' – a reflection of 'us' versus 'them', forest hills versus non-forest plains, underdeveloped local versus advanced outsider or subsistence peasantry versus production-for-profit industry. As the pioneer of the Chipko movement, C. P. Bhatt, rightly says:

> On the one hand, forests are central for maintaining a balance in the physical environment. On the other hand, the focal point of the ecological system – the human being – is directly dependent on the forests. Various products such as firewood and timber, grass, vegetables, honey, medical herbs and fruits come from forests, and agriculture and animal husbandry are also dependent on forests. For these reasons, the ecological balance and traditional human relations with forests are so intertwined that it is difficult to view them separately.
>
> (Bhatt 1990)

Though the DGSM was formed in the 1960s, yet earlier it could not launch a movement; actually it was not a typical NGO where there is a formal organisation, with defined roles and functions, and paid staff to carry out those tasks as per the directions of the donors (though 'resource mobilisation' theory would have expected so). However, there are three strands in the Chipko movement with their perception of the causes being different (Gadgil and Guha 1994; Omvedt 1984; Guha 1989; Akula 1995):

(a) The Crusading Gandhian (deep ecologist) strand, of Sundarlal Bahuguna, blamed the modern industrial (western) civilisation where man is the 'Butcher of Nature' (cited in Guha 1989). Here the ecological problem was seen as a 'moral problem', originating in the values of materialism and consumerism. They cited relevant examples of ecological sustainability from Hindu scriptures to generate consciousness among the local people.
(b) The second ecological Marxist strand, Uttarakhand Sangharsh Vahini (USV) blamed capitalism for the ecological degradation of the forest hills of the then State of Uttar Pradesh, now Uttarakhand. To them, unequal access to natural resources, rather than values, was the main cause of the ecological problem, i.e. in a stratified society, the rich exploit nature optimally in pursuit of profit while the poor do so in order to survive.
(c) The 'Appropriate Technologists' (AT), led by Chandi Prasad Bhatt, blamed the development paradigm that was biased towards the city and the technology of the big industry which were against local, small-scale economic, and ecological self-reliance. In a pragmatic way, it tried to have a working synthesis between agriculture and industry, big and small units, Eastern and Western technological traditions, practically closer to Gandhism, especially to his philosophy of reconstruction.

These three strands got support from different intellectuals: while Shiva (1989), Bandyopadhyay and Shiva (1987), and others were supporting the first, Omvedt (1984) was supporting the second, and Agarwal and Narain (1982) were supporting the third. Actually, all of these facets (cultural values, economic production system, and appropriate technology, respectively) are interrelated. That is why despite the differences, all of them joined together in the Chipko movement against the destruction of the local forests.

Silent Valley

The cause of this movement was different from that of the Chipko movement. In this proposed hydroelectric multipurpose dam, there was no question of displacement of human beings involved since it was an uninhabited forest area. So it was only the direct ecological ground on which it was resisted by the people. First, the scientists found three mammal species endangered – the lion-tailed macaque (*Macaca silenus*), tiger, and Nilgiri Tahr (*Hemitragus hylocrius*) – as well as seven mammal species vulnerable – Nilgiri or John's Langur (*Presbytis johnii*), Panther, small Indian civet, sloth bear, Asiatic elephant, Gaur, Indian pangolin; the approximate population of macaque was estimated as 200, of tiger 20–40, and of Nilgiri Tahr 75–125 (Balakrishnan 1984). Second, the project would have submerged and devastated 530 ha of the undisturbed and rare evergreen rainforest, hence the 'silence of the valley' would have been broken badly (Balakrishnan 1984, Singh et al. 1984). Apart from the possibility of the spreading of diseases like malaria, increasing soil erosion, and siltation, there would have been indirect effects like cutting of wood for fuel purpose by the 3,000 dam construction workers for seven years (18,396 tons of wood estimated), opening the area to new settlers and poachers, failure of regeneration of trees, etc. (Singh et al. 1984).

Save Narmada

In the early phase of the movement, NBA protested against the inadequate compensation for the land acquired and resettlement of the displaced people on a large scale (about 2,00,000). Later the activists broadened their protest because of the following major costs: first, *social costs* – displacement; second, *ecological costs* – the destruction of flora and fauna, which have also been the source of livelihood of a large section of people, soil erosion, microclimatic changes, siltation and sedimentation, seismicity, water-logging, and salinity; third, *financial costs* – huge expenditure without giving, on an average, cheaper rate of irrigation; fourth, *economic costs* – loss of agricultural land/soil fertility – are borne by the poor peasants in Madhya Pradesh while benefits would be cornered by the rich farmers and industrialists in Gujarat (Sanghvi 1994; Kothari, A. 1989).

Composition: Participants

We have attempted to answer the question about the participants who directly participate and/or support ecological movements in terms of gender, class, age, and ethnicity/community (Sharma 2016). These participants may be broadly put into two categories: (a) activist participants and (b) non-activist participants, i.e. supporters, who may be from the press, experts, international NGOs, political parties, or other organisations/institutions.

Chipko

In the Chipko movement, both men and women participated. In the beginning, most of the activists were men at Mandal and Phata villages, but in the later phase (at Reni) the movement was sustained by the women who have been playing a major role in the family as well as economic life in the Garhwal region. There exists a 'money-order economy', as Guha (1989) calls it, i.e. there is a massive out-migration of adult males (for employment) who send a part of their savings to their families regularly. Therefore, the fact of proportionally more women in the later phase should not be construed as a sign of its being a feminist movement as Shiva (1989), enthusiastically but wrongly, depicts. Only, as an exception, at Dungri-Paintoli (Guha 1989), there was a conflict between men and women regarding the control of forest resources. However, even there, the political power of the village elites, through the village council (Panchayat), was more significant than the gender aspect. Further, both young and old actively participated in the Chipko movement. Since in the Garhwal region the peasantry has been less differentiated, there was no polarisation in class terms, hence internal class contradiction was not manifest. Rather there the issue of underdeveloped 'insider' (as user and defender of forests) versus developed 'outsider' (as exploiter) was the main contradiction. Therefore, almost the entire peasantry comprising the subaltern groups of all communities participated in it. Omvedt's (1984, 1865) contention of Chipko's base in low caste peasantry was not true not only because the two most important leaders of the Chipko movement (S. L. Bahuguna and C. P. Bhatt) were from the upper caste (Brahmin) but also because almost all caste groups participated in the Chipko movement, since anything relating to caste difference/conflict was not an issue there at all. The primary concern of all was the 'critical life issues' of all the local people irrespective of caste and gender. As far as leadership is concerned, it was totally inside the leadership reflected in the community-based grassroots organisation, namely DGSM, which was not a typical NGO where the functionaries work in a formal hierarchical organisation, are regularly paid salaries, and are provided other perks and amenities in lieu of their (assigned) works by the donors. DGSM and USV have internal democracy more than the ascetic charismatic Bahuguna (who ran Silyara Ashram) who liked master–disciple relationship between leadership and activists (Guha 1989). USV is more radical in this regard – it opposes the leadership principle or 'movement bureaucracy' as such (Omvedt 1984).

Among non-activist participants (supporters), there were mainly two segments: the press and scientists/experts. The press played a positive role – especially the local press – in highlighting the social–ecological problems as well as the movement actions in the region. Similarly, most of the experts, too, took a stand in the interest of the masses. For instance, the Virendra Kumar Committee, appointed by the then State Government of Uttar Pradesh, established a link between the deforestation and floods in Alaknanda and emphasised the ecological sensitivity of the Reni forest in Chamoli (Bhatt 1990).

Silent Valley

In the Silent Valley movement, the participants were mainly adult men. This was basically a middle-class movement led by a Marxist organisation called Kerala Sastra Sahitya Parishad (KSSP – a 'science to the people' movement organisation of mainly teachers and students). The 'science to the people' movement was launched by the school and college teachers and students to generate 'scientific temper' by popularising science through the demonstration, both empirically and logically, against the dogmas and superstitions prevalent among the masses. So in the Silent Valley movement, the main activist participants were the teachers and students. They also got support from international environmental NGOs, as well as experts (both within the government and outside). They successfully expanded their support base by involving a group of 'counter-experts' – botanists, zoologists, and economists – who could prove that, on the one hand, hydroelectric project would badly affect the endangered mammals' species and rare evergreen forests, and, on the other hand, an additional power generation could be obtained by setting up the thermal power units at other locations, or by improving the efficiency of the transmission system or through alternative renewable energy resources like solar energy (Sethi 1993; Swaminathan 1983; Gupta 1988). When this movement got a massive support from international NGOs like World Wide Fund for Nature (WWF), and International Union for the Conservation of Nature (IUCN), especially for saving the endangered lion-tailed macaque, the movement became internationalised. Thus, this was a unique alliance of leftist intellectuals/teachers and green (wildlife) conservationists. But while the KSSP focused on the techno-economic appraisal of energy-generating alternatives, its allies like the green conservationists emphasised on the need for the conservation of rare plants and animals (Gadgil and Guha 1994). An expert committee under the chairmanship of Dr M. S. Swaminathan, the then Secretary in the Ministry of Agriculture, Government of India, questioned the propriety of costs, expressed concern over the submergence of forests of rare plants and animals, and pinpointed whether alternative methods of power generation (solar energy, etc.), employment, and irrigation could easily replace the proposed project, hence 'development without destruction' (Swaminathan 1983). Thus, it is quite clear that even an expert from the government, who had a far-sighted vision, could help achieve the goal of a purely ecological movement. The scientific views of the experts, on the one hand, gave credibility to the movement, and, on the other hand, it also provided an alternative mode of development. There was no centralised leadership in the movement, yet M. P. Parmeshwaran, as the President of the KSSP, was the most prominent leader. Finally, almost the entire press at all levels (local, national, and international) was favourable to the ecological cause. It is interesting that though the Communist Party of India (Marxist) was opposing the movement, some of its intellectuals like Govind Pillai, the editor of the Party's newspaper, *Deshabhimani*, were actively supporting the movement.

Save Narmada

This movement had participants from all age groups, with almost equal participation of both sexes. The major participants were the victims themselves – the oustees of SSP and other dams on the river Narmada in Madhya Pradesh (Central India). The Narmada Bachao Andolan, to a large extent, successfully united the marginalised tribals like Bhils and Bhilalas, on the one hand, and the upper caste landowners like Patidars, on the other, in Nimar region transcending the negative aspects of caste, class, and party differences (Baviskar 1997). The mother organisation was Narmada Bachao Andolan (NBA – Save the Narmada Movement) under the leadership of a woman activist Medha Patkar (a PhD degree holder who, in 1991, was given the international 'Right Livelihood Award'). The NBA did not believe in individual leadership, rather it believed in collective leadership. Its support from the press, like the Economic and Political Weekly, was with full commitment. Dedicated activist-journalists and intellectuals (Smithu Kothari, Ashish Kothari, Krishna Kumar) strongly supported the movement. On the other hand, the vernacular press, especially the Gujarati language press, as well as a section of the national press and some journalist – intellectuals (like Tavleen Singh 1998), were against the movement. Regarding the experts' support for the movement, their reaction was mixed. However, a five-member Review Committee of Experts (Jayant Patil, a member of the Planning Commission of India, L. C. Jain, Ramaswamy Iyer, and two others) on SSP constituted by the Ministry of Water Resources, Government of India, in August 1993, did substantially confirm in its report (July 1994) the concerns shown by the movement:

First, it doubted the government's estimates of water availability in the Narmada (28 million acre feet (MAF) as mentioned by Narmada Water Disputes Tribunal in 1979) as estimated water availability, by many engineers and the Government of Madhya Pradesh, was only 23 MAF.

Second, it criticised the proposed plan of growing sugarcane in water-stressed Gujarat as it would require more water for irrigation.

Third, it recommended an increase in the allocation of water to the needy Kutch and Saurashtra regions.

Fourth, it asked for a detailed plan for the drinking water needs and the publishing of the full list of 8,214 villages and 135 urban centres that were to be provided drinking water facility.

Fifth, it recognised that the SSP dam would require a height of only 436 ft (about 132.9 m), and an additional 19 ft (about 5.79 m) for power generation; if this 19 ft (about 5.79 m) is slashed, the adverse impact would be reduced to a large extent, and power could be generated from other sources.

Sixth, it pinpointed that the project work was started even before the approval was given by the Planning Commission of India in 1988.

Finally, it also found that action plans were incomplete, and it also doubted the advisability of releasing the forest lands for compensatory afforestation (cited in Sanghvi 1995).

Contents (Goals)

Chipko

The Chipko activists had mainly six goal-oriented issues in their struggle (Bhatt 1990):

First, only specific trees and vegetation should be grown so that appropriate needs for fertiliser, soil, water, and energy can be met; '*only people living near forests* can establish a practical and harmonious relationship with the forest'.

Second, areas prone to and affected by landslides and soil erosion or where forest areas were crucial for the conservation of water resources should be identified and *reserved*.

Third, 'the minimum needs of the people living next to the forests who have been customarily using them for *survival* and their village economy should be established'.

Fourth, the contractor system should be completely stopped in forest conservation, development, and exploitation, and instead people living in the forest areas should be organised for this.

Fifth, trees of use and utility to the local villagers should be planted near the villages so that the villagers need not go to the reserved areas.

Finally, based on minor forest produces, *village industries* should be set up to enable the local population to get gainful employment and thus reduce migration from villages to plains/urban areas.

Thus, the Appropriate Technologists (AT) group wanted, at the local level, a ban on tree felling in sensitive areas as well as a judicious extraction for local use only, aimed at employment generation through ecologically sound technology; at global level, it desired an 'alternative path of industrialisation – with political and economic decentralisation based on technologies that promote self-reliance, social control and ecological stability' (Guha 1989). The Crusading Gandhian of the Chipko movement, S. L. Bahuguna, however, had a different goal perception. At local level, his goal was a *total ban on green felling* (both by the insiders and outsiders) and forests to revert to villages; at global level, he desired to return to pre-industrial economy (Guha 1989). Further, while the crusading Gandhian, S. L. Bahuguna, had tried to link Chipko with national and international ecological concerns, DGSM had turned from struggle to 'reconstruction' work at grassroots level: (a) afforestation in the villages of upper Alaknanda valley; e.g. in 1978, they planted 1,00,000 trees in 480 ha through people's participation and its survival rate was 75% as against 14% in government afforestation then; (b) soil conservation measures in landslide-ridden areas, plugging of gullies, small check dams, and fast-growing grass planting; (c) energy-saving devices – fuel-efficient cooking stoves and biogas plants; (d) eco-development camps for consciousness raising (Agarwal and Narain 1982; 1985; Gadgil and Guha 1994; Bhatt 1990). The Uttarakhand Sangharsh Vahini (USV) desired people's participation, ban on alcoholism, and also structural change in the economy. Thus, a right to participation by the local people in the use and control of natural resources was the common thread between these three strands of the Chipko movement.

Silent Valley

The immediate goal of this movement was the ecological preservation of rare plants and endangered animals, while the long-term goal was an alternative method of energy generation (thermal and/or solar) with the least ecological exploitation, i.e. '*development without destruction*', as M. S. Swaminathan (1983) called it.

Save Narmada

In the initial phase of the movement, the main issue was an adequate compensation to the displaced persons; later the issue became rehabilitation of the displaced persons; again the issue shifted to local opposition to the dam due to several kinds of losses; and, in the last phase then the issue was an '*alternative development paradigm*' (Roy and Sen 1992; Gadgil and Guha 1994). This alternative paradigm of development began with the 'right to information', then 'planning from below' at the community level, for a 'self-reliant and self-confident', not self-sufficient, community (Patkar cited in Roy and Sen 1992). This paradigmatic shift in the movement's goal was due to three reasons:

First, because the government was not publishing correct data about costs and benefits, e.g. which villages were to be submerged and which villages and urban centres would get drinking water.

Second, the assurances and promises were found to be unreliable; e.g. earlier the State government had declared to give 'land for land' to the displaced persons but, on investigation, it was found that adequate land was not available for redistribution.

Third, the movement activists had been learning in the course of their struggle as to why such kind of development – large dam – was causing massive ecological, cultural, economic, and social problems, and whether there was any alternative to those woes.

The paradigm of alternative development in NBA was to achieve 'democratic environmental socialism', not in a dogmatic way, 'not just an irrigation system or an agricultural pattern but the lifestyle as a whole', and for this 'democratic environmental socialism' Patkar wanted a '*combination of green and red values and ideas*' (Patkar cited in Roy and Sen 1992).

212 Ecological Movements in India

Mobilisation Process

The mobilisation process has three aspects (Sharma 2016):

(a) Mobilisation of a movement in various phases (spatio-temporal dimension as well as the direction of a movement
(b) Level or scale of action
(c) Strategies and tactics adopted by a movement (how to achieve goals which change over time)

The first aspect having been explained earlier (see the section 'Chipko Movement, Garhwal Uttarakhand, North India'), the second and third aspects need to be analysed. Strategies of an ecological movement have four components:

(i) *Confrontational action* (struggle)
(ii) *Institutional action* (public interest litigation in courts, or question-raising in the legislature, or lobbying the executive)
(iii) *Communicative action* for consciousness raising (oral, e.g. singing songs or reciting poems or discussions/discourses from religious texts; or songs/poems written by others or by the activists themselves and coverage by the media; demonstrative actions, e.g. plays, success stories, models)
(iv) *Reconstructive action* (afforestation, soil conservation measures, alternative energy sources/devices like solar energy)

Chipko

Regarding the level or scale of action, the Chipko movement's three strands had different perceptions. Appropriate Technologists (DGSM) preferred an ecological action at a micro-level, e.g. a group of villages, for the viability of an alternative development model. The crusading Gandhians took nationwide lecture tours, long Padyatras (footmarches), and were close 'to think globally and act locally', while the Appropriate Technologists acted locally and thought locally too, and the Marxist strand worked in the intermediate range at the State level, as KSSP did in Silent Valley. The Chipko activists took confrontational actions on several occasions besides 'Satyagraha' (peaceful protest). Among these, demonstrations, road blockades, hunger strikes, foot marches, hugging the tree, etc. were major methods. Further, they also used communicative action. For instance, Ghanshyam Shailani, a local folk poet, composed poems concerning ecological damages and therein suggested a way out from the felling of trees: To hug the trees; the word 'Chipko' was used by him for the first time in a local ecological context (Guha 1989; Shiva 1989). Folk songs and reciting poems along with discourse from ancient religious texts were also used for the purpose (Bandyopadhya and Shiva 1987). S. L. Bahuguna and C. P. Bhatt wrote articles in popular newspapers and magazines to mobilise the masses. Television and radio could not be used then because they were under State control, and also because of the non-existence of private television channels then. Activists also briefed journalists about the collective action proposed and taken. In addition, the Appropriate Technologists group also organised eco-development camps (Bhatt 1990). The Marxist strand (USV) mainly used mass demonstrations, rallies, and foot marches.

Silent Valley

It adopted a number of strategies/tactics like debates, lobbying through the press, parliamentary and expert committees, seminars, demonstrations, court litigation (though the Kerala High Court dismissed the petition seeking relief for restraining the government from constructing the project), networking with international environmental NGOs (WWF, IUCN), etc. (Sethi 1993; Oza 1981). Various private experts like G. M. Oza suggested declaring the entire Silent Valley as a World Heritage Site (Oza 1981), while the government expert M. S. Swaminathan suggested developing an area of 39,000 ha, comprising Silent Valley forests, New Amarambalam reserve forests, Kundas forests, and Attapadi reserve forests, into a 'National Rain Forest Biosphere Preserve' (Swaminathan 1983).

Save Narmada

It has also used many strategies/tactics like delegation, demonstration, court litigation, rallies, discussion in seminars/workshops, hunger strikes, *dharna* (sit-in protest), *gherao* (picketing officials/politicians), road blockade, struggle march, and sometimes sabotaging public properties, especially the pillars fixed by the project officials and some construction materials of the SSP. Thus, both violent and non-violent strategies were adopted. This movement had not taken up any reconstructive activities like providing schooling and health services to the displaced people. As Medha Patkar believes that 'the State has the responsibility and the people can only organise themselves to get those things for themselves .... The entire system functions only on paper – the schools and dispensaries are run on paper – and we say we don't want them' (cited in Roy and Sen 1992, 283).

Their strategies have been very wide and comprehensive from organising women with veils in rural areas to mobilising journalists, experts, lawyers, other intellectuals, and people's representatives at state and national levels to contacts with Asia Desk and Senior Vice President of the World Bank (Roy and Sen 1992). Though Patkar admits that 'violent strategies seldom produce results. Instead, the strength that is inherent in the non-violent strategies can shake our system', yet she feels a need for a combination of both (Roy and Sen 1992, 297). But Patkar also clarifies:

> Attacking the unjust system, attacking the targets and vested interests has to be a part of our strategies, but we have not attacked them with weapons at any time and I think only that constitutes violence. I don't think that one can change the world through a small armed revolution.
>
> (Patkar cited in Roy and Sen 1992, 298)

Regarding the scale of action, this movement tried to work at the grassroots level, but at the same time, it did not want to be isolated from the outside world because the reality is complex and inter-linked at different levels in different dimensions. Hence, it attempted to see the linkages – picking up issues 'rooted in micro reality' and 'linking it to the total economy' (Patkar cited in Roy and Sen 1992).

Control

The State, an opponent class, a private agency (contractor, trader, or industrialist), or a political party may try to control a social movement. Hence, in the following, we analyse these three movements from this angle.

Chipko

The Chipko movement was attempted to be controlled by both the State and the contractors. The contractor, who was prevented from felling of 300 ash trees in Mandal village by the local villagers in March–April 1973, and again prevented from felling of trees at Phata forest, made a third attempt in December 1973 to fell trees at Phata. On resistance by Chipko activists, in order to disintegrate them, the contractor spread a rumour that the DGSM wants a bribe! To counter such a nasty allegation, the movement activists called a meeting of local villagers in the evening and the contractor, too, was invited to attend it. The contractor tried to intimidate the local people with threats and insults but the villagers did not budge from their stand, and ultimately the contractor retreated (Bhatt 1990). This attempt to disintegrate the villagers (activists and supporters) was a kind of '*demobilisation by disintegration*' (Sharma 2016). Another type of demobilisation of the Chipko movement occurred in early 1974. In the Reni village of Chamoli district, the Forest Department had auctioned 250 trees despite the resistance of the activists. As an attempt to demobilise the movement, the then State government of Uttar Pradesh suddenly announced to pay compensation for lands acquired earlier for military purposes (land was acquired during the Indo-China war in the early 1960s but compensation was not paid till then), and so the adult males went to collect it. The same day, when they went to meet the officials of the Forest Department concerned with Reni, the Chipko activists were detained at Gopeshwar (district headquarters) and Joshimath (nearby town). Subsequently, on the other hand, the contractors and forest officials went to the Reni forest in a bus to fell the auctioned trees! But, on their way to the forest, they met with stiff resistance by the village women who, led by a 50-year-old illiterate woman named Gaura Devi, blocked their way to the forest; thus 2,500 trees were saved (by hugging the trees) from being felled (Bhatt 1990). This tactic of the state government to segregate the local protesting women from men is another kind of demobilisation called '*demobilisation by division*' (Sharma: 2016).

Silent Valley

Firstly, since it was a project expected to give hydroelectricity and irrigation, all political parties (including leftists) of the State of Kerala supported it. In fact, they passed a unanimous resolution in the Kerala State Legislature to support the project and expressed their concern over its delay.

Secondly, an all-party delegation from Kerala met the then Prime Minister of India, Shri Morarji Desai, on 4 April 1978 and pressed for an early clearance of the project. Shri Morarji Desai replied to the then Chief Minister of Kerala that, in view of the State government's commitment to comply with the safeguards suggested by the NCEPC Task Force, he had no objection to its implementation (Swaminathan 1983).

Thirdly, local political leaders of different political parties formed a Silent Valley Scheme Protection Committee under the Presidentship of A. K. A. Rahiman. He emphasised mainly the supposed benefits like 240 MW of power required to meet the energy needs by 1982, irrigation for raising additional crops in 10,000 ha in Palghat and Malappuram districts, and creation of employment for 3,000 persons in seven years' period of construction (Swaminathan 1983, 116). Thus, a kind of '*contra-movement*' (Sharma 2016) was launched by the *political class* (politicians) and their followers in Kerala. However, there was nothing like repression of the activists of this movement.

Save Narmada

Since most of the water and power would be used by the people and industrialists of Gujarat State while the costs would be borne by the people of Madhya Pradesh, there was maximum opposition to the Sardar Sarovar Project works in Madhya Pradesh. The pro-dam State government of Gujarat, urban people, some religious trusts, and NGOs (ARCH Vahini), who publicised SSP as the 'lifeline of Gujarat' or Gujarat's 'identity', had actually launched an organised *'contra-movement'* (Sharma 2016).

First, the most vocal was the upper class of rich farmers who were planning the cultivation of cash crops like sugarcane and paddy there.

Second was the political class of almost all major political parties who accused NBA as 'anti-national', 'anti-India', and 'anti-development', and even charged it as an agent of the CIA, of the Naxalites and even of Pakistan (Sanghvi 1994, 537)! This shows how notions of 'nation, nationalism, and nationalist' are parochialised and voices of dissent are branded negatively by tarnishing their image.

Third was the rich class of urban industrialists who would get maximum power (electricity) from the project. Gujarat State tended to get political mileage out of this project in electoral politics, and the central government did not want to annoy the Gujarat politicians and industrialists, hence the dilly-dallying went on for years. The activists were harassed, tortured, and repressed by the Gujarat police in various ways – threats, assault and battery, lodging of false criminal cases, and indiscriminate arrests (with or without charges) of protesters (including women and children). For instance, in January 1993, 200 policemen ransacked local tribals' houses and beat up the local tribals and fired many rounds in Anjanvara village in the Jhabua district in Madhya Pradesh, and, on the second occasion, arrested 21 villagers who were participants in the movement (Gadgil and Guha 1995). This was a *'destabilisation by repression'* (Sharma 2016). While the NBA demanded the drastic lowering of the height of the Sardar Sarovar dam, the Gujarat government made a pact with Rajasthan government to demand an increase in its height up to 100 m despite non-compliance with the resettlement and environmental requirements; the government also alleged that the dam-supporters had no regard for the laws of the land or even the Indian constitution (Sanghvi 1994), not to say of morality and social justice. The Supreme Court of India first stayed the construction works of SSP but later allowed the SSP authorities to raise the height of the dam beyond 90 m. When the NBA activists and supporters criticised the Supreme Court's judgment by holding a people's court, the Supreme Court of India issued a notice of contempt of court against them, and, ultimately Medha Patkar and a radical literary writer Arundhati Roy (who vehemently supported NBA) were sent to jail for the said contempt of the court by the Supreme Court of India!

Outcome

Chipko

The question now arises as to whether these movements have succeeded or failed in achieving their short-term or long-term goals. It is now very clear that the Chipko movement achieved a grand success in preventing the felling of thousands of trees when in 1981 a 15-year ban on felling of trees in the hill forests of Uttar Pradesh was declared by the then Prime Minister of India, Shrimati Indira Gandhi. However,

216 Ecological Movements in India

its outcome was more in its spread in length and breadth in India, especially as the Appiko movement later in the Uttara Kannada district of Karnataka (South India). Panduranga Hegde in the Western Ghats started the Appiko movement in 1983 to resist forest destruction resulting in the decline of natural forest produces, soil erosion drying up of water resources, etc. Hence men, women, and children in Salkani village hugged the trees, in 1983 to resist the contractors trying to fell these trees. In the Shimoga district of Karnataka, the youth took pledge not to allow the contractors to fell trees. The Appiko activists non-violently mobilised the people through street plays, foot marches, and folk songs/dances to raise mass awareness. They promoted afforestation on the degraded lands as well as introduced alternative energy sources to reduce the pressure on forests. Parisar Samrakshan Kendra (Environment Conservation Centre) played a significant role therein. So it spread to whole South India. Similarly, Rakshasutra movement in Uttarkashi and Tehri Garhwal districts of Uttarakhand in 1995–1996, too, was influenced by the Chipko movement. In fact, Uttar Pradesh Forest Corporation (UPFC) was created for better management of forest resources but since 1993 it connived with the timber traders and contractors and destroyed Chaurangikhal and Rayala forests substantially. UPFC was penalised for illegal activities and had to pay Rs 1.6 crores (16 millions) to Forest Department, UP Government. Hence, men and women of these two districts launched Rakshasutra (Safety Thread) movement in July 1995, when they learnt that labourers from Kashmir were brought to cut Rayala forest – they foot marched 15 km to reach the hills to resist the cutting of trees. The contractors' labourers had cut down three trees by then, but fled away after seeing the movement activists. Women tied threads to 2,500 trees in order to save these. In December 1995, Himalayan Environment Education Society launched a similar movement to protect the Chaurangikhal, Vedali, and Haruntal forests. In February 1996, women of Uttarkashi, Satyali, and Lakvarakha also participated in this movement. The activists gave a memorandum to the district and forest officers to stop cutting trees and punish the criminals cutting the trees. The National Women Commission wrote to the district and police officers to support the movement of women in this regard in the Tehri Garhwal region. On 8 March 1996 (International Women's Day), Save the Forest Day was celebrated. Movement activists fully stopped the felling of trees in the region and, later, the High Court of Allahabad decided in their favour on 12 December 1996. Similarly, in Uttarakhand, new issues like alcoholism (led by USV) or resistance to Tehri dam construction on the Bhagirathi river (led by V. D. Saklani and actively supported by S. L. Bahuguna) or opposition to Vishnuprayag dam on the Alaknanda river, close to the Valley of Flowers (led by C. P. Bhatt), were strongly taken up by the Chipko activists. Further, it broadened from a peasant movement ('private face') to ecological sustainability ('public profile') (R. Guha). Further, it also turned to reconstruction in a significant way – thus providing a new knowledge system and an alternative development paradigm primarily rooted in Indian tradition and culture. Thus, unlike the post-modernists who only talk of deconstruction of 'grand narratives', AT strand of Chipko movement believed and acted upon the 'reconstruction' too.

Silent Valley

The Silent Valley movement, a truly ecological one, fully succeeded in achieving its goal. It sensitised the political class as well as the masses about the ills of the mainstream development model based on big dam technology; it also opened a new debate about 'development without destruction' and ecological sustainability. However, it succeeded more because of the charismatic leadership of the then Prime Minister of India, Shrimati Indira Gandhi, who desired to enhance her image in the international environmental community (D'Monte 1985 cited in Gadgil and Guha 1995).

Save Narmada

Save Narmada movement continued for decades, especially from 1987 to 2017, when India's Prime Minister Shri Narendra Modi dedicated it to the nation on 17 September 2017, even then Medha Patkar, along with her thousands of activists and supporters, protested against its construction because, even by then, all the displaced persons were not rehabilitated. It did not succeed, in the narrow sense of the term, in achieving its declared goals like abandoning the dam or even lowering its height, but it has been fairly successful in conscientisation of not only the common people, NGOs, and professionals in India but also those in the whole world (including economists of the World Bank that stopped funding). It also compelled the national government of India to declare a national rehabilitation and resettlement policy for the displaced persons for the first time; it has indirectly helped the revival of the traditional tribal community's self-rule at the village level, going beyond the Panchayati Raj (village council) in Scheduled Areas of tribal majority. Hence, the Government of India first formulated 'The National Rehabilitation and Resettlement Policy, 2003' and then followed it by 'The National Rehabilitation and Resettlement Policy, 2007' which was notified on 31 October 2007. The comprehensive national rehabilitation and resettlement policy was given statutory backing by the Government of India through the enactment of 'The Right to Fair Compensation and Transparency in Land Acquisition, Rehabilitation and Resettlement Act, 2013' (RFCT LARR Act, 2013), which came into effect on 1 January 2014. The RFCT LARR Act, 2013 provides compensation of four times of market rate in rural areas and two times the market rate in urban areas. With its fair publicity in the international press and support from Northern NGOs, the NBA was able to pressure the World Bank to withdraw its financing in 1993. However, taking support from the Northern NGOs was also looked down upon by a section of the people, press, intellectuals, and other activists in India as having undesired (negative) consequences for the sovereignty of the nation (Sethi 1993). Maybe this fact consolidated the claim of the pro-dam people, rich farmers, industrialists, and the State government of Gujarat that the movement is influenced by the undesired foreign agencies, which have been destabilising developing countries in one way or the other, as well as influenced by the Naxalites. However, such an allegation has not been proved. It has rightly stood for an *alternative development paradigm*.

Thus, it is crystal clear that all six hypotheses proposed earlier are proven. A comparison of the three movements may be seen in Table 5.4.

Conclusion

All these major ecological movements stood against the neo-liberal/neo-classical development paradigm of 'trickle down' or planning 'from above' and provided 'bottom-up' models of inclusion, equity, justice, and sustainability. However, it is crystal clear

Table 5.4 A Comparison of Chipko, Silent Valley, and Save Narmada Movements (India)

Aspect \ Movement	Chipko	Silent Valley	Save Narmada
1. Causes			
Proximate causes	Denial of access to trees to a local cooperative, but large auction of trees to industry (AT)	Threat to three endangered animal species; threat to rare evergreen rain forests	Inadequate compensation; lack of adequate rehabilitation of displaced persons; loss of agricultural lands/forests as sources of livelihood
General causes	Floods caused by deforestation; bias towards city and big technology; top-down State planning (AT); unequal access to means of production (USV-MS); materialistic values; 'a moral problem' (CG)	Top-down development model bringing destruction	Large-scale displacement of 2,00,000 people in 243 villages; wrong planning (top-down) based on big dam technology; ecological damage; soil erosion, siltation, water-logging, and salinity; avoidable huge expenditure; small peasants of MP to suffer while rich farmers/industrialists and urbanites of Gujarat to benefit; loss of cultural identity
2. Composition (Participation)			
Main organisation	DGSM, USV, Silyara Ashram	KSSP	NBA
Leadership	From within	From outside	From outside
Main leader	C. P. Bhatt, S. L. Bahuguna	M. P. Parmeshwaran	Baba Amte, Later Medha Patkar
Sex	Male, female	Male (mostly)	Male, female
Age	All age groups	Adults	All age groups
Ethnicity/community	All castes/communities	All castes and religions	All castes and religions
Class	Lower, lower middle class	Middle class	Lower, lower middle class
Supporters	Local/national press, experts, no support from any political party; leftist USV (MS) had active mobiliser	Local/national/ international press, experts, no support from any political party, supported by communist press 'Deshabhimani'	Mixed response from the national press; a hostile local Gujarati press; but supported by experts
3. Goals			
Short term	Local people's access to forests; abolition of contractor system	To preserve rare plants and endangered animal species	Adequate compensation; rehabilitation and resettlement of displaced people; protecting sources of livelihood

Long term	Ecological reconstruction through soil conservation, energy-saving devices, and village forestry; establishment of forest-based small industries for local use (AT); pre-capitalist mode of production, ban on cutting trees (CG); more space for local people vis-à-vis the State; redistribution of economic and political power, as well as the removal of illegal liquor trade and unregulated mining (USV-MS)-bottom-up planning	Development without destruction; bottom-up planning	Right to information; alternative paradigm of development from below; democratic – ecological socialism; more space for people by reducing State's hegemony; bottom-up planning

4. Mobilisation process

Level of Action	AT At local level CG 'Thinking globally, acting locally' MS At hill region/State level	State level	Inter-state, national

Mobilisation Strategies

(a) Institutional	AT No lobbying CG Lobbying with high officials/politicians MS Against lobbying	Court litigation, Questions in Parliament	Court litigation, Questions in Legislature (Centre/State), delegation to meet policy-makers
(b) Confrontational	AT Hugging of trees, road blockade, demonstration CG Foot-march, hunger strike, demonstration MS Strike, road blockade, demonstration	Demonstration	Demonstration, hunger strike, rally, sit-in, struggle March, gherao, road blockade, sabotaging project, *Jal Satyagrah*
(c) Communicative	AT Folk songs, eco-development camps (discussion and action), writing articles, meetings, press coverage CG Reciting religious texts, writing articles, meetings, walking tours, press coverage MS Ideological consciousness – raising, meetings	Press coverage, writing articles, plays, and folksongs	Press coverage, writing articles, songs, plays
(d) Reconstructive	AT Soil conservation, afforestation, use of alternative energy sources CG and MS Had no such strategies	'Science to the people' campaign	No belief in providing social services (like schooling and health services), rather terms it as state's responsibilities

5. Control

(a) State	Demobilisation by division	____	Demobilisation by repression
(b) Opponent class	____	Contra-movement	Contra-movement

(Continued)

Table 5.4 (Continued)

Aspect \ Movement	Chipko	Silent Valley	Save Narmada
(c) Private agency – Contractor/ Industrialist/Trader	Demobilisation by disintegration	———	———
(d) Political party	———	———	De-organisation
6. Outcome			
(a) Positive	Check on deforestation; reconstruction; spread to other regions – Appiko movement in Karnataka; Rakshasutra movement in other parts of Uttarakhand in 1995–1996 taking up more issues like alcoholism, Tehri dam, and Vishnuprayag dam (in Uttarakhand); alternative development paradigm; right to self-determination	Check on loss of rare evergreen rainforests and endangered animal species; conscientisation for 'development without destruction'	A national rehabilitation and resettlement policy; mass conscientisation for a right to information; an alternative development paradigm; providing an opportunity for Indian NGOs to present a united front
(b) Negative	———	———	Taking support from international governmental and non-governmental agencies created suspicion about it

Note: AT, Appropriate Technologists; CG, Crusading Gandhian; MS, Marxist Strand.

Source: Based on Sharma (2016).

that while the 'Silent Valley' movement was a 'purely ecological' movement, the other two movements (Chipko and NBA) were 'social ecological' movements, which are concerned with people's livelihoods and socio-cultural space ('critical life issues'). As Medha Patkar (2019), the main leader of NBA, observes, a people's movement emerges when the system and the society are shaken on some issues; there should be continuity, longevity, people's force, and local, national, or global vision. However, in a people's movement, there should also be construction or creation along with the struggle – both on field and legal fronts. People should have right to rehabilitation and right on the water, forest, hills, and lands, she adds. There should be no rift between the development and natural continuity; hence, the issue of environmental protection is brought to the political agenda.

If we look at the ecological movements in India, there were three major ecological movements since the 1970s: The Chipko movement in Uttarakhand, the Silent Valley movement in Kerala, and the Narmada Bachao Andolan (Save Narmada) movement in Madhya Pradesh. These movements had their specific context, genesis, formation, and consolidation. Undoubtedly no movement is spontaneous because it requires various phases of events and thought processes for fixing its main goals, participants (both activists and supporters), particular and relevant kinds of strategies, and tactics of mobilisation process because its causes are situation-specific. Further, in a socially economically heterogeneous and democratic (politically) society, as exists in India, there are various factions (in a village itself), ideologies, political affiliations, socio-cultural belief systems, economic strata, etc., who conceive and perceive a specific phenomenon with a different perspective and orientation. Thus while, on the one hand, some individuals in a collective formation visit certain policy decisions, schemes, or projects because of they being anti-people, on the other hand, the establishment (political, administrative, media) itself or its followers or partners use various strategies to oppose such a protest movement by dividing the movement activists on flimsy grounds, or by misinforming/dis-informing/rumour-mongering about them indirectly or by confronting them or controlling them directly in the name of 'nation', 'development', 'innovation', or 'collective benefit'. Such contra-movement agents sabotage the genuine people's movement and declare or brand the movement leaders and activists as 'the traitors', 'the anti-nationals', 'urban naxals', or 'the nation's enemies'! Further, the grand politician-bureaucrat-contractor-industrialist-supplier-criminal nexus harasses and tortures them physically, mentally, or legally – they are finally booked under serious sections of the Indian Penal Code (1860), or even under the Unlawful Activities (Prevention) Act, 1967 (as amended by Amendment Act of 2019) for 'committing' heinous crimes by misusing the powers at the behest of the establishment. And the so-called 'fourth estate' (media) often tarnishes their public images and reputations by distorting the facts and the untruth or half-truth or 'post-truth' ad nauseam with more intensity and frequency – and thus Goebbels has the last laugh!

However, the determined movement leaders like C. P. Bhatt and Sundarlal Bahuguna (of the Chipko movement), M. P. Parmeshwaran (of the Silent Valley movement), and Baba Amte and Medha Patkar (of the Narmada Bachao Andolan) were not deterred from the obstacles created by the contra-movement, and the local people, activists, supporters, NGOs (both national and international), and the international media gave them company and empathised with them in letter and spirit. That is why C. P.

Bhatt was awarded the Ramon Magsaysay Award (Asian Nobel Prize) and Medha Patkar was awarded the Right Livelihood Award (Alternative Nobel Prize) – as well as other awards – in recognition of their stewardship and strong vision for an alternative development paradigm that was finally appreciated both nationally and globally. All these ecological movements stood up (in varying degrees) for bottom-up planning, for local people's participation, and development without destruction. From these movements, the younger generation has learnt a lot and has been committed to intra-generational and inter-generational equity and justice. In 2021, the Madhya Pradesh Government decided to cut 2.15 lakh trees (of Mahua, Jamun, Sagaun-teak-Bahera, etc.) in Baxwaha forest (382 ha) in Chhatarpur district in order to mine for diamond underground. Various wildlifessw would also lose their habitat. But, taking inspiration from the Chipko movement, local people and some NGOs are hugging those trees and tying 'Raksha Sutras' (protection threads) on those trees. The Appiko movement was launched in Karnataka, and Rakshasutra movement in other parts of Uttarakhand too, on similar lines. Central government has not given forest and environmental clearances so far for diamond mining in Baxwaha forest (MP) till mid-2022.

However, one has to understand global environmental politics, especially the difference between environmentalism in the North and the South. We will discuss it in the next chapter.

Points for Discussion

(1) Why did the people launch the Chipko movement? How many strands or streams did exist in the Chipko movement?
(2) In what respects was Silent Valley a 'purely ecological' movement while the Chipko and Save Narmada (NBA) movements were 'social ecological'?
(3) What was the composition of the Chipko, Silent Valley, and Save Narmada movements in terms of gender, class, and community? Examine critically.
(4) Compare the goals and outcomes of the Chipko, Silent Valley, and Save Narmada movements.
(5) How did the major ecological movements, launched in India, change the development paradigm?

Chapter 6

Global Environmental Politics

Introduction

During the 1960s and thereafter, the concern for environmental protection increased globally both at the level of ideas, notions, and thought, on the one hand, and at the level of action, on the other. Environmentalism emerged as the buzz word for global environmental politics showing its concern for the protection of the biosphere (commonly called nature) that consists of various entities like the atmosphere, oceans, lakes, ponds, rivers, soil, plant life system, animals, birds, fish, humans, and other minor organisms. Over the decades, if not centuries, we humans have used and abused various natural assets for the so-called material prosperity. And these anthropogenic activities have devastating consequences for not only various nations separately but also for the world as a whole because most of the ecological problems are, by their nature, inter-related, inter-dependent, and trans-boundary. For instance, the oceanic tides, winds, tsunamis, earthquakes, the flow of large river systems, migration of birds (from Siberia to tropical countries like India during winter), rainforests, etc. have trans-boundary effects on humans of different nations. Thus, at first, we create various environmental problems and then make a hue and cry about their solutions; hence, we begin environmental (or green) movements against such problems (which turn into challenges with the passage of time). Broadly speaking, traditional agrarian societies conceived of nature as alive, active, and nurturing just like a mother nurtures her children. These societies usually thought of the universe as an organic whole comprising both animate and inanimate entities; finally, these respected nature and practised various ethical restrictions on the use of natural assets. However, on the other hand, with the onset of the industrial revolution in Europe (first and later in the USA and Canada), the very view of nature changed fundamentally, as Francis Bacon, Isaac Newton, and their fellow modern scientists conceived of various forms of nature as passive and inert and of instrumental value only in order to fulfil various human needs, wants, and comforts. Modernisation (in essence 'westernisation') promoted individualism that, in turn, promoted the control and private ownership of various natural and other assets for productive and profitable purposes. Thus, capitalist modernity pushed for taming, controlling, and conquering nature and making human life independent of nature itself (Merchant 1990).

Unfortunately, the policymakers in most of the developing countries follow suit the mode of development (modernisation) in the West in order to 'catch up'. Here the main motto is to exploit natural assets optimally and with high speed. Western

DOI: 10.4324/9781003336211-6

224　Global Environmental Politics

extractors' notion of nature was disliked and even questioned long back when Chief Seattle of the Sugumish Indians in North America perceptively commented on the land-grabbing by the white settlers in 1855.

The Whiteman

> is a stranger who comes in the night and takes from the land whatever he needs. The earth is not his friend but his enemy, and when he has conquered it, he moves on. He kidnaps the earth from its children. His appetite will devour the earth and leave behind a desert. If all the beasts were gone, we would die from a great loneliness of the spirit, for whatever happens to the beasts, happens also to us. All things are connected. Whatever befalls the earth, befalls the children of the earth.
>
> (Cited in Cohen & Kennedy, 2000, p. 323)

In the late twentieth century and the early twenty-first century, most of the policy-planners, industrialists, and economists have further become uncertain about nature. For instance, commercialisation of nature for more and more profit through the use of more chemical inputs has resulted in 'refashioning' of agriculture and thus 'denaturalisation of foods' (Goodman and Redclift 1991). Now the so-called modern 'food system' has political–economic concentration, environmental degradation, and trade disputes. These are 'the outcome of industrial capital to control "nature" in agricultural production and food manufacturing and the interest of the state in pursuing cheap food policies' (Goodman and Redclift 1991, p. XV). Further, such forces are attempting to 'disarticulate' the agriculture of developing countries like India. Goodman and Redclift (1991) are of the firm view that the loss of indigenous food security is responsible for rural poverty, rural–urban migration, and increased dependence on exotic grain and animal protein primarily due to the penetration of western transnational corporations (TNCs); the application of 'agri-genetics' has created the following major problems:

(a)　Focus on production growth rather than on low-input, production sustainability
(b)　Corporate appropriation of agri-biotechnologies
(c)　Privatisation of the biosphere

Thus, their conclusion is that 'commoditisation' and 'industrialisation' of food production, processing, and consumption have fundamentally transformed what, how, and where we eat. Commoditisation of food implies the increased consumption of consumer durable goods for the home and increased consumption of processed foods. On the other hand, they point out that the natural organic cycle has been broken in rural society and instead it has integrated with industrial processes. Thus rural society has left its own values because 'agrifood' system has been industrialised in a 'fordist-productionist' model:

(a)　Mechanisation
(b)　'Agri-chemicals'
(c)　Genetics and hybrids
(d)　Farm research and crass profit

One more dimension is extracting pleasure from nature, hence the notion of 'eco-tourism', 'rural tourism', or 'agro-tourism' has come up. Therefore, more demand for more parks, wilderness, organic foods, eco-friendly holidays in hills/wildlife areas, etc. This aspect signifies as formal 're-naturalisation' (though often not much substance). How, when, and by whom such issues were raised globally may be seen in the next section.

Rethinking for and Return to Nature in the North

It is now widely accepted that in eighteenth and nineteenth century Europe, various romantic poets like S. T. Coleridge (*The Rime of the Ancient Mariner*) and William Wordsworth (*Daffodils*) drew people's attention to nature (forests and rural areas) by not only appreciating nature's beauty, but also by publishing jointly *Lyrical Ballads*, and that was the beginning of the romantic era in British English poetry (in 1798). Later Robert Southey joined them and the trio residing in Lake District became known as the 'Lake poets'! Other famous romantic poets were P. B. Shelley (*Ode to the West Wind*), John Keats (*To Autumn*), Victor Hugo (*Tomorrow, at Dawn*), and Lord Byron (*Childe Harold's Pilgrimage*). Romantic poetry movement was, to a large extent, a creative and emotional reaction to the industrial revolution and the so-called scientific rationalisation/utilisation of nature. Later some environmental organisations were formed to emphasise the conservation issues like the preservation of wilderness and protecting animals and birds. For instance, such organisations launched campaigns not to buy leather/fur coats/hats. In this regard, the oldest international organisation, concerned with birds, was established in 1922 – International Committee for Bird Protection. Four active major green international NGOs are as follows:

(a)   IUCN, 1948
(b)   World Wide Fund for Nature (WWF), 1961
(c)   Friends of the Earth, 1969
(d)   Greenpeace International, 1971

This may be seen more vividly with details in Table 6.1.

Thus, these four international environmental non-governmental organisations have spread their environmental actions in different ways to many developed and developing countries. These have inspired and supported many local and national groups to launch environmental movements against prevailing environmental threats and challenges. Comparatively speaking, Greenpeace International has the widest organisational networks globally and has been more pro-active and aggressive in directly launching environmental protest movements globally. The oldest international environmental organisation is the International Union for Conservation of Nature (IUCN) that has membership of both states/government agencies (208) and a huge number of NGOs/indigenous people's organisations (12,000) with 15,000 plus experts in six commissions, and it has members in more than 160 countries. However, at national level, the American Society for Prevention of Cruelty to Animals (1866) and Sierra Club (by John Muir in the USA in 1892) were the oldest environmental NGOs in the USA.

The mobilisation for nature at global level, popularly called environmentalism, claims to work for the entire humankind because it engages in 'transcultural discourse'

Table 6.1 Leading Green International NGOs

Sl. No.	Name	Year of Foundation	Affiliated National Groups and Worldwide Membership	Campaigns and Initiatives
1.	IUCN (International Union for Conservation of Nature) (HQ, Gland, Switzerland)	1948	208 States/govt. agencies and 12,000 NGOs/indigenous people's organisations as members in 160 countries, plus 15,000+ experts	Since 1963, IUCN publishes the Red List of endangered/threatened/ vulnerable species in the world. IUCN Red List (2018) includes 96,951 species of which 26,840 are threatened with extinction – Dodo, Redheaded duck, Asian Cheetah, Sea Cow, Pink Zebra are found extinct animals by IUCN; its experts dedicated to species' survival, environmental law, protected areas, social and economic policy, ecosystem management, and education and communication; it divides species into nine categories: Not Evaluated, Data Deficient, Least Concern, Near Threatened Vulnerable, Endangered, Extinct in the Wild, and Extinct
2.	World Wide Fund for Nature (HQ, Gland, Switzerland)	1961	In 100 plus countries with 5 million plus supporters; 90 plus offices in the world	Cooperates with local groups and governments on conservation projects in 96 countries, its first successful 'debt-for-nature swap' took place in 1987 in Equador
3.	Friends of the Earth International (HQ, Amsterdam, Netherlands)	1969	77 national member groups in 74 countries in 2020; about 2 million members and 5,000 local activist groups	In 1992 a campaign was launched 'the Mahogany's Murder' project to boycott Brazilian rainforest timber; by 1994 six largest Do It Yourself (DIY) chains in the UK agreed to stop selling Mahogany and its imports fell by 68% during 1992–1996. Its motto is to mobilise, resist, and transform
4.	Greenpeace International (HQ, Amsterdam, Netherlands)	1971	Operating in 55 countries, 3 million paid-up members in 158 countries; has 27 independent national/ regional organisations	In the early 1970s campaigned against US and French nuclear testing; in 1975 confronted USSR whaling fleet; campaigned in 1991 to turn Antarctica into a world wilderness park – works for greener and more peaceful world

Source: Prepared by the authors based on google.com (30.5.2020).

(Kay Milton's terms 1996); K. Milton rightly observes that both capitalist and socialist economies were of the view that 'nature' was to be exploited for human benefit, and environmental problems, if any, could be solved by the technology. But since the 1980s, the impact of non-technological factors (especially political and economic factors) on the environment has also been duly recognised. For the first time, it was accepted by the policymakers in both developed and developing countries that economic/financial policies of national governments as well as international funding organisations were responsible for environmental degradation, e.g. British Government's fiscal policies caused the loss of significant habitat to commercial forestry in North Scotland; similarly common agricultural policy had serious environmental impacts in the European Union. Further, the emergence of 'green consumerism' did determine the processes of manufacturing various products; Kay Milton (1996, 5–6) makes a perceptive remark in this regard:

> It is not simply technology that determines the human impact on environment, but a combination of technology with economic values, ethical standards, political ideologies, religious conventions, practical knowledge, the assumptions on which all these things are based and the activities that are generated by them.

Hence, he calls for all specialists to join together. Earlier M. Nicholson (1987, 193), too, rightly appealed for such joining for

> permanently arresting the deterioration in the functioning of the biosphere as a viable life support system for the earth. The time limit must permit the biosphere to recover its equilibrium, and to renew its vigour sufficiently to enable human, animal and plant life to continue to flourish into the indefinite future.

However, in the early 1960s, two classic books inspired many people globally for environmental activism: *Our Synthetic Environment*, by a green anarchist, Murray Bookchin (under pseudonym Lewis Herber), was published in 1962 and criticised the effects of modern technology, polluted air, food with pesticide residues, contaminated water, milk contaminated by strontium-90, and filthy cities. After six months Rachel Carson's *Silent Spring* (1962) brought out the 'ecological inter-connectedness' by contending that due to excessive use of pesticides, including DDT through aerial spraying, various species of birds died, hence even the spring season was 'silent' (without chirping of birds). These pesticides are actually 'biocides' (life killers) and are much deadlier than their manufacturers admit; she also found the presence of such poisonous chemical 'biocides' in human foods causing cancer. Thus, in her findings, such chemical biocides were harmful to humans, animals/birds, and planets, rebounded massively after spraying, and, finally, many insects/pests were developing resistance to such 'biocides' in accelerating speed and pattern like an arms race. Hence, her prudent way is to forego such 'easy' solutions and return to the 'road less travelled by' – that nature exists to serve humanity in a better way. Rachel Carson rightly observed: 'The modern world worships the gods of speed and quantity, and of the quick and easy profit, and out of this idolatry monstrous evils have arisen'. In fact, she argues that the nature has been reduced to factory-like organisation for faster and more economic returns – this is the main reason for most of the ecological problems.

228   Global Environmental Politics

In the early 1970s, the Club of Rome commissioned a study (1972) 'The Limits to Growth' (Meadows et al. 1972) that focused on the finite nature of natural resources, hence the limits to growth. Further, the famous journal *The Ecologist* also published *Blueprint for Survival* (Edward Goldsmith et al. 1972) that argued for adopting strategies like restrained consumption and population control. Thus, these two reports highlighted the ecological crises and hence the question of the very survival of humankind. In 1972, Barbara Ward and Rene Dubos wrote a book called *Only One Earth*, exploring the constraints on development – probably this facilitated (at the Stockholm) UN Conference on Human Environment (1972). They showed with evidence how 'there exists a single unified system from one end of the cosmos to the other' – from the orderly macrocosm of the universe to the equally orderly microcosm of the atom and the gene. But now, disturbing the subtle balances of the biosphere, which evolved over millennia, humans have developed their new 'technosphere'. Hence there is a need for restoring the earth to the 'health, beauty and variety'. Later, in 1979, Theodore Roszak wrote a book *Person/Planet* and questioned the idea of materialistic growth leading to various social, economic, and environmental costs. T. Roszak, the proponent of counter-culture, opposed radically the urban industrial society, and stood for the *equation of personal and planetary needs and rights*. He was the founder of the eco-psychology movement. In 1982, Erik Eckholm wrote *Down to Earth* and argued for protecting the world's environment. In 1987, Brundtland Commission's report, 'Our Common Future', highlighted the *sustainable development* for both intra- and inter-generational equity. Thus during the 1960s, 1970s, and 1980s, ecological thought had different themes of crisis, hence different solutions were suggested by three major schools (Table 6.2).

Global environmentalism further got a boost due to following major environmental disasters, if not catastrophes.

(a) In 1979, on 28 March, Three Mile Island (USA) accident occurred due to the partial meltdown of reactor (No. 2) of Three Mile Island Nuclear Generating

Table 6.2 A Comparison of Three Ecological Schools about Ecological Crisis (1960s–1980s)

Period	Ecological Crisis Identified by Ecological Schools	Causes of Concern	Solutions Suggested
1960s	Crisis of participation (thesis)	More pesticides (biocides), nuclear plants, toxic waste dumps, urban industrial pollution	More equitable distribution of environmental 'goods' and 'bads'; environmental resource as a human resource to be used more efficiently and equitably
1970s	Crisis of survival (anti-thesis)	Exponential growth of consumption and population	Population control, reduction in consumption, resource rationing, and govt intervention necessary
1980s	Crisis of culture and character (synthesis)	Industrialism, culture, income disparity, large technology, all systems of domination	Decentralisation, population reduction, distributive justice, revitalisation of civil society

Source: Sharma (2016).

Station in Dauphin County, Pennsylvania (USA), and subsequent radiation leak. It was a nuclear accident with wider consequences: both a mechanical and human failure. Due to the partial meltdown, there was a release of radioactive gases and radioactive iodine into the environment. Its total cleanup cost was $1 billion (during 1979–1993). Anti-nuclear movement activists highlighted the regional health effects of it, but officially there was a statistically non-significant increase in the rate of cancer in and around the area.

(b) In 1984 (2–3 December), the Bhopal gas tragedy occurred due to the leakage of methane gas from Union Carbide Co. (later Dow Co) – a US multinational company. It caused the deaths of 16,000 people and affected 6 lakhs of people. It mainly occurred because of the casual and callous attitude of the management as earlier there were minor incidents but management did not pay heed to those. It did not install adequate safety measures, and, as per the movement activists (Bhopal Gas Peedit Udyog Sangathan), the State government, central government, and even the Supreme Court of India did not fix the full responsibility and onus on its owner. Its owner fled away due to the laxity and connivance of the then State government of Madhya Pradesh. In 2013, Bhopal gas victims reiterated their demand for payment of Rs 5,00,000 to each victim (instead of Rs 25,000).

(c) In 1986, on 26 April, Chernobyl nuclear power plant disaster took place in Ukraine (former USSR) – less than 100 deaths were directly attributed to it, but subsequently it led to 4,000–16,000 cancer deaths of people (in Ukraine, Belarus, and Russia) exposed to the highest level of radiation in the long term. There were also birth defects in the new-born babies. Total damage caused by this disaster was $235 billion. In addition, it also spread radioactive material to large parts of Europe.

(d) In 1989, on March 24, at Prince William Sound, Alaska (USA), from Exxon Valdez oil tanker there was a huge oil spill – then the temperature was 43°C. Crude oil quantity spilled there was 10.8 million US gallons (37,000 MT) for some days, hence caused huge environmental damage as the region is a habitat of salmon, sea otters, seals, and seabirds; it affected 2,100 km of coastline of which 320 km was heavily or moderately oiled; location was not easily accessible.

(e) Fukushima Daiichi nuclear accident occurred in 2011, on March 11, in north Japan, on its Pacific coast. It was a major accident (level 7) caused by Tohoku earthquake and tsunami. Three of six reactors at the plant sustained serious core damage and released hydrogen and radioactive materials. Plants' design basis for tsunamis was inadequate; multi-unit interactions complicated the accident response; there were no adequate procedures and training for managing water levels and pressures in reactors; lack of clarity of roles and responsibilities at the onsite emergency response centre; failure to transmit information and instructions in an accurate and timely manner. It caused only one death instantly from radiation and 2,202 deaths (1,368 deaths related to nuclear power plant) from evacuation, and 37 persons had physical injuries.

In fact, environmentalism became a big political issue in many developed countries in the 1980s. For instance, opinion poll data show that in the UK in 1989, 18 million people confirmed as 'environmentally conscious shoppers' (Garner 1996), while in the USA in 1987–1988 the public opinion supporting environment was 'all time' high in

230 Global Environmental Politics

17 years (Portney 1992). But people's interest in environment rises and falls due to issue-attention cycle that has five phases (A Downs 1972; W. Solesbury 1976):

(a) Public attention rises due to some environmental crisis or event.
(b) The public demand action in that regard.
(c) The government or its administrative agency responds.
(d) Public attention declines.
(e) The environmental issue slides off the political agenda, till the next such crisis or event happens.

But the social reality is more complex than such rise and fall. That is why scholars like Burke (1995) focus on the distinction between 'salience' and 'latency' (appears similar to Robert K. Merton's concept of 'manifest' and 'latent' functions); to Burke, though environment moves up and down on the immediate political agenda ('salience') there is constant long-term increase in people's concern for it ('latency'). Further the green parties did succeed more at local elections in North and Western Europe and green consumerism in terms of ethical shopping (cruelty free cosmetics), organic farming and green products like chlorofluorocarbon (CFC)-free refrigerators, lead-free petrol, and recycled paper led to more consciousness (Gray 1997). In addition, the environmental movement has spread quite fast in most of the nations of the world, more so in the UK wherein the membership of eight biggest environmental groups increased 15-fold during 1971–1989 – from 0.218 million to 3.147 million. Hence, Buttel and Taylor (1994) rightly observe that environmental NGOs have replaced social justice NGOs in combating developmentalism. In fact, due to the rise of environmental consciousness among the people, the politicians (including conservative Margaret Thatcher of the UK and Richard Nixon of the USA) and political parties in the developed countries positively responded in three ways (Gray 1997):

(a) By espousing green rhetoric
(b) By greening their parties
(c) By enacting green policies, e.g. European Commission enacted 300 environ-mental laws binding on member-states; in the UK, in 1989 National Rivers Authority was created, and Mrs Margaret Thatcher hosted international con-ference on ozone layer depletion and recommended for placing of environment G7 in Economic Summit agenda, and also producing white paper leading to Environment Protection Act 1990. In the UK, the power of local bodies was significantly enhanced, hence they produced environmental audits. They also pre-pared local Agenda 21, following the Earth Summit at Rio in 1992. In the USA, on the other hand, Clean Air Act 1970, National Environment Protection Act 1970, establishing Environment Protection Agency and making Environmental Impact Assessment (EIA) mandatory, Clean Water Act 1972 (imposing national standards on effluent discharges) were enacted earlier. In Japan, 14 environmental laws were enacted in 1970, and setting up of Environment Agency in 1971, and Chemical Substances Control Act. Thus, the USA and Japan were ahead of others in environmental policies and laws by 1980. Globally too, many international conventions and summits were held – mainly Rio Earth Summit (1992), Kyoto Protocol (1997), Montreal Protocol on Ozone Layer Depletion (1987), Paris

Agreement (2015), CITES (Convention of International Trade in Endangered Species of Wild Fauna and Flora) 1973, Moratorium on Commercial Whaling adopted by International Whaling Commission in 1986, MDGs (2000–2015) Sustainable Development Goals (2015–2030), etc. However, there is a huge gap between word and deed, in environmental issues like climate change at local, national, regional, and global levels. Yet Montreal Protocol and CITES have been a success globally. The USA, one of the most emitters of carbons (per capita), withdrew from Paris Agreement (2015) during President Donald Trump's regime, Yet, his successor Joe Biden, later in 2021, rejoined it and assured to comply the targets set in Paris Agreement.

However, due to the trans-boundary nature of many environmental problems, these affect one and all – e.g. climate change, global warming, acid rain, ozone depletion, air pollution, and loss of biodiversity. But some western economists like Lester Thurow (1980) thought environmentalism as an interest of upper middle class only:

> If you look at the countries that are interested in environmentalism, or at the individuals who support environmentalism within each country, one is struck by the extent to which environmentalism is an interest of the upper middle class. Poor countries and poor individuals simply are not interested
>
> (p. 104–105).

But, this observation is not correct, especially in the context of developing countries. Yes, it is not homogeneous, rather it has many strands and streams. While most of neo-liberal economists saw environmentalism as a 'post-materialist' phenomenon, most of Marxists saw it as 'a bourgeois deviation from the class struggle' (seeing 'Man, the Sovereign of Nature'), rightly remarks Ramchandra Guha (2006). He also compares environmentalism with three other great movements of modern age:

(i) The *democratic* movement that gave political voice to the common people.
(ii) The *socialist* movement that desired fruits of economic development to be distributed equitably.
(iii) The *feminist* movement that demanded all economic and political rights to be given to women as enjoyed by men.

Ramchandra Guha (2006) also talks of four types of chauvinism prevailing in the study of environmentalism:

(i) Disciplinary chauvinism (mostly by economics)
(ii) Religious chauvinism (one's faith is privileged)
(iii) National chauvinism (one's nation is privileged, e.g. USA)
(iv) Ideological chauvinism (e.g. Marxist, deep ecology)

Hence both environmental theorists and activists should be duly conscious of avoiding such chauvinisms in thought, word, and deed.

Crisis of Consumerism and the Hazardous Wastes

Now it is very much visible in most of the developed countries, especially the USA, that due to more energy-intensive and pollution-generating activities by the 'consuming class', termed by Alan Durning (1992), in his book, *How Much Is Enough: The Consumer Society and the Future of Earth*, environment is badly affected and as a result there are various distortions in both the developed countries and the developing countries. According to him, while the consumer society has been highly harming the world, it has not given people 'a sense of fulfilment' because they 'suffer from social, psychological, and spiritual hungers'. Further, if our earth suffers because of people having little or too much, five questions arise (Durning 1992):

(a) How much is enough?
(b) What level of consumption can the planet support?
(c) When does having more things cease to add appreciably to human life?
(d) Is it possible for all people of the world to live comfortably without bringing on the decline of earth's natural health?
(e) Is there a level of living above poverty and subsistence but below the consumer life style a level of sufficiency?

He clearly mentioned in the early 1990s that, on an average, a person in an industrialised country consumes three times as much fresh water, ten times as much energy, and 19 times as much aluminium as a person in a developing country like India does. In the USA, average per capita consumption is much more than other industrialised countries.

Second, in industrialised countries, burned fuels emit three-fourth of sulphur and nitrogen oxides causing acid rain; their factories generate most of the world's hazardous wastes.

Third, industrialised countries' military facilities have built more than 99% of the world's nuclear warheads; their atomic power plants have generated more than 96% of the world's radioactive waste.

Fourth, their air conditioners, aerosol sprays, refrigerators, aeroplanes, and factories emit about 90% of chlorofluorocarbons (CFCs) depleting ozone layer.

He, therefore, suggests that we have to realise that *more is not always better*, and the only option is to reduce the high consumption level of consumer society by 'taming consumerism' as the developing world's population cannot economically achieve/afford the level of the former, and the former's level cannot be 'politically possible, morally defensible or ecologically sufficient'. He, therefore, argues to focus on three aspects of daily life:

(a) What we eat and drink
(b) How we get around
(c) Things we buy and use

In his view, the ultimate goal of reform is to bring about convergence of the rich, middle class, and the poor – not raising the poor and middle-income groups into the consumer class. The consumer class would contribute super-efficient technology and the middle class would give local produce and clean drinking water. But the class contradictions cannot be resolved so easily.

Unfortunately, the 'consumer class' continuously and increasingly throws (partly due to the new 'use and throw' technology) in bulk the wastes (solid, semi-solid, and liquid), toxins and gases into lakes, ponds, rivers, seas, oceans, landfill sites, and the open atmosphere itself (considering these as 'empty'). Ulrich Beck (1992), therefore, makes some important points in this regard:

(a) The changing nature of society's relation to production and distribution is related to the environmental impact as a totalising, globalising economy, based on scientific technical knowledge; it is more central to social organisation and conflict.

(b) He does not agree with the concept of industrial nations as 'post-modern', because these are in a transition between 'industrial society' and 'risk society' – the latter is the 'morning after' the industrial night (mare) before, with 'reflexive modernisation' as the 'hair of the dog', a necessary but not a sufficient condition for dealing with risk.

(c) The contemporary crisis is not of modernity but 'within' modernity – the question is how to make the best of modernity – to move from the *semi-modernity of industrial to modernity proper.*

(d) Society fights the 'devil of hunger' with the 'Beelzebub of multiplying risks'; in industrial society, the driving force is 'I am hungry', while in risk society it is 'I am afraid'; in risk society, therefore, people have replaced the politics of voting with *faith in progress.*

(e) In earlier class-based societies, only the proletariat was victimised, but in present risk society all the groups – both the poor and the rich – are threatened and are victims – the perpetrators become victims through the 'boomerang effect'; actually use of chemical fertilisers, pesticides, and weedicides results into the decline of soil fertility, soil erosion as well as the extinction of various plants and animals' species. Further, their residues in food grains, vegetables, and fruits lead to serious health hazards including cancer. Risk and class positions overlap on a national and international scale. Hence, a social theory has to take into account the roles played by science, technology, and cultural constructions of risk. The greatest of risks, the risk-infested society carries, is the indifference to the risk aspects of individual, sectional, and global actions.

However, we do not agree with the basic assumptions of U. Beck because the social class differentiation, exploitative character, and the luxurious lifestyle of the upper class make them more and more rapacious and inhumane, while, on the other hand, the poor are not capable of even fulfilling their basic needs, especially due to the unequal access to natural assets and more so, due to the privatisation and/or statisation of the common property (natural) resources. Here Oscar Wild's words make sense: 'We live in an age when unnecessary things are our only necessities'.

The culture of consumerism creates vast amounts of wastes (Box 6.1).

Box 6.1: Consumerism Generates More Wastes

(1) In the whole world the 'consumer class' numbered 1.1 billion in the early 1990s – mostly in advanced industrial countries but also one-fifth of the wealthy people in the developing countries. This number seems to have doubled by 2020.

(2) They use maximum energy per capita due to the use of various electronic gadgets, aeroplanes, aerosols, private cars, air conditioners, and air-conditioned buildings. They throw away the goods with the change of fashion.

(3) The poorest 1.1 billion people in the world then lived in mud-mortar and wooden houses, travelled on foot, ate root crops/beans/lentils, and used unsafe drinking water; they did not have durable assets.

(4) Both groups have suffered from global pollution – the global poor people due to the survival necessity and poverty, not greed (luxury), have no choice but pressure on the already fragile eco-systems (semi-deserts and steep hill sides). The rich people suffer due to over-consumption because of greed and luxury.

(5) The middle-income group of 3.3 billion people ate mostly vegetables and cereals well-fed and they have clean drinking water. They have used public transport and bicycles, and they have durable possessions.

(6) The 'throwaway economy' generates a large quantity of wastes – including the release of toxic poisons from farmlands and factories – more than 70,000 synthetic materials created by chemical industries; emission of greenhouse gases; dumping of outdated household goods; e.g. in the USA 3,200 kg of waste is generated for every 100 kg of manufactured products – 32 times! Disposal of such wastes is costly and damages environments.

(7) In the early 1990s, packaging created almost half of municipal wastes; it accounted for 40% of paper consumption in Germany, about two-fifths of plastics produced in the USA was used for packaging; each US consumer spent $225 on packaging – including 140 million cubic metres of styrofoam used for wrapping snacks like peanuts.

(8) Regarding discarding, annually Germans reject 5 million domestic appliances, US citizens throw away 7.5 million TV sets; the Japanese people use 30 million disposable cameras. This number has multiplied now.

(9) Advertising fosters consumerism – total global spending on advertising rose from $39 billion in 1950 to $495 billion by 1990; it has grown many times now.

(10) In the early 1990s, 14 billion mail order catalogues and 38 billion additional adverts were sent through the US mail system every year.

Source: Modified Durning (1992) & Baird (1997, cited in Cohen & Kennedy 2000).

The present state, and terms of international trade, based on *unequal ecological exchange*, goes against the developing countries economically, technologically, geopolitically or strategically, and environmentally. For instance, the developed countries usually transfer only the obsolete technologies to the developing countries which emit more carbon (but less productivity); further, many developed countries and also lateral institutions like World Bank, International Monetary Fund (IMF), Asian Development Bank, etc. continue to provide 'tied aid' with various conditionalities like the purchase of particular machines/equipment/weapons. Worst of all, many developing countries of Africa, Asia (including India), and Latin America accept the dumping of electronic and mechanical wastes (e.g. breaking of old ships) in some way or another. Similarly in order to get foreign exchange for debt repayment, developing countries often cut down their invaluable rainforests – that, in turn, leads to reduction in the capacity of the earth to absorb carbon dioxide. Top 20 countries, used as dumping grounds for world's wastes, are provided in Table 6.3.

In 2012, China generated 11.1 million tons of waste and the USA did 10 million tons but an average American citizen generated 29.5 kg of waste in a year and an average Chinese generated 5 kg of waste then; in the UK it was 21 kg per person (The guardian.com, 3 June 2020). Hence, Sustainable Development Goals (SDG) target 12.5 kg (to reduce waste) is not achievable. India's 80% of 1.5 lakh MT daily garbage remains exposed, untreated; sometimes India's highly polluting unprocessed solid waste in dump sites reaches 300 million (or 3 crore) MT; as per National Green Tribunal (NGT), over 28 million tons of waste lay at Bhalaswa, Ghazipur, and Okhla landfill sites. As per European Environmental Agency, between 2,50,000 tons and 3 million tons of used (actually non-functional) electrical products are shipped out of EU annually, mostly to Africa and Asia – 'these goods may subsequently be processed in dangerous and inefficient conditions, harming the health of local people and damaging the environment'. As per an MIT study, the US discarded 258.2 million computer monitors, TVs, and mobile phones in 2010 of which only 66% was recycled, and 120 million mobile phones were shipped to Hong Kong, Latin America, and the Caribbean. Unfortunately, 53.6 million tons of e-waste was dumped in 2019 in the whole world but only 17.4% was recycled. The e-waste management has four stages – generation, collection, segregation, and treatment or disposal. In India, 95% of e-waste is managed informally but under unhealthy working conditions. Yet it has less operating costs and no overheads nor administrative requirements. The failure to recycle is undoubtedly leading to the shortage of rare earth minerals to make future generations of electronic equipment. If at such rate and volume of various forms of wastes continue to be generated, mostly by the developed countries, there will be a grand environmental crisis globally, and their dumping will be vigorously resisted by the local communities and people's movements in the South. E-waste may be drastically reduced if electronic products are made safer, more durable, repairable, and recyclable by using less toxic materials. The electronic manufacturers must re-use recyclable materials and must change their cornucopian view (that the resources are plenty).

Milestones in Global Environmentalism

Since the onset of rapid industrialisation in Western Europe and North America (especially the USA) in the nineteenth century, there have emerged many environmental problems. People became sensitive and conscious of various sorts of cruelty to different animals, hence Society for Prevention of Cruelty to Animals (SPCA) was established in 1824 in England. It was concerned with (a) the prevention of cruelty to animal cases, (b) rehabilitation centres for victim animals, and (c) shelter homes for maltreated and

236 Global Environmental Politics

Table 6.3 Top 20 Countries Used as Dumping Grounds for the World's Trash

Sl. No.	Countries	Type of Wastes Dumped and Sources	Remarks
1.	Ghana (Africa)	Waste from all over the world 'go to die'	Hardest hit region is Agbogbloshie (former wetland) – health hazards for those picking up valuables from obsolete technology
2.	The Philippines	Manila port receives e-waste from North America and Europe	People protested to 50 containers of Canada supposed to have plastic for recycling but dirty diapers along plastic were found – 'Canada Take Back Your Waste' slogan in protest
3.	Nigeria	Port city of Lagos receives from Europe 15 shipping containers of e-waste daily; it receives toxic wastes also from Italy	European e-waste arrives illegally and sold as 'second hand goods' – health hazards
4.	Somalia	Due to Tsunami surge in 2004, hazardous waste containers arrived on the shores of southern Somalia	Some western countries' firms took advantage of Somalia's lack of a functioning govt to dump wastes off its coast for years; hydrogen peroxide toxic waste and radioactive wastes were also found in central and southern Somalia
5.	China	China manufactures electronic devices and many of such devices are shipped back as e-waste	Guiyu town is a biggest dumping hub in China
6.	India	European wastes including metals, textiles, and tires arrive in India, along with illegal e-waste	At Alang shipyard, half of all the ships salvaged in the world are sent for recycling – dismantling is done dangerously by labourers
7.	Vietnam	Its IT industry generates huge e-waste; and also illegal import of e-waste from other countries including 340 tons of electronic parts and thousands of computer screens, laptops, chargers, etc.	'Rubbish metropolis' of Minh Khai village receives waste from Vietnam itself, and Europe and Asia
8.	Pakistan	More than 5,00,000 used computers are sent to Pakistan every year from Singapore, USA, and European countries	Only 15–40% of computers are in a usable condition, the rest are recycled in extremely hazardous conditions
9.	Bangladesh	Dumping of plastic, asbestos, defective steel, waste oil, lead waste, and used batteries from many countries; it is second largest country in ship breaking in the world	83% of child workers are exposed to toxic substances related to e-waste recycling – 15% of such children die every year
10.	Ivory Coast	500 MT of toxic waste from Europe were offloaded illegally at 14 dump sites around Abidjan	Sites close to water supply and food growing fields caused death of 8 persons and more than 80,000 took medical treatment

(Continued)

Global Environmental Politics 237

Table 6.3 (Continued)

Sl. No.	Countries	Type of Wastes Dumped and Sources	Remarks
11.	Indonesia	Exporting scrap metal to Indonesia is legal with some conditions but illegal waste dumping continues; in 2012 it sent back 1,800 tons of contaminated waste to many dumping countries including the UK	In Jakarta 5,00,000 scavengers make a living, searching through hazardous e-waste
12.	Kenya	E-waste recycling facility is located in Nairobi to handle 15,000 tons of e-waste each year; annually e-waste in Kenya is 11,400 tons from refrigerators, 2,800 tons from TVs, 2,500 tons from personal computers, 500 tons from printers, and 150 tons from mobile phones	Health hazards
13.	Guinea	From the USA in the 1980s, 15,000 tons of waste was dumped in the island of Kassa, near its capital, but it contained a dangerous mixture of heavy metals and toxic dioxins; it still receives wastes	Finally, it was returned to the USA and there it was buried in a landfill
14.	Haiti	4,000 tons of toxic incinerator ash aboard the Vessel, Khian Sea, were dumped on a beach near Gonaives when Honduras, Bermuda, and Dominican Republic refused it. The ship left quietly in the night dumping the ash, though fertiliser was promised	Illegal dumping of toxic materials continues
15.	Mexico	During the 1980s, Mexico was the dumping ground for hazardous waste coming from the USA	Health hazards
16.	Zimbabwe	In the 1980s, it was also the recipient of toxic waste shown as commercial cleaning fluid	Waste was brought using grants from USAID
17.	Guinea-Bissau	Upto 3 million tons of toxic waste were sent from Switzerland, the UK, and the USA; an agreement was signed for the burial of 15 million tons of toxic waste sent over by European pharmaceutical companies and tanneries	People protested against the contract, hence it was repudiated
18.	Lebanon	In 1987, Radhost ship sailed to Venezuela carrying 2,400 tons of industrial waste, but it was turned away by authorities and finally it delivered its toxic cargo to Lebanon	Local media reported it and people campaigned to return the toxic waste to Italy

(Continued)

238 Global Environmental Politics

Table 6.3 (Continued)

Sl. No.	Countries	Type of Wastes Dumped and Sources	Remarks
19.	South Africa	In 1989, 120 drums of waste was dumped in South Africa from the USA, it contained sludge laced with mercury	Illegal dumping of waste
20.	Sweden	Sweden recycles 99% of wastes and only 1% of garbage is dumped. Hence, it welcomes trash from other countries, especially Europe for its waste-to-energy programme	No health hazard; well-planned processing of waste for energy done

Source: When on Earth, in Atchuup.com.

unwanted animals. Various SPCA organisations function independently all over the world, having separate governing bodies, financial management, methods of awareness generation, etc. There is no affiliation of any kind anywhere in the world. In India, various SPCAs function in major cities like Chennai, Cuttack, Kolkata, Kollam, Lucknow, Mumbai, Noida, Punjab, Thane, Vishakhapatnam, Jamshedpur, etc. The American Society for Prevention of Cruelty to Animals (ASPCA) was founded on 10 April 1866 in New York City by the philanthropist and diplomat Henry Bergh. It has more than 1.2 million members. Actually, Henry Bergh was appointed by Abraham Lincoln, US President, as a diplomat to the Russian court of Czar Alexander II; there he saw horrific scenes of torture of the horses by the peasant drivers. Later, in June 1865, he visited the Royal Society for Prevention of Cruelty to Animals (London); on his return, he argued in Central Hall, New York on 8 February 1866 for such 'a moral question', 'a matter purely of conscience'. The New York State legislature passed a charter for ASPCA on 19 April 1866. In 1867, ASPCA operated the US first ambulance for horses.

Second, in 1872, the world's first national park, Yellowstone National Park (Idaho, Montana & Wyoming, USA) was established.

Third, the British Royal Society for the Protection of Birds was founded in 1889.

Fourth, in 1892, John Muir founded Sierra Club (USA) that was concerned with the protection of natural environment, extinction of various species, and loss of wild species for industrial/infrastructural development. It shifted its goal from pro-wilderness (preservation) to the support for 'wise use' of natural resources. It has 3 million members and supporters. It lobbies for green politics and was active to get Clean Air Act, Clean Water Act, and Endangered Species Act enacted in the USA.

Fifth, in 1948, the World Conservation Union, later the International Union for Conservation of Nature and Natural Resources, was founded in Gland (Switzerland).

Sixth, in 1961, the World Wide Fund for Nature (WWF) was founded as a charitable trust in Morges, Switzerland, for conservation, research, and restoration of the natural environment.

Seventh, Friends of the Earth was founded in 1969 as an anti-nuclear organisation by David Brower, Donald Aitken, and Jerry Mander (in San Francisco) to challenge the Sierra Club and its moderate groups. It has 2 million activists in 74 countries. It works for changing the perceptions of people, mass media, and the policy-makers (governments), on the one hand, and for sustainability and socio-economic equity between and within societies. Further, it works hard for the empowerment of the indigenous people, local communities, and women for genuine participation in decision-making. For instance, it helped the local people launch the Sarawak Movement in East Malaysia in the late twentieth century. Its head quarter (HQ) is in Amsterdam, Netherlands.

Eighth, Greenpeace International was founded in 1971 in Vancouver, Canada, to oppose US nuclear testing in Alaska. It directly confronts governments and private corporations, though non-violently. It addresses issues of climate change, defends oceans, protects old forests, animals, and plants; it also works for disarmament and peace and acts for nuclear-free world. It vigorously campaigns for sustainable agriculture, rejects genetically modified foods, protects biodiversity, and encourages socially responsible agriculture. Its networks function in 55 countries on different continents, but it does not accept donations from governments or the private corporate sector – it just depends on individual contributions and foundation grants. Various efforts for environmentalism by NGOs may be called 'bottom-up environmentalism'.

On the other hand, the United Nations and its organisational bodies as well as several national governments have also contributed to the 'formal' environmentalism, which may be called 'top-down global environmentalism'. This may be seen in Table 6.4.

Comparison between Environmentalisms of Developed and Developing Countries

Before we discuss the range and depth of environmentalism of developed and developing countries, it seems quite pertinent to pinpoint the ideological divisions between the national governments of the developed (North) and developing countries (South), as in 1992 at the Earth Summit in Rio and at the Kyoto Protocol (1997) such fundamental differences and divisions became manifest, especially on the issue of global warming. The following divisions are notable (Cohen & Kennedy 2000):

(a) The North uses 70% of the world's energy, three-fourth of its metals, three-fifths of its food, and 85% of its wood, though its population is only one-fourth of the world. The USA tops the list, with less than 5% of the world population, it consumes 25% of its energy, highest propensity for air travel and uses 31% of the world's total vehicles; on the other hand, India has world's 17% of population, but consumes only 3% of the world's energy. Hence the developing countries rightly hold the North more culpable for most of the environmental problems.

(b) Due to unequal world trade and cheap labour from the South, the North enjoys a high standard of living. The South does not get the due price for its agricultural raw materials (due to unequal ecological exchange) and since the 1970s has transferred a huge amount of funds in debt repayment to North's governments and multilateral banks.

(c) In view of (a) and (b) above, the South desires that the North should subsidise the cleaner technologies to developing countries so that their industrialisation may not be polluting. Further, the North should reduce its consumption level and also

240 Global Environmental Politics

Table 6.4 Major Environmental Events Organised by the UN

Sl. No.	Year	Institutional Body	Events and Activities
1.	1972 (5–16 June)	UN Conference on Human Environment	At Stockholm, Sweden; 113 countries' political representatives (including India's Prime Minister, Smt. Indira Gandhi) and over 250 NGOs took part; UNGA voted for forming United Nations Environment Programme (UNEP)
2.	1983	UNGA decided to form World Commission on Environment and Development (WCED)	Norway's Prime Minister Gro Harlem Brundtland was made Chairperson of WCED; its report 'Our Common Future' (1987) floated the concept of 'sustainable development'
3.	1985 (22 March)	Vienna Convention for the Protection of the Ozone Layer	To protect health, the right mechanism for funding, and scientific panels; 197 nations ratified it; its effective date was 22 September 1988 – multilateral fund aided thousands of projects in 150 countries, preventing the usage of 2,50,000 tons of ozone-depleting chemicals
4.	1987 (16 September)	Montreal Protocol on Substances that Deplete the Ozone Layer	To regulate the production and use of chemicals that deplete earth's ozone layer; initially signed by only 46 countries but now by about 200 signatories. It came into force on 1 January 1989 – it had good results in preventing ozone layer depletion
5.	1988	Inter-Governmental Panel on Climate Change (IPCC) established	To assess human-induced climate change, with substantive scientific evidence
6.	1991 (4 October)	Protocol on Environmental Protection to the Antarctic Treaty (Madrid Protocol)	At Madrid, it was signed on 4 October 1991 but came into force on 14 January 1998; it designated Antarctica as a 'natural reserve, devoted to peace and science'
7.	1992	UN Earth Summit (on global environmental issues)	In Rio de Janerio, govt. delegates from 178 countries, more than 5,000 journalists, representatives from 9,000 green NGOs/ INGOs attended many declarations on biodiversity and climate change – not binding commitment; Agenda 21 offered guidelines to implement sustainable development
8.	1997 (December)	Kyoto Protocol (Japan) to UNFCC	Govt. officials from 159 nations, 10,000 journalists, green activists, and industrialists participated; industrialised nations agreed to cut six green house gas emissions by 5.5% (on 1990 levels) by 2012; but varying targets for different countries – 'common but differentiated responsibility' was the major policy decision – industrialised nations to meet their targets primarily through national measures, through three market-based

(Continued)

Table 6.4 (Continued)

Sl. No.	Year	Institutional Body	Events and Activities
			mechanisms: (i) International Emissions Trading, (ii) Clean Development Mechanism (CDM), and (iii) Joint Implementation (JI) – these stimulate green investment and help to meet the target in a cost-effective way. Later US Congress did not ratify it. Kyoto Protocol was adopted in Kyoto on 11 December 1997 and came into force on 16 February 2005
9.	2002 (September)	UN World Summit on Sustainable Development at Johannesburg	Over 180 nations and 100 heads of States attended it; it brought together States, civil society, and private sector to formulate a new deal on sustainable development. Recognised that benefits and costs of globalisation are unevenly distributed, hence to increase access to basic needs to be achieved by 2030 like clean water, sanitation, adequate shelter, renewable energy, health care, food security & protection of biodiversity, to fight against chronic hunger, malnutrition, foreign occupation, armed conflicts, illicit drug problems, organised crime, corruption, natural resources, illicit arms trafficking, trafficking in persons, terrorism, intolerance incitement to hatred, xenophobia, endemic, communicable & chronic diseases, especially HIV/AIDS, malaria & TB; women's empowerment, more effective democratic, and accountable international & multilateral institutions
10.	2012 (21–22 June)	UN Conference on Sustainable Development, at Rio de Janeiro (Brazil) Rio + 20	Two broad themes were (i) a green economy in the context of sustainable development and poverty eradication; and (ii) institutional framework for SD. It was decided to develop a set of SDGs by improving MDGs. Second, to establish an inter-governmental process under General Assembly to prepare options on strategy for SD financing. Third, to establish a commission on SD – in 2013, it was replaced by High-level Political Forum on SD. Fourth, to strengthen UNEP. Finally, its outcome document 'the future we want' (with 283 principles), called 'our common vision'
11.	2015	UN Sustainable Development Goals	UN declared 17 SDGs and 169 targets to be achieved by 2030

(*Continued*)

242 Global Environmental Politics

Table 6.4 (Continued)

Sl. No.	Year	Institutional Body	Events and Activities
12.	2015 (December)	UN Conference on Climate Change (Paris Agreement)	Extension of 2012 target upto 2030 – keeping global temperature rise in the twenty-first century well below 2°C above pre-industrial levels and try to limit it further to 1.5°C; all nations to put forward Nationally Determined Contributions (NDCs); transparency of action; came in force on 4 November 2016; signed by 195 signatories on 22 April 2016; but US President, Donald Trump, announced to withdraw from it by November 2020; developed countries agreed to support developing countries financially & technologically; targets CO_2 reduction by 20%; increasing renewable energy market share by 20% & increase energy efficiency by 20%; new US President Joe Biden (2018–2022) supported and joined it
13.	2018	World Sustainable Development Summit (New Delhi)	Theme – 'Partnerships for a Resilient Planet'

Source: Authors.

should compensate the South for slowing down the rate of deforestation and for preserving biodiversity.

On the other hand, the North points out the huge population of some developing countries which have a cumulative effect on energy consumption – especially China and India (the emerging economies). As per one estimate, the developing countries would need to spend $625 billion annually and the North is expected to provide 20% of this amount – but the North has not done so far. However, due to the cornering of the benefits of industrial revolution by the North, resulting in huge carbon emission, it was by and large agreed in the Kyoto Protocol (1997) for 'common but differentiated responsibility' (CDR).

Not only because of such ideological divisions, but also because of some historical-colonial reasons, the very notion of environmentalism varies in developed and developing countries. For instance, western conventional wisdom often considers environmentalism as an elitist action (consequence of affluence) by upper middle class, and that the poor countries and poor individuals are not interested in such actions, and the North Americans (especially the USA) just want a cleaner environment and more enjoyable goods and services (Thurow 1980 cited in Gadgil and Guha 2007). Environmentalism in the USA was 'not a throwback to the primitive, but an integral part of the modern standard of living as people sought to add new "amenity" and aesthetic goals and desires to their earlier preoccupation with necessities and conveniences' (S. P. Hays, 1982, cited in Gadgil & Guha 2007, 421). A British journalist further adds that it is 'safe to assume that when everyone turns environmental, prosperity has truly arrived. Greenness is the ultimate luxury of the consumer society' (C. Moore 1989, cited in Gadgil & Guha 2007, 421). Thus, in 'post-industrial' society, the very notion of environmentalism is lived by the expansion of leisure opportunities, signifying a 'post-material' worldview (Inglehart 1977). Thus this is an 'ecology of affluence' (Gadgil & Guha 2007). On the other hand, developing countries (like India, Kenya, Malaysia, Brazil, the Philippines) and their lower and lower-middle classes have shown their genuine interest in environmental movements, especially for their *critical life issues* (Sharma 2016). A comparison of environmentalism of developed world (represented by the USA) and that of developing countries (represented by India) has been made here (Gadgil & Guha 2007):

(a) North America and Western Europe, in the initial stage of industrialisation, did face the problem of land degradation, deforestation, dust bowls, etc., but genuine scientific resource management and technological substitution could manage these issues. But, in the second phase (after World War II), air and water pollution and the destruction of wilderness have appeared on the public agenda in the West. Thus, there is a shift from 'environmental sustainability' (protection of soil and forest produce) to 'environmental quality' (clean air and water, protection of habitats, etc.). On the contrary, India has been facing simultaneously problems of land and resource depletion, pollution, and loss of biodiversity. Colonial exploitation of natural resources and even planned development after the independence of India, assert Gadgil and Guha (2007), have linkage to such a situation.

(b) In India and other developing countries, environmentalism originates from various conflicts over livelihood and productive resources between competing constituencies – peasantry and industry. On the other hand, the environmental conflicts in the West have come from the threats to health and leisure options. But, in both cases, for the environmental destruction, the State agencies and private corporations are by and large responsible. In developed industrialised nations, the 'quality' of life issues (environmental protection) have taken the place of economic conflicts 'as the motivating factor behind collective action', and, on the other hand, in developing countries 'environmental conflict is, for the most part, only another form of economic conflict' (Gadgil & Guha 2007, 423). Similarly, in Ecuador, in 1999, a local group of blacks, FUNDECOL, circulated a message from a black woman about 'environmental racism' as those dependent on mangroves are suffering:

> they want to humiliate us because we are black, because we are poor .... Now we are struggling for something which is ours, our ecosystem, but not because we are professional ecologists but because we must remain alive because if the mangroves disappear, a whole people disappears ... we shall eat garbage in the outskirts of the city of Esmeraldas or in Guayaquil, we shall become prostitutes ... I shall fight to give them (children) a better life than I have had ... when the govt gives them (Camaroneros Shrimp ponds owners) the lands, will they put up big 'Private Property' signs, will they even kill us with the blessing, of the President?
> (Martinez-Alier 2002)

244   Global Environmental Politics

(c) Regarding tactics of protest, in India, direct action (e.g. hugging the tree in Chipko Andolan, demonstrations, attacks on government property, sit-in, foot march, and Jal Satyagraha (standing in deep water in Narmada Bachao Andolan) have been prevailing; on the other hand, in the USA and other western nations indirect action like litigation in court, media publicity, and political lobbying has been prevailing. However, in the recent past, there is a changing trend. Environmental movements in India have added public interest litigation (PIL) as a supplement to direct popular protest. Similarly, in America and other western nations, the radical environmental NGOs (like Greenpeace International) have taken direct action too (like the spiking of trees) along with or in place of indirect political lobbying, litigation, and media publicity.

(d) In the USA and other western countries, scientists like Rachel Carson, Barry Commoner, Paul Ehrlich, Garret Hardin, and Meadows helped bring ecological concern to a wider public; on the other hand, natural scientists and social scientists in India have not played any key role in such concerns; rather journalists, Gandhians, and environmental activists have played a more significant role. Further, unlike the US society, Indian society does not give much recognition to scientists, nor does it command any 'moral authority' here.

(e) Most significantly, environmental degradation is a very serious issue in developing countries like India, especially because of affecting the life chances of the people directly. On the other hand, in western countries, environmentalism is primarily concerned with health and natural habitats valued for science, aesthetics, and leisure. As a result, in the USA, the ecological movement has run parallel to the consumer society 'without questioning its socio-ecological basis'. But, in developing nations, since the questions of subsistence and survival are involved, people question both the culture of consumerism and the model of unbridled economic development. The environmentalism of developing countries undoubtedly questions the western lifestyles and economic priorities based on the affluence and massive use of energy in daily life-causing huge environmental degradation globally.

A comparative paradigm, for better appreciation, may be seen in Table 6.5.

Ramchandra Guha (2006) rightly remarks about two waves historically in Indian environmentalism: (i) the first wave from the early twentieth century to the outbreak of World War II; (ii) after India's independence, on 15 August 1947, the second age of ecological innocence, catching up with the West through industrialisation began – hence environmental concerns were relegated to the background; but in the early 1970s environmental concerns re-emerged as a vocal new social movement. But since 1973 (when Chipko movement began), the environmental movement in India has passed through three broad phases (Guha 2006):

(a) Age of the 'struggle to be heard' – activists throughout the 1970s made environmental problems (soil erosion, deforestation, water shortage, etc.) visible but the state remained silent.

(b) In the 1980s, the second phase found the support of media, scientific studies, and investigative reportage; hence the State slowly responded to concrete evidences of the 'social costs of environmental abuse'; consequently, the Department of Environment was created both at the centre and States; alarmists by over-stating

Global Environmental Politics   245

Table 6.5 A Comparison of Environmentalism in the North and the South

Aspects	Environmentalism in the North (USA)	Environmentalism in the South (India)
(a) Concerns	(i) In first phase – land degradation, deforestation, dust bowl (ii) In second phase (after World War II) – environmental quality – clean air and water, protection of habitats	Land and resource depletion, loss of biodiversity, pollution
(b) Origin of movements	Threats to health and leisure options	Conflicts over livelihood and productive resources between peasantry and industry
(c) Tactics of protest (action)	Indirect action (public interest litigation, political lobbying, media publicity) but a new trend of direct action also	Direct action ('hugging tree' in Chipko) demonstrations, foot march, attacks on govt. property, sit-in, Jal Satyagraha (standing in water in Narmada Bachao Andolan) but a new trend of PIL also
(d) Inspiring agencies	Scientists (Rachel Carson, Barry Commoner, Paul Ehrlich, G. Hardin, Meadows, etc.); played significant role; they have moral authority	Scientists played no role; they have no moral authority; Gandhians and environmental activists played a major role
(e) Attitude towards consumerism, industrialism, and capitalism	Not questioning socio-ecological basis of consumerism, industrialism, and capitalism	Questioning of consumerism and pattern of energy use (western lifestyle) and inequality – poverty-enhancing industrialism and capitalism (crusading Gandhians in Chipko, KSSP in Silent Valley, Uttarakhand Sangharsh Vahini (Marxist Strand) in Chipko, and Narmada Bachao Andolan)

Source: After Gadgil and Guha (2007).

the problems to ensure policy-making sensitive to both nature and human beings simultaneously.

(c) To make the public and politicians responsive to their agenda, environmentalists often over-stated the gravity of environmental problems, hence were blamed for being alarmists (predicting deaths, diseases, collapse of national sovereignty, and decline of Indian culture due to prevailing situation). This, along with major structural transformation in global geopolitics, brought by the fall of the Berlin wall, resulted into green backlash in the 1990s; due to the fall of the Soviet Union the propagandists of free market over-emphasised the single American economic model of production and consumption but the environmentalists opposed the globalisation of consumerism because an average American citizen consumes energy equal to 50 Indians, hence that model is not sustainable. Despite the backlash against activist environmentalists like Medha Patkar (NBA), they succeeded in bringing the 'victims of development' to the centre stage; but such activist environmentalists are usually opposed by most politicians, industrialists, and elite urban consumers. What Andre Siegfried (1932, cited in Guha 2006, 220) observed about the US grand design came into practice: 'The United States is presiding at a general reorganisation of the ways of living throughout the world'.

Needless to say that neo-Malthusians like Paul Ehrlich and G. Hardin over-emphasise on the population explosion in developing countries like India, China, Bangladesh, Indonesia, but, on the other hand, the grave reality of over-consumption of energy by industrialised countries led by the USA is more responsible for the carbon emissions leading to climate change and other global environmental problems (hence opposed by the environmentalists). In this context, Ramchandra Guha's (2006, 226–227) perceptive remarks are notable:

> The truth about America (USA) is that it is at once deeply democratic and instinctively imperialist ... the clearest connection between democracy at home and imperialism abroad is provided by the American consumer economy, its apparently insatiable greed for the resources of other lands ... the USA imports well over 50 percent of the oil it consumes.

At present the industrialised countries of Europe and North America consume three-fourths of global energy and resources, hence the green theorists (Werner Hulsberg 1988, cited in Guha 2006) and activists demand the cancellation of the international debt, based on unequal ecological exchange, the banning of the trade in such products that damage ecosystems, and for the free migration of people from the developing countries to the developed countries. Long back in December 1928, Mahatma Gandhi, criticising the West had remarked: 'to make India like England and America is to find some other races and places of the earth for exploitation', and 'the distinguishing characteristic of modern civilisation is an indefinite multiplicity of wants', while ancient civilisations had an 'imperative restriction upon, and a strict regulating of, these wants' (cited in Guha, 2006, 231). Gandhiji and his true followers in Sabarmati Ashram followed the principle of wantlessness and 'simple living and high thinking' in thought, word, and deed. Environmentalists have rightly criticised the prevailing development, on the line of liberalisation, privatisation, and globalisation, in India and other developing countries, because it has destroyed nature and did not remove poverty, rather divided the population into three socio-ecological classes (Guha 2006):

(i) Omnivores – who use and abuse natural assets.
(ii) Ecosystem people – who prudently use but also protect natural assets.
(iii) Ecological refugees (living in slums in urban areas) – who over-use natural assets due to their compulsions of survival and face most environmental problems like lack of safe water, lack of sanitation, and pollution.

Generic Modes of Action in Indian Environmentalism

According to Gadgil and Guha (2007), there are various ideological trends in Indian environmentalism and three generic modes of action too. These generic modes are discussed as follows:

Global Environmental Politics 247

(a) Struggle mode
(b) Publicity mode
(c) Restoration mode

(a) Struggle Mode: As far as *struggle mode* is concerned, three significant environmental struggles in early modern India were as follows:

  (i) Movement by the Bishnois of Rajasthan was launched against the felling of Khejri trees, ordered by the then Raja of Jodhpur (Rajasthan) in September 1730. In the desert region of Rajasthan, Khejri leaves were traditionally used as fodder for the cattle. When the Bishnois strongly protested against the order, the mercenaries of Raja ignored them who 'hugged the trees'. Even then king's soldiers killed hundreds of the protesters. Later Raja of Jodhpur atoned and withdrew it (see Box 6.2) and banned felling of Khejri trees in Bishnoi-dominated villages.

  (ii) In 1921, there was a Kumaun forest movement (Uttarakhand) against the British Government's forest policy depriving the local communities of taking various forest produces from the forest. In fact, in 1864, Forest Act was enacted by the British Government, which declared the State ownership of forests, and deprived the traditional rights of local communities.

  (iii) In 1921, there was Mulshi Satyagraha against the dam construction, near Pune in the then Bombay State, by the Tata industrial house. During the first dam constructed near Lonavala, farmers were not paid compensation, hence they opposed the second phase of the project. Then Government of Bombay State promulgated an ordinance to the effect that the Tata group may acquire lands after paying compensation. Hence, this resulted in two factions – landlords were willing to take compensation while the tenants opposed it. Ultimately the Tata group paid compensation for the submerged lands but did not proceed with the third phase. This dam was supposed to generate electricity to light up the latrines of Bombay Chawls (dwelling units of industrial workers). Actually, the local peasants were not consulted for the project; the local farmers opposed it due to fertile land, natural facility for irrigation, and a nominal amount of revenue being paid. The submerged areas included ancestral homes, place of worship, and agricultural lands.

Box 6.2: Save Khejri Tree Movement by the Bisnois in Rajasthan (India)

In the recorded history of Indian environmentalism, the Save Khejri movement of Bishnois, led by Shrimati Amrita Devi, in September 1730, is the first glorious movement. Bishnoi community considers Khejri (acacia) trees as sacred for long due to the preachings of its Hindu Sect leader Guru Jambeshwar (Jambhoji: 1451–1536) who thought that if the Khejri and other trees are protected, and animals are also protected, the local Bishnoi community may survive against the onslaught of droughts, frequently occurring in the desert of Rajasthan. Hence, he formed 29 guiding postulates including a ban on the cutting of any green tree and killing of any animal or bird (in harmony with local ecology). The Bishnoi community followed his guiding principles, and their lands turned into a green forest. In 2019, there were 9,60,000 followers of Bishnoi sect in north and central India. Historically speaking, in September 1730, King of Jodhpur. Abhai Singh ordered his Minister Giridhar Das to cut several trees for the burning of bricks for building of a new palace. He went to Khejrali village to cut green Khejri trees. The villagers protested against the tree felling, but the functionaries of the King did not stop. Then Amrita Devi (a mother of three daughters from the Bishnoi community) led a large crowd from the community and hugged the sacred Khejri trees, declaring: 'A chopped head is cheaper than a felled tree'! Thus 363 Bishnoi volunteers, who had embraced the Khejri trees, were killed by the mercenaries of the then King of Jodhpur, Abhai Singh. Later the King came to know of such killings, he apologised and as atonement for such a massacre, he ordered a ban on felling of green Khejri trees in Bishnoi villages. Every year Bishnois assemble on Dashmi, Shukla Paksha of Bhadrapad (Hindi month) to commemorate the sacrifice of their ancestors to save the sacred trees.

(b) Later in 1753, King of Bikaner, Anup Singh, also banned felling of green trees in Bishnoi-dominated villages.
(c) In 1754, Maharaja Ajit Singh also banned felling of green trees in his area.
(d) In 1858, Man Singh, King of Jodhpur, too, banned felling of green Khejri trees.
(e) In 1900, King Takht Singh, along with ban on felling of green trees, also ordered ban on the slaughter of any animals in Bishnoi-dominated villages.

Three of 29 principles of Guru Jambeshwar are directly related to ecological sustainability:

(i) Do not cut green trees, save the environment.
(ii) Provide shelters for abandoned animals to avoid them from being slaughtered in abattoirs.
(iii) Do not eat meat, always remain pure vegetarian.
(iv) Do not sterilise bulls.
(v) Be merciful to all living beings and love them.
(vi) Obey ideal rules of life: modesty, patience (satisfaction), and cleanliness.

Subsequently, the Government of India instituted the Amrita Devi Wildlife Protection Award for village communities for the protection of forests and wildlife. The Government of India declared the day of massacre (11 September) as National Forest Martyr's Day. Though in 1973, in Uttarakhand, local people embraced trees and launched Chipko Andolan (Hug the tree movement), yet both movements have no linkage – a gap of about 250 years exists between these two movements.

Source: Rajagopalan (2009); Nepal (2009); Wikipedia.org.

They also feared displacement and forced migration. In the 1950s, four major dam projects in India – Bhakhra Nangal (Punjab), Hirakud dam (Odisha), Tungbhadra dam

Global Environmental Politics 249

(Andhra Pradesh), and Rihand dam (UP) – displaced tens of thousands of local people but they did not protest because they were preached of 'nation-building' through dams (J. Nehru called dams 'modern temples') but later people were disillusioned with these projects. Hence, they protested in Koel Karo, Narmada, Silent Valley, Tehri dam, and other dam-building projects on major rivers.

(b) Publicity Mode: In most of the environmental movements, media has played an important role. In Chipko movement (Uttarakhand) and Narmada Bachao Andolan (MP, Gujarat, Maharashtra, and Rajasthan), two main leaders Sundarlal Bahuguna (in one strand) and Baba Amte (later Medha Patkar), respectively, used the mass media significantly. In the Silent Valley movement (Kerala), Kerala Sastra Sahitya Parishad (KSSP) used media plus street corner plays as well as folk songs befitting the purpose. In the Chipko movement too, folk songs and recitals from Hindu scriptures were used. In 1982–1983 Sundarlal Bahuguna (Chipko) took an environmental 'Padyatra' (foot march) covering 4,000 km – from Kashmir to Kohima, and it got much publicity. In the second strand of Chipko, Chandi Prasad Bhatt organised 'eco-development camps' and got trees planted, on the one hand, and generated awareness among local people. Similarly, Save the Western Ghats' Footmarch in 1987–1988 (2,500 km) was well organised : (i) to study environmental degradation and its social consequences; (ii) to activate local groups to prevent further ecological deterioration; and (iii) to generate mass opinion on such issues. Similarly, the National Fisherfolk Forum, in April 1989, organised Kanya Kumari March with a slogan 'Protect Waters, Protect Life'. It had five objectives:

(i) To generate mass awareness of the link between water and life, and encourage people to protect water.
(ii) To form a network of all people concerned with this issue.
(iii) To pressure government for sustainable water use policy and to democratise and strengthen the existing water management agencies.
(iv) To assess the damage already done, identify the problem areas for detailed study, and to rejuvenate water resources.
(v) To revive traditional water conservation practices and regenerative fishing technologies (Gadgil & Guha 2007).

On the last day of that March (May Day 1989), a government bus deliberately disrupted the marchers, leading to the police firing – many persons were killed there and the rally was abruptly called off. But it highlighted the significance of water issue.

On 10 March 2010, the Ministry of Environment, Forests and Climate Change (Government of India (GOI)) constituted Western Ghats Ecology Expert Panel under the chairmanship of Madhav Gadgil, along with other experts. It classified in its report in 2011 the entire hill range of 142 talukas as an ecologically sensitive area – zone 1, zone 2, and zone 3. In zone 1, all development activities like mining, thermal power plant, dam-building, etc. were debarred. It refused environmental clearance of Athirapally of Kerala and Gundia of Karnataka for hydel power projects (falling in zone 1). Madhav Gadgil committee also recommended that changes be brought by following 'bottom-up' approach (from gram sabhas to higher authorities), and decentralisation of power to local bodies. But GOI did not accept Gadgil committee's report and formed another Kasturirangan committee that recommended 37% of total areas of the Western Ghats under the ecologically sensitive area (ESA); it recommended a total ban on mining but allowed thermal and hydropower projects only after a detailed study. It distinguished between cultural and natural landscapes – cultural landscape (of agricultural lands, human settlements, and human-oriented plantations) forms 58% of Western Ghats. On the other hand, 90% of the natural landscapes were brought under ESA. Highly polluting Red Industries were banned. It excluded farmer-inhabited regions from the ESA. But the Kasturirangan report's recommendation of the use of remote sensing/aerial survey to demarcate zonal lands ignored ground reality, hence it caused errors. Finally, human settlements were excluded from the ESA. Consequently, in 2019, there was a huge landslide in Puthumala, and in August 2020 again there was a huge landslide in Pettimudi (Rajamala regions in Idukki) – both in Kerala – and the latter killed 52 persons (plus 20 persons were feared to be trapped under the debris). At this Madhav Gadgil told *Times Now Digital*:

> High rainfall and steep slopes which are often witnessed in these areas, are susceptible to landslides … land slides are under check in areas with intact natural vegetation because of the binding of the soil by the roots. However, any disturbances of natural vegetation render a locality with high rainfall and steep slopes susceptible to landslides. Such disturbances may include the construction of buildings and roads, quarrying or mining.
>
> (Karindalam 2020)

Therefore, various expert environmentalists in India have criticised the government for not accepting the report of Madhav Gadgil Expert Panel and not acting accordingly. It is also notable that during 1–20 August 2018, Kerala received 771 mm rainfall and 341 landslides took place in 10 districts of Kerala – 143 landslides in Idukki alone; such devastating floods/landslides affected 54 lakh people, displaced 14 lakh people, and killed 433 persons (during 22 May–29 August 2018). It was the worst ever flood since 1924 – 42% excess of the normal average rainfall.

(c) Ecological Restoration Mode: In order to have ecological restoration or rehabilitation, in preference to struggle or publicity, various environmental movements/groups have adopted the following modes (Godgil & Guha 2007):

(i) Constructive work or re-construction (Gandhian way – in Chipko movement by Chandi Prasad Bhatt)
(ii) Small dam construction (on cooperative basis in Khanapur, Sangli, Maharashtra)
(iii) Inspiration from religious reforms
(iv) Inspiration from international relief organisations
(v) Inspiration from a catalyst agent (Anna Hazare in Ralegan Siddhi, Ahmadnagar Maharashtra; he was engaged in water storage/recharging, anti-corruption, afforestation anti-liquor, dairy, etc. through self-help)

The Appropriate Technologist strand of Chipko movement (Uttarakhand) was involved in afforestation, soil conservation, small check dams, and promoting of fuel-efficient cooking stoves and biogas plants. After the submerging of the small town of Harsud (MP), various movement organisations and activists like Baba Amte, Dr B. D. Sharma, Medha Patkar, formed Jan Vikas Andolan (People's Development Movement) in September 1989, against the 'destructive development' with four objectives (Gadgil and Guha 2007):

(i)   To coordinate 'collective action against environmentally destructive policies and practices'
(ii) To provide 'national solidarity' to these struggles
(iii) To 'mobilise wider public opinion for new development path'
(iv) To work for an alternative vision 'ecologically sustainable and socially just' for India's future

Thus, generic modes of environmental movements in India are broad-based and are encompassing many local movements for a national 'movement identity' and integrated mass action.

Ideological Orientations in Indian Environmentalism

Despite a lot of contention about the ideological orientation of environmental movements in India, there have been three ideological perspectives or orientations in Indian environmentalism (Gadgil and Guha 2007):

(a) *Crusading Gandhian Perspective*, by and large, rejects the modern (western) way of life (rationalism) and is instead inclined to moral/religious orientation to everyday life. It perceives environmental problems first as a moral problem, rooted in the basic ideology of materialism and consumerism that separates humans from nature, on the one hand, and promotes wasteful lifestyles, on the other. Hence, they want to return to the pre-colonial/pre-capitalist (indigenous) village society, as in their view, the traditional village society had social and ecological harmony but that was disturbed and distorted by the modern systems of land, water, and forest management during the British rule (Forest Act of 1864 brought Indian forests under the British Govt ownership). They also dream of implementing ideals of Gandhi's notion of 'Ramrajya' (rule of Lord Rama) – happiness of people in harmony with nature. Sunderlal Bahuguna of Chipko movement took this ideological perspective by invoking Hindu scriptures showing nature–society harmony.
(b) *Ecological Marxist Perspective* usually perceives environmental problems basically ingrained in political and economic structures and processes, e.g. unequal access to natural resources (not the values) ultimately generates ecological problems like land degradation, water scarcity, loss of biodiversity, deforestation, and so on. The upper class destroys nature by its unbridled exploitation for more and more profit, while the poor people take/use natural resources for their survival and subsistence. The Ecological Marxists want economically a just and egalitarian society through redistribution of economic/natural assets and political power for ensuring ecological sustainability, and hence resolving social and ecological conflicts

in everyday life. Such eco-Marxists were the activists of Kerala Sastra Sahitya Parishad (KSSP) who were at the forefront of Silent Valley movement in Kerala. They had widened their notion of 'science to the people' to include environmental protection too. They differed from the crusading Gandhians in two ways: (i) they opposed the regressive traditions and upheld modern science and modernity; and (ii) they used confrontational mode of struggle much more than the consensual mode of resistance. Further, Chipko's Marxist Strand, Uttarakhand Sangharsh Vahini (USV), too, took an ecological Marxist perspective in terms of banning alcoholism and mining and also ensuring land redistribution for social justice.

(c) *Appropriate Technology*: Unlike the ideological extreme of the crusading Gandhians and the political extreme of ecological Marxists, the Appropriate Technology group takes the middle-of-the-road perspective. This perspective believes in a functional synthesis between agriculture and industry, big and small units, and modern and traditional technology. Influenced by western socialism, it focuses on constructive works (as Mahatma Gandhi had emphasised). They have pragmatically contributed to the 'diffusion of resource-conserving, labour-intensive, and socially liberating technologies'. They thus do not question the very prevailing economic and political systems (as ecological Marxists do).

Thus, the most celebrated environmental movement in India, the Chipko movement, has all these three ideological strands: Sunderlal Bahuguna represented the crusading Gandhian strand, Chandi Prasad Bhatt represented the Appropriate Technology strand, and Uttarakhand Sangharsh Vahini (collective leadership) represented the ecological Marxist strand. They launched massive popular movements against the profit-making commercial forestry, illegal and unregulated mining, and illicit liquor trade.

We may have a comparative glimpse of ideological orientations and generic modes in Indian environmentalism in Table 6.6.

Further, in developing countries, sometimes social conflicts have also ecological content while at other times ecological conflicts have also social conflicts because often the poor people try to retain the ecological system services for their livelihood that are threatened by the State or the market system. For instance, in 1991, Hugo Blanco, a former peasant activist in Peru, described the environmentalism of the North as the 'cult of wilderness', having concern for the disappearing blue whales and pandas, while the 'environmentalism of the poor', on the other hand, is concerned with 'how to get their daily bread', e.g. fighting against the pollution of water from mining or trying to prevent strip-mining in valley or Amazonia people defending their forests against depredation or fighting against water pollution in the beaches, hence it is the 'environmentalism of the livelihood' (Martinez-Alier 2002). However, in the USA, though the extension of 'cult of wilderness' is the mainstream, the minorities also carry on their 'environmental justice' movements regarding local issues, though at a small scale. She is of the firm view that ecological conflicts may arise because of different values and/or different interests. For instance:

(i) Some people want to preserve mangroves as they appreciate their ecological and aesthetic values.

(ii) Others want to preserve them because they live from them (role as coastal defence and as fish breeding grounds).

Table 6.6 Ideological Orientations and Generic Modes in Indian Environmentalism

Ideological Orientations	Perceptions and Priorities	Generic Modes		
		Struggle	Publicity	Restoration
(a) Crusading Gandhian (Sundarlal Bahuguna, Sylwara Ashram strand in Chipko)	• Rejection of modern (western) way of life and life style • Moral/religious inclination • Return to pre-colonial/pre-capitalist village society – questions capitalism • To implement ideals of Gandhi's Ramrajya (happiness in harmony with nature)	Direct but peaceful satyagraha	More	No
(b) Ecological Marxist (KSSP in the Silent Valley; Uttarakhand Sangharsh Vahini strand in Chipko)	• Environmental problems ingrained in political and economic structures and processes – unequal access to natural resources, hence questions capitalism • Solution of environmental problems lies in redistribution of economic assets and political power • Upholding of modern science and modernity and rejection of regressive traditions	Direct – confrontational mode	More	No
(c) Appropriate Technology (Chandi Prasad Bhatt, Dashauli Gram Swaraj Mandal, in Chipko)	• Middle-of-the-road perspective between ideological extreme of crusading Gandhians and political extreme of ecological Marxists • Balance between agriculture and industry, big and small units, modern and traditional technology • Diffusion of resource-conserving, labour-intensive, and socially liberating technologies • They do not question the prevailing economic and political systems	Direct – peaceful satyagraha (hug the tree)	Less	Top priority on afforestation, development eco-camps, improved stoves, small check dams, soil conservation

Source: After Gadgil and Guha (2007).

(iii) Others ((ii) category) appeal to the sense of culture sacredness, etc.

Thus, there is 'the existence of value pluralism', hence cost–benefit analysis is one of several points of view', and 'a reflection of real power structures' (Martinez-Alier 2002). In this regard, O'Connor and Spash (1999) rightly talk of the clash in standards of valuation when the languages of environmental justice, or indigenous territorial rights, or environmental security, or sacredness are used against the monetary valuation of environmental burdens. Hence, non-compensatory multicriteria decision aids or participatory methods of conflict resolution are appropriate for such a situation. Further, in most of the environmental movements in developing countries, women have overwhelmingly participated as they are, more or less, dependent on common property (natural) resources, hence their better awareness and respect for community harmony and solidarity (Agarwal 1992). Further the 'unequal ecological trade' – the South-to-North flow of raw materials (including energy, especially oil and gas) is growing in volume, hence the North is benefitting more. There is a need to incorporate the ecological and social costs of raw materials in their prices. Due to unequal ecological trade, bio-piracy, damage from toxic exports, and the disproportionate use of carbon sinks and reservoirs, there results an ecological debt from the North to the South. Agarwal and Narain (1991), therefore, contrast 'livelihood emissions' (South) versus 'luxury emissions' (North); this has been raised in various fora of the United Nations, but it has not become the immediate concern of environmental movements in developing countries. But Oil-watch has emerged as a South–South cooperation network of resistance with members from more than 50 countries in Africa, Russia, and Central Asia; it was formed after killing of Ken Saro-Wiwa and his activists in November 1995 in Nigeria – its concerns are biodiversity conservation and resource degradation in extraction areas, local air, soil and water pollution, loss of forests, violation of human rights and indigenous territorial rights, and links between global climate change and the increased consumption of fossil fuels.

Thus, the environmentalism of the North and of the South differ in letter and spirit in terms of attitude to life, need versus greed, and individual affluence versus societal sustainability.

Eco-philosophy and Momentum of Environmentalism

Environmental issues exist at many levels, often being trans-boundary, but some issues are, for the time being, only local or national, hence at local, national, and global levels both in thought and action they play their respective role. Hence following nine categories may emerge if we classify such environmental thought and action at different levels (Table 6.7).

In environmental/ecological thought as well as action, there are various models/perspectives. For instance, Arne Naess, in 1973, made two broad perspectives on ecology: 'deep ecology' and 'shallow ecology'. Deep ecology believes in radically experiencing a lived sense of identification with other entities of the earth, i.e. empathy and compassion for non-humans (animals, birds, and plants), hence it is a radical eco-centric perspective. On the other hand, shallow ecology takes a reformist view of environmental problems and suggests only temporary and functional technical-managerial solutions. Arne Naess postulated eight 'eco-sophical' premises of deep ecology (1989, cited in Sharma, 2016):

Global Environmental Politics 255

Table 6.7 Environmental Thought and Action at Different Levels

Category	Thought (at Three Levels)	Action (at Three Levels)
(1)	Local	Local (Ganga Mukti Andolan, Bihar; Chilka Bachao Andolan, Odisha; Appiko movement, Karnataka)
(2)	Local	National (Anti-Chico Dam, the Philippines)
(3)	Local	Global (Silent Valley, Kerala)
(4)	National	Local (Chipko, Uttarakhand)
(5)	National	National (Amazonia Rubber tappers, Brazil)
(6)	National	Global (Narmada Bachao Andolan, India)
(7)	Global	Local (Sarawak movement, East Malaysia, crusading Gandhians in Chipko)
(8)	Global	National (Earth First, USA)
(9)	Global	Global (Anti-nuclear test, Alaska (USA) opposed by Greenpeace International)

Source: Authors.

(1) The flourishing of both human and non-human life on earth has *intrinsic value*-value in itself, not the merely instrumental value of usefulness/utility.
(2) Richness and *diversity of all life forms* are values in themselves and mutually contribute to the flourishing of both human and non-human life on earth.
(3) Humans have *no right to reduce this richness of diversity*, except to satisfy basic needs.
(4) Human interference in non-human nature is *excessive* at present.
(5) For the flourishing of human life, culture, and non-human life, there is a need to *substantially decrease in human population*.
(6) *Change in policies* is needed for a significant change in conditions for a better life.
(7) The ideological change is primarily of *appreciating 'life quality'*, not adhering to a high standard of living.
(8) Those who believe in the above-mentioned premises (1 to 7), should *participate, directly or indirectly, for bringing the necessary change*.

In this regard, it is clarified that Mahatma Gandhi believed in the ideals of such deep ecology, much before this idea conceptually came into being. His two famous statements in this regard are as follows:

(a) 'Earth has everything to fulfil the needs of everybody, but not the greed of anybody'.
(b) 'Be the change, you want to be'.

Now we may compare the major tenets of deep ecology and shallow ecology in Table 6.8.

As indicated earlier, most of the western environmentalism took the shallow ecology perspective, except a few NGOs like Greenpeace International that took the deep ecology perspective, e.g. by opposing the nuclear test in Alaska (USA). Further, radical

256 Global Environmental Politics

Table 6.8 A Comparison of Deep Ecology and Shallow Ecology

Sl. No.	Tenets	Deep Ecology	Shallow Ecology
1.	Science and Technology	Questions western science and technology, stands for process-oriented new science of nature	Believes in western science and technology for 'modernisation'
2.	Outlook	Spiritual outlook to integrate self into greater self	Economic outlook, limited to production and reproduction
3.	Ethic	New ecological ethic of egalitarianism	Anthropocentric view – nature exists for utilisation by humans
4.	View of nature	Intrinsic value of nature	Instrumental value of nature
5.	View on environmental problems	Radical view, environmental problems rooted in the socio-economic–political structures and cultures	Reformist view – technical-managerial solutions
6.	Diversity of life forms	Diversity contributes to flourishing of both human and non-human life on earth – *empathy* towards non-humans	At the most, limited sympathy towards non-humans – but *human supremacy* is at the centre
7.	Population control	Human population needs to be controlled for the flourishing of non-humans	Human population to be controlled to ensure affluence, otherwise more population will lead to poverty
8.	Change for sustainability	Significant change in policies requires basic change in economic, technological, and ideological structures	Only incremental change will lead to better life and high standard of living
9.	Participation for change	Obligation to participate for change (movement action) directly or indirectly	Not necessary

Source: Authors based on Arne Naess (1989, cited in Sharma 2016).

NGO, Earth First (US), and animal liberation movements like People for Ethical Treatment of Animals (PETA) also take a deep ecology perspective. Ecofeminists like Rosemary R. Ruether, Maria Mies, and Vandana Shiva, more or less, also take a deep ecology perspective as does eco-anarchist (an extreme view) Murray Bookchin who wants to overthrow capitalism (the root cause of ecological problems). M. Bookchin proposes workers' management of economy based on reciprocity with nature, use of eco-technology and eco-friendly renewable sources of energy, decentralisation of industries, use of recycled and recyclable raw materials, integration of town and village, and also integration of agriculture and industry. On the other hand, some environmental movements in India like Chipko (crusading Gandhians) and Silent Valley have taken a deep ecology perspective. While Chipko movement was primarily a *social ecological* movement (having a concern for the rainforests of Uttarakhand as well as the livelihood of the local people who get various forest produces from these forests in everyday life), the Silent Valley movement in Kerala was a *purely ecological* movement (concerned only with the protection of rare species of lion-tailed macaque and rare rain forests). Ramchandra Guha (2006), however, had criticised deep ecology perspective on four counts in 1989 itself:

(i) The anthropocentric–biocentric distinction was of little help in understanding the dynamics of ecological degradation.
(ii) The most challenging environmental problems globally were over-consumption and militarism but deep ecology has ignored both.
(iii) Deep ecology was, in essence, an elaboration of the American Wilderness movement.
(iv) In other cultures, 'radical' environmentalism expressed itself very differently.

Ramchandra Guha (2006) later admitted that his criticism of American Wilderness (making national parks) was perhaps a kind of his *national chauvinism*. But at the same time, he elaborates that while American environmentalists were driving thousands of miles in a polluting automobile to enjoy 'unspoilt wilderness', Chipko movement leaders and activists like Chandi Prasad Bhatt integrated their lives with their work: 'Deep ecology tended to ignore inequalities within human society. While the Gandhian Greens, I knew, worked among and for the poor' (2006, 27).

Two more streams of environmentalism are 'green capitalism' and 'eco-socialism'.

Green Capitalism is led by Thomas Friedman, Daniel C. Esty, Michael Braungart, Andrew S. Winsto, William Mc Donough, etc. They provide free market-based solutions to almost all environmental problems. They, by and large, believe that since natural resources are limited, their scarcity will lead to costly products, hence the industries, businesses, and technologists should find out: Whatever is produced, should be produced 'more (quantity) with less' raw materials, energy, water, and also go for the re-use, recycle, etc. Thus, they are basically reformist and close to a shallow ecology perspective.

Eco-socialists, like William Morris, John Bellamy Foster, Paul Burkett, etc. criticise green capitalists as neo-liberals' 'old wine in new bottle'. They have an alternative vision of an economy and society fully in harmony with nature. They directly question various negative social, economic, political, and cultural systems (like capitalism, patriarchy, racism, fossil fuel-based economy) and their processes, and want to replace these with ecologically sustainable and socially just and equitable alternative systems and processes – Stateless and classless society indeed! They are not as green as 'deep ecology (green-green) but they forge a balance between red and green perspectives, hence they are called 'red-green'. Some sociologists and political scientists want an alliance of 'red-green' and 'green-green' perspectives for a broad radical and progressive environmentalism.

Any environmental movement's critical analysis and understanding depends on multiple dimensions (Sharma 2016): *context* (where the socio-economic–political background), *causes* (why proximate causes or long-term causes), *composition* (who do participate – men, women, and/or children, the elite or the lower classes or both, political party or non-party formations), *goals* (what to be achieved – to prevent deforestation, to prevent submergence due to dam construction, to prevent opening of ranches or for ecological sustainability/reconstruction, creating a space for forest

258 Global Environmental Politics

dwellers and indigenous people, or development without destruction or redistribution of economic and political power, etc.), mode of *mobilisation process* (how – direct or indirect, mass media/social media, court litigation, lobbying straight fight/confrontation, etc.), *control* (threats and challenges to the movement from outside, but sometimes from within too), and the final *outcome* (results of the movement, going beyond output by realising the alternative vision of development in practice). This is discussed in detail in Chapter 5.

Participation in environmental action is a very significant aspect; it has various aspects like levels, methods, social norms, the type of leadership, and the agenda (issues). The competence of both an individual and a community (or higher form of a society including a nation) in an environmental movement needs 'both presence and voice, and the former does not guarantee the latter' (Bina Agarwal 2010, 172). In fact, effective participation requires a higher level (to be more active). Bina Agarwal (2010) presents a comprehensive typology of participation, in the specific context of 'gender and green governance', yet it is quite relevant in the context of environmentalism, particularly in developing countries like India (Table 6.9).

Thus from Table 6.9, it is crystal clear that the first and second forms of participation are just formal, hence these have no consequences for the environmental movement as well as the society as a whole. Further, the third and fourth forms are moderate or intermediate participation. Finally, the fifth and sixth forms are real and effective participation which has a positive contribution to the movement as well as the society. In fact, the last one (interactive) is a form of empowering the local people, in letter and spirit, hence it is adjudged as the best form, and the highest level of participation in an environmental movement, especially in developing countries.

Is there any theoretical perspective on global environmentalism? We will discuss it in the next section.

Theorising Politics of Global Environmentalism

Various theories of convergence are used regarding global environmentalism. The major ones are as follows (Jasanoff 2004).

Table 6.9 A Typology of Participation in Environmental Action

Sl. No.	Form/Level of Participation	Characteristic Features
(1)	Nominal participation	Membership in the group
(2)	Passive participation	Being informed of decisions ex-post facto, or attending meetings and listening in on decision-making, without speaking up
(3)	Consultative participation	Being asked an opinion in specific matters without guarantee of influencing decisions
(4)	Activity-specific participation	Being asked to (or voluntarily to) undertake specific tasks
(5)	Active participation	Expressing opinions whether or not solicited, or taking initiatives of other sorts
(6)	Interactive (empowering participation)	Having voice and influence in the group's decisions; holding positions as office bearers

Source: Bina Agarwal (2010).

(i) **Positivist Theory of Technological Determinism:** It considers social change occurring either due to visible exogenous events (new scientific research findings, new crises, new disasters, etc.) or due to the work or private entrepreneurs or institutions promoting private or scientific agenda into 'wider political causes'. It explains why some environmental issues get prominence over others on social and political agendas. John Kingdon (1984) has used this approach; similarly American technological determinists like M. R. Smith and Leo Marx (1994) have also used it.

(ii) **Transnational Alliance Theory of Human Agency:** Here environmental policy change is attributed to the works of *'epistemic communities'*. To Haas (1990), 'epistemic communities' primary actors are bureaucrats, scientists, technocrats, and experts who provide policy-relevant knowledge. They are 'goal-seaking actors', 'loss impeded by institutional rigidities, and disciplinary blinders' (1990). They are politically empowered due to their authoritative knowledge and are motivated by 'shared, causal and principal beliefs'. Thus, they provide relevant but apolitical services to the political system including its policymakers. Its second sub-stream focuses on the role of advocacy coalition frameworks in policy-making (Sabatier 1999). Its third sub-stream takes a broader view that the power to mobilise the people lies not only in the hands of the policy elites but also with wider networks of activist citizens. For instance, Margaret Keck and Kathryn Sikkink (1998) suggest that NGOs often succeed in spreading transnational social agendas like environmentalism through a mix of communicative strategies.

(iii) **Theory of Environmental Discourse:** Here the discourse is considered more significant than human actors in consensus-building on global environmental issues. For instance, Karen Litfin (1994) found that various nations joined hands on Montreal Protocol on ozone-depleting chemicals, through a common discourse of prevention, before the establishment of cause–effect relationship between the discharge of CFC gases and the destruction of the stratospheric ozone layer. On the other hand, Maarten Hajer (1995), regarding European politics of acid rain, argues that actors' cognitive and moral sensibilities are made by their own discourses. They are free to interpret various pieces of evidence and facts in specific ways. Thus, the publicity of a particular discourse (e.g. precaution) promotes consensus and agreement on earlier disputed claims.

But Brcyer (1993 cited in Jasnoff 2004) criticised the linear model of policy formulation (knowledge-centred policy by actor-centred) because the citizens/common people are considered passive, to be led by the charismatic leaders and the untoward events! The general public's view is given due weight only when it differs from experts' views, e.g. resistance to nuclear power in the 1970s, resistance to environmental chemicals in the 1980s, and resistance to agricultural biotechnology in the 1990s. But, experts' explanations have often been rejected with new questions: whose knowledge was considered significant? Further, due to the high level of specialisation when a specialised branch's expert does not know the other branch, why should people believe what one domain expert says? The second and third theories, at least, recognise the role of human agency in social context and, without human actors to mobilise others in wider areas, the experts' claims about environmental problems would have no social–political meaning.

(iv) **Theory of Environmental Rights, Justice, and Equity:** The first, second, and third theories are grounded in the western industrialised world; hence, they are hardly applicable to the South. As Sheila Jasanoff (2004) remarks: 'The space-based view of Earth remains a uniquely American achievement born of a conjunction of wealth, pride, and insecurity, and a culture of separating scheme from values, not shared by most other nations' (2004, 49). She agrees with Anil Agarwal and Sunita Narain's (1991) 'Global Warming in an Unequal World', arguing that the grounds on which the North fixes responsibility for greenhouse gas emission was wrong as it did not take into account history (colonialism) and equity. Talking of 'environmental colonialism', they proposed to divide the basket of greenhouse gases into 'subsistence emission' and 'luxury emissions' – the former by the poor to meet their basic needs while the latter by the wealthy for high levels of consumption for pleasure, comforts, and fashionable lifestyles. A person from the developing world first thinks of 'critical life issues' (e.g. livelihood) and joins an environmental movement, while one from the industrialised world calls for the protection of trees/forests for absorbing greenhouse gases (Sharma 2016). Thus, people in developing countries face problems of unsafe drinking water, polluted air, lack of sanitation, degraded land, etc. leading to less production and productivity, ill effects of the green revolution, unequal ecological exchange, etc., hence they join the environmental movement(s) as a local form of peasant movement (Gadgil & Guha 2007). Therefore, the fourth theory of environmental rights, justice, and equity, to be seen in a historical (colonial) context, is more contextual and appropriate to developing countries. Hence, in this theory, the cause of environmental action arises due to the deprivation of critical life issues, primarily the sources of livelihood (land, water, minerals, and forests) but also due to the loss of cultural heritage that is inter-connected with land, water reservoir, and forests and mangroves, especially among Hindus as well as the tribals.

A comparison of such four theories may be seen in Table 6.10.

Conclusion

It is now crystal clear that there is no uniform environmentalism between the developed and developing countries – rather among the developed countries, some North European countries, especially Scandinavian countries like Sweden, Iceland, Finland, Denmark, and Norway, are by and large in favour of radical ecological politics for several decades. India has a very rich history of radical environmental/ecological political movements – among whom the Chipko movement is the most celebrated environmental movement, not only in India but also in the whole world. The major features of environmentalism in developing countries like India are the concern for '*critical life issues*' (livelihood, land degradation, water, sanitation, etc.), direct action, mass support of men, women, and children, and alternative vision of development. However, the Silent Valley movement in Kerala was a 'pure ecological movement' (concerned with rare animal species – lion-tailed macaque – and rare rainforests (plants), while the Chipko movement, Narmada Bachao Andolan (movement), Anti-Tehri dam movement, Koel Karo movement, etc. were 'social ecological' movements, primarily concerned with 'critical life issues' of the local people. On the other hand, the elitist

Global Environmental Politics   261

Table 6.10 A Comparison of Theories of Politics of Global Environmentalism

Sl. No.	Theory	Major Cause of Environmental Action and Social Change	Role of Human Agency	Main Theorists/Proponents
(1)	Positivist theory of technological determinism	Exogenous events (new scientific findings, crisis, disaster) or work of entrepreneurial individuals/ institutions	No role	John Kingdon, M. R. Smith, and Leo Marx
(2)	Transnational alliance of theory of human agency	(a) 'Epistemic communities' (b) Advocacy coalition (c) Networks of activist citizens (actor-centred)	Major role	(a) Haas (b) Sabatier (c) Margaret Keck & Kathryn Sikkink
(3)	Theory of environmental discourse	Consensus-building discourse (knowledge-centred)	Major role	Karen Litfin and Maarten Hajer
(4)	Theory of environmental rights, justice, and equity	Anti-people nature of polity/ state, deprivation of livelihood, basic needs & cultural heritage	Major role	Anil Agarwal & Sunita Narain, Madhav Gadgil & Ramchandra Guha, Subhash Sharma, Sheila Jasanoff

Source: Modified after Sheila Jasanoff (2004).

environmentalism in developed countries is more concerned with issues like raising the standard of living and the culture of consumerism (resulting in pollution, loss of rain-forest, and ozone depletion). Further, in the North, *movement organisation*, hierarchy, salary, etc. are given more importance than actual movement action while in the South *action* connected with people's lives and livelihoods is more important than the movement organisation, and there is hardly any hierarchy among leaders and activists, and no salary, etc. is paid to the activists (as in the North).

Now the question arises how climate change is affecting people's lives every day, and hence what measures are being taken locally, nationally, and globally. This aspect will be dealt with in the next chapter.

Points for Discussion

(i)   How does environmentalism in India differ from that in developed countries?
(ii)   What are major theories of global environmentalism? Please compare these.
(iii)   Discuss and compare at least three major international environmental NGOs.
(iv)   How does the dumping of the wastes, in which developing countries, affect them?
(v)   Compare deep ecology and shallow ecology perspectives critically.
(vi)   Discuss major environmental events organised by the United Nations since the 1970s.

Chapter 7

Climate Change and Society

Introduction

Climate change is a change in the space and time distribution of weather patterns or conditions or properties of a region or some regions or the entire earth for a relatively long period. Climate change connotes three things: (i) deviation from mean magnitudes; (ii) phase difference from mean periodicity; or (iii) altered frequency of occurrences. It is caused by (a) natural processes like biotic processes, variation in Earth's orbit, variation in albedo or reflexivity of the oceans and continents, continental drift and mountain-building, variation in solar mediation on earth, glacier melting, floods, volcanic eruptions and plate tectonics, or (b) anthropogenic activities like deforestation, burning of crop residues, use of fossil fuel and high energy consumption through machines and electronic gadgets (use of air conditioners, aeroplanes, refrigerators, vacuum cleaners, industrial machines, etc.). While the term 'global warming' means a specific increase in surface temperature due to human activities, the term 'climate change' is very comprehensive and includes global warming (as one of many causes of climate change) as well as other changes in weather patterns/conditions (wind, precipitation, length of seasons, and extreme weather conditions/events) resulting in more emissions of greenhouse gases due to both human activities and natural processes. Climate change manifests in many ways:

(i)   Global warming (including heat waves)
(ii)   Melting of glaciers
(iii) Long spell of drought
(iv)   Long spell of floods
(v)   Heavy floods in a very short duration
(vi)   Ocean acidification
(vii) Forest fires
(viii) Loss of seasonal crops (due to changing pattern of rain)
(ix)   Migration of species (and shifting disease vectors)
(x)   Loss of biodiversity

Many natural scientists have found in their research that there are internal and external forcing mechanisms for climate change – internal forcing mechanisms are natural processes within the climate system (e.g. thermohaline circulation) while external forcing mechanisms may be either natural (e.g. changes in solar output) or anthropogenic

DOI: 10.4324/9781003336211-7

(human activities leading to more emission of greenhouse gases). In fact, our concerns are as follows:

(i) How is climate change created by the political, social, and economic systems in interaction with the natural systems?
(ii) How do these systems try to prevent it?
(iii) How do they cope with its effects (mitigation, adaptation, dislocation/displacement, or permanent deprivation)?

Now there is no scope for sceptics as climate change has emerged as a real problem. The year 2020 was the hottest year and 2016 was the second hottest year and 2019 was the third hottest year in the recorded history of climate till 2020, and January 2020 was the hottest month in 141 years since January 1880 (since then monthly record is available). During January–July 2019, the global temperature was 1.71° above the twentieth-century average of 56.9°. Thus, in the twenty-first century, during 2005–2020 there were ten warmest years on the earth. In Delhi, the maximum temperature on 25 February 2021 was recorded at 33.2°C – the highest day temperature recorded in February in the previous 15 years (on 25 February 2006 at 34.1°C) as per India Meteorological Department (IMD) (TOI, 26/2/21). In Delhi, February 2021 was recorded as the second warmest February in 120 years and March 2021 was recorded as the warmest March in 11 years (since 2010), as per IMD (Gandhioik 2021). There are three categories of nations in terms of per capita carbon emission in the world:

(a) There are 60 countries with an average per capita GDP of $1,768 that emit up to 2.3 tonnes of carbon per capita.
(b) There are 74 countries with an average per capita GDP of $3,058 that emit up to 4.5 tonnes of carbon per capita.
(c) There are 13 countries with an average per capita GDP of $33,700 that emit above 10 tonnes of carbon per capita (as per the World Bank, 2014).

The New Normal: The most popular buzzword for the year 2017 was 'toxic'. That is, the pollution level became intolerable in different parts of the world in both range and depth. But worse was the catastrophe of the weather. Hence, we tend to agree with the Centre for Science and Environment (CSE) environmentalist Sunita Narain (2018, 9).

The 'person of the year' is the weather. And the face of 2017 is the farmer, who has borne the brunt of weather changes, from unseasonal rain to drought. This is the year, when India has had drought at the time of flood. And flood at the time of drought.

This cryptic remark underlines the intensity, range, and depth of the vagaries of nature in the whole world in general and in India in particular. This is widely known as 'climate change' that is, the consequence of both natural phenomena and anthropogenic activities. It has devastating conservancies:

farmers take loans, sow seeds, invest in labour into growing their crops and then comes a devastating weather event – too much rain; too cold or too hot or of a freak hailstorm. Something. But it is a killer. This is not an anomaly. The anomaly is the new normal.

(Narain 2018, 9)

264 Climate Change and Society

The canvas of the new normal is very vast (Box 7.1).

Box 7.1: The New Normal due to Climate Change

Due to global warming (autumn temperatures in northern latitudes have risen 1.1°C) and climate change, there will emerge 'the new normal' in the following ways:

(a) There will be a substantial decline in the polar bear population.

(b) The colour brown will be the new black.

(c) Organic cotton will be the new leather.

(d) Autumn will be the new spring; shorter fall and winter, longer growing season – (leaves are coming up sooner than earlier and falling later).

(e) Shorter ski season.

(f) Amazon forest could become a desert by 2100 AD.

(g) There will be an impending Ice Age.

(h) Global warming would also affect autumn colour if it brought cold and rainy weather, severe drought, or early forest to the fall season.

(i) Autumn in northern Europe may get shorter, while Costa Rica may see a fall season it previously never had.

(j) Mount Everest has slid from an elevation of 5,320 m to 5,280 m (since 1953 when Edmund Hillary and Tenzing Norgay reached the highest peak), and continues to sink; glacial lakes may unleash a massive flood; out of 9,000 glacial lakes in the Himalayas, 200 such lakes face glacial outburst floods (like atomic bomb); a flood there in 1985 created a torrent of 10 million cubic metres of water; a village with a local power station was swept away with some people and debris ending up 55 miles away! Himalayan glaciers have 40% of the world's fresh water, feed nine large rivers and provide one-sixth of the world's drinking water, there is a huge trash (1,00,000 pounds) at Mt Everest, including abandoned equipment like oxygen cylinders; 'the term *development* is simply a euphemism that covers up how projects like road building may destroy Everest's extremely fragile ecology'. Tibet's Buddhists call Everest a holy place – an object of 'pride, affection and reverence'; Mt Everest straddles the border between Nepal and Tibet, hence both have the responsibility to clean up and conserve it.

Source: Science.howstuffworks.com.

Major Consequences: As per a study in 2019 by Stanford University researchers (*Hindustan Times*, 24 April 2019), Climate Change has the following major consequences:

First, global warming arising from human activities increased economic inequalities between countries by 25% during 1961–2010 – rich countries became richer, poor countries became poorer.

Second, climate change reduced India's growth by 31%, Sudan's growth by 36%, Nigeria's growth by 29%, Indonesia's growth by 27%, and Brazil's growth by 25%.

Third, a UN report also points out the link between climate change and biodiversity loss – earth is home to 8–10 million distinct species but 25% of animal and plant species are being crowded, eaten, or poisoned out of existence, much of it is due to the way humans utilise Earth's natural resources that deplete the planet faster than it can regenerate (*Hindustan Times*, 24 April 2019). But, on the other hand, research at the University of Arizona estimated in 2017 about two billion living species on earth (*The Quarterly Review of Biology*, September 2017).

Fourth, new projections by the 'Inter-Governmental Panel on Climate Change' (IPCC) reveal that there would be a threefold increase in the number of people vulnerable to sea level rise, and in India 3.6 crore people in coastal areas are at risk. In the past decade, the sea level in India rose by 3–6 cm and globally the average rise is 3 cm. As a climate scientist, Roxy Mathew Koll (2019) observes:

> If we take the worst case scenario, that will mean a half-metre rise by 2050 and up to 1 metre by 2100 … we will see greater and prolonged flooding, covering large parts of cities like Mumbai where many areas are already low-lying.

Thus, it transpires that the rise in ocean/sea level will spread to wider areas differently (as the sea level is non-uniform) and water will remain there for a longer period, but it will occur gradually. While in Europe melting of ice and glaciers will result from global warming, on the contrary, in India, there will be an expansion of water volume due to ocean warming. In addition, says Roxy Mathew Koll (2019), while in Europe the population is less, hence their exposure to global warming will be less and they already have dikes and other measures to prevent flooding, Asian (including India) and African nations have more population, low-level land, and lack of appropriate measures regarding flooding, hence they will be more vulnerable (Box 7.2).

Box 7.2: Climate-induced Migration in South Asia

(1) Various climate disasters in South Asia may lead to the forced migration of 37 million people by 2030 if the present trend and the situation continue. Of this 27.5 million will be from India, 3.5 million from Sri Lanka, 2 million from Bangladesh, 1.2 million from Pakistan, and 3.1 lakh from Nepal.

(2) By 2050, in total, 62 million people will be forced to migrate in South Asia – 45.5 million from India, 11.5 million from Sri Lanka, 3.3 million from Bangladesh, 1.9 million from Pakistan, and 5.5 lakh from Nepal.

(3) This estimate is based on a study by three organisations Action Aid, Climate Action Network South Asia, and Bread for World; this research was led by Bryan Jones; this study was released on International Migrants Day (18 December 2020).

(4) In the present scenario, global temperature will rise by 3.2°C by 2100; but if global temperature is restricted to 2°C only (by 2100) due to the best efforts of the high emitters of greenhouse gases (GHGs), then only 34 million people will be forced to migrate by 2050.

(5) This study also noted McKinsey Global Institute's finding that slow-onset climate impacts could cause nations in South Asia to lose about 2% of their GDP by 2050 'without strong mitigation and adaptation measures'.

Source: Vishwamohan (2020).

Fifth, climate change will have serious consequences on the health of humans and animals, e.g. heat waves/stress and high humidity will result in skin cancer, malaria, dengue, diarrhoea.

Sixth, as per the Global Climate Risk Index (2020), Japan ranks first, the Philippines second, Germany third, Madagascar fourth, and India fifth in climate risk in the world. India had the highest number of deaths, in 2018, due to the south-west monsoon.

Seventh, according to the Climate Change Performance Index (CCPI), no country in the world could rank 1, 2, or 3 places as none could achieve the target of reducing emission below 2%; Sweden ranked 4th (at the top of all) among all 57 countries and EU/India ranked 9th while China ranked 30th; the performance of the USA and other industrialised countries was bad. Against its voluntary target of India to reduce emission intensity by 35%, by 2030, India reduced it by 21% by 2019 by reducing coal consumption, better management of waste, and promoting renewable energy.

Finally, during 1750–2011, the atmospheric concentration of nitrous oxide (N_2O) rose by 160% and carbon dioxide (CO_2) by 40% (Prabhu 2014).

Oceans/seas will also have critical impacts of climate change (Box 7.3), though there are some sceptics.

Box 7.3: Oceans and the Impact of Climate Change

As per an assessment by renowned scientists of the world in 2015, five oceans and various seas cover 70% of the earth's surface; further 40% of the world population live in low-lying areas within 100 km from the seas/oceans. As per the Food and Agriculture Organisation (FAO) estimate, more than 10% of the global population is engaged in fisheries, especially in coastal areas. Climate change affects the oceans/seas in different ways, thus may be categorised into three broad categories (Ossewaarde 2018):

(a) *Climate-Related Impacts*: Oceans have become acidic due to the absorption of a huge amount of carbon dioxide. Due to the phenomenon of 'global warming', the temperature of the oceans has also increased and these have expanded. Consequently, the oxygen level of oceans has dropped in several places. This has an enormous impact on the death and damage of coral reefs – e.g. there has now emerged the dead zone in the Gulf of Mexico. Further, sea level rise will submerge the Maldives, parts of Indonesia, Bangladesh, India, Mauritius, etc.

(b) *Pollution by Land-Based Sources*: Pollution from land-based sources including discharge of sewage, run-off agricultural nutrients, etc. lead to a huge growth of algae

Climate Change and Society   267

> in oceans. Second, untreated industrial/municipal effluents, particles of chemical pesticides and herbicides, and hydrocarbon pollutants from shipping and oil industries, radioactive wastes from nuclear tests/accidents, etc. damage marine life to a large extent. Third, plastic wastes are also piled in the oceans, and these affect marine birds and other creatures (fish, prawn, etc.) in different ways.
>
> (c) **Over-fishing**: As per the FAO estimate, 70% of world fish is exploited in a non-sustainable way. Most of the coastal countries/regions do not properly monitor and regulate the code of conduct for responsible fisheries. This affects the resilience of the oceans. On the other hand, the above-mentioned tendencies affect the livelihood of coastal people in the world. Further, their pollution spoils the visits of tourists who desire to see aesthetic and clean ocean life. Finally, the anthropogenic pollution of oceans boomerangs on the humans themselves as the changes in the chemistry of oceans lead to climate change that finally affects the pattern of rainfall – this affects agricultural crops, etc.
>
> Source: Based on Martin J. Ossewaarde (2018).

Sceptics about Climate Change: In different parts of the world and at different periods of time, there have been sceptics about climate change. They may be categorised into six categories (Dryzek et al. 2013):

(a) Those who deny that climate change could ever exist.
(b) Those who believe it could exist, but is not caused by humans.
(c) Those who believe it exists, but does not do any substantial damage.
(d) Those who believe it exists, is damaging, but nothing can be done about it.
(e) Those who think that something should be done but not by themselves.
(f) Those who are unsure about what is happening and require convincing.

But the fact is that climate change is occurring all over the world, though varying in degree, range, and intensity as supported by the huge pieces of evidence collected by IPCC. Three scientific links in this regard are as follows:

(a) The concentration of GHGs in earth's atmosphere is directly linked to the average global temperature on Earth.
(b) The concentration has been rising continuously, and the mean global temperature along with it, since the industrial revolution.
(c) Carbon dioxide (CO_2) accounts for two-thirds of GHGs, and is largely the product of burning fossil fuels. According to the World Resources Institute, the climate changed over the 2010s in six ways (see Box 7.4).

Box 7.4: Climate Change Over the 2010s in Six Ways

(1) **Carbon Dioxide Emissions from Fossil Fuels Grew 10%**: During 2010–2018 CO_2 emissions from fossil fuels and cement grew from 32,000 to 36,500 Gt CO_2

emission, as per climate watch; even if nations fully implement their current climate commitment, even then we will face an emission gap in 2030 of 32 Gt CO_2 emission (about twice the size of China and India's combined emissions) for the 1.5°C goal, and 15 Gt CO_2 emission for the 2°C goal.

(2) **Global Average Temperature Increased**: In 2010 global average temperature was 0.88°C above pre-industrial levels but increased to about 1.1°C, as per WMO/UK met office. The world will face severe climate impacts: (a) under 1.5°C of warming, the Arctic will have one sea-ice free summer every 100 years; at 2°C of warming, frequency will increase to at least and such summer every 10 years.

(3) **Concentrations of Carbon Dioxide in Atmosphere Exceeded 400 PPM**: In 2010, carbon dioxide concentrations at the Mauna Loa observatory in Hawaii (US) were 390 PPM on average but by 2018 it went beyond 400 PPM threshold – reaching 408 PPM (pre-industrial concentrations of carbon dioxide were 280 PPM). While during 2000–2009 average growth rate of carbon dioxide concentrations at Mauna Loa observatory was 2 PPM, per year during 2010–2018 it grew to 2.4 PPM per year and in 2018 its growth rate was 2.9 PPM per year.

(4) **Sea Rose More Than 1.6 inches**: Global mean sea level rise was about 3.3 mm per year (0.13 inch per year) during 1993–2018 but during 2010–2018 it grew to about 4.4 mm per year (0.17 inch/year), rising almost 2 inches overall (2010–2018). In 2018, the global mean sea level was the highest, as per NASA. Hence eight of the world's 10 largest cities, located in coastal areas, will have frequent high-tide floods, hence devastating risks to homes, habitat, and infrastructures.

(5) **Ice Reached Record Lows**: Rate of September sea ice decline has been 13% per decade relative to the 1981–2010 average but during 2010–2018 Arctic sea-ice minimum reached its lowest level since 1880. Ice sheets in Green Land and Antarctica have also been losing mass – more loss during 2010–2018. Glacier melt grew from 460 mm of liquid water in the 1990s to 500 mm in the 2000s and to 850 mm in 2010–2018. Ice loss will lead to sea rise and change ocean's surface reflexivity, exposing dark waters that absorb more solar radiation.

(6) **Extreme Weather Became More Frequent and Severe**: Probability of drought has increased substantially in the Mediterranean region due to human-caused emissions; human-induced warming resulted in increased frequency, intensity, and amount of heavy precipitation events. In 2017, Hurricanes Harvey, Irma, Jose, and Maria reached unprecedented rapid intensification — Irma set a record for sustained winds of 185 miles per hour, and Harvey and Maria had record rainfall.

Source: wri.org/blog/2019/12/6-wa.

Challenges and Major Global Events of Climate Change

Challenges of Climate Change: National and Global Scene

Climate change has widespread effects in different arenas of human, animal, and botanical lives. For instance, the production of food grains may decline by one-third due to climate change, bringing both drought and floods, and much of the agriculture in developing countries (up to 50%) is usually dependent on rains for most of the Kharif crops. Over the years, India has been experiencing more rain in a short span and/or less rain in a long span. For instance, in Bihar in 2019, there was almost drought in most of the districts from June to mid-September but in the last leg of September there were torrential rains, resulting into the worst floods and water-logging in many urban areas, especially in Patna. Similarly, there were massive floods in Mumbai in July 2005 (killing 1,094 persons and damaging property worth crores of rupees), in Kedarnath in 2013 (killing 10,000+ persons), in Kerala in 2018 (killing 470 persons and damaging properties worth Rs 40,000 crores and in 2019 also (killing 121 persons, damaging thousands of houses and affecting 2 lakh persons who were shifted to relief camps), and in Chennai in 2015 (killing 422 persons there and 81 in Andhra Pradesh, and damaging more than Rs 200 billion resulting into the huge loss of lives of humans, animals, natural resources, and properties. All these floods were caused by man-made factors – encroachment of lands/nullahs/rivers/lakes/ponds.

Second, climate change and sea level rise would wipe out 50% of the world's beaches (1,32,000 km of coastline) by 2100 CE (assuming carbon emission continues unabated) and the loss of more than 40,000 km of coastline by 2050. Beaches usually act as the first line of defence from coastal storms and floods, and in their absence, the impact of extreme weather events would be higher, as per a study by Michalis Vousdoukas and others (2020) at European Commission's Joint Research Centre. Further, Australia could be hit the hardest, with about 15,000 km of white beach coastline to be washed away by 2100, followed by Canada, Chile, the USA, Mexico, China, Russia, Argentina, India, and Brazil. Further, the storms and hurricanes (e.g. Katrina in 2005, Hurricane Sandy in 2012) have increased, especially the Atlantic storm per year is now double what it was 100 years ago with huge economic, social, and environmental costs.

Third, climate change has enhanced the Australian summer by a month, compared to the mid-twentieth century and bushfire in 2019–2020 burnt about 12 million hectares of bushland, killing 33 persons and one billion native animals, remarks Richi Merzian (2020) of Australian Institute. So is the case in India in terms of extreme weather events, e.g. in April–May 2002 heat wave killed in India about 1,000 persons; in Europe it killed 35,000 to 50,000 persons, especially in France, Germany, and other European countries; in 2010 heat wave killed 15,000 persons in Russia; heat wave in Chicago in July 1995 killed 739 persons (cited in Bell 2016). Due to global warming, the Amazon rainforest could turn into Savannah in 2070, and 1.5°C of atmosphere warming above pre-industrial levels would doom 90% of the world's shallow water corals and 2°C rise would spell their almost total demise (*TOI*, 12.3.20).

Fourth, 13 towns across India, Nepal, Pakistan, and Bangladesh in the Hindu Kush Himalayan region are facing increased water insecurity, i.e., 'urban Himalaya is running dry' despite being in the region of high water availability – India (Mussoorie, Devprayag, Singtam, Kalimpong, Darjeeling), Nepal (Kathmandu, Bharatpur, Tansen, Damauli), Pakistan (Murree, Havelian), and Bangladesh (Sylhet, Chittagong); Shimla was not a part of this study by Anjal Prakash et.al. (published in the journal *Water*) but it, too, is facing the worst water crisis due to poor water governance, lack of urban planning, and climate-related risks – the study noted that the inter-linkages of water availability, water supply systems, and rapid urbanisation and consequent increase in demand (due to increase in population and tourist inflow) were leading to water insecurity in these 13 towns (Vishwa Mohan 2020a). Further, as per the Ministry of Earth Sciences, seven States of Uttar Pradesh (UP), Bihar, West Bengal, Himachal Pradesh, Arunachal Pradesh, Meghalaya, and Nagaland have significantly decreasing trend in annual rainfall in the last 30 years and the number of 'dry days' (daily rainfall of 2.5 mm or less) in UP, Bihar, West Bengal, Delhi, and some southern States are increasing, hence there the situation for agriculture and water recharge is 'alarming' (Viswa Mohan 2020b).

Fifth, there has also been a connection between the frequent wet spells in North India's winter in 2019–2020 and unusually a record freeze in the Arctic region; Arctic sea ice cover was at a 10-year high in 2019–2020 winter – 'polar vortex' (anti-clockwise wind circulation) being the strongest in 2019–2020 caused the Arctic freeze and kept the Arctic cold trapped within the polar region. This, supported by other global factors, led to 20 western disturbances (WD) against the normal WD of 10–14; this resulted in a wet season without temperature extremes (Bhattacharya 2020). This fact clearly shows the inter-connectedness of the local and the global climate change.

Sixth, various new diseases like novel Corona virus (COVID-19) have links to global climate change and deforestation – the natural habitat of bats – as about 60% of 300 new pathogens originate in the bodies of animals:

> We cut down the forests that the bats live in, they don't just go away – they come and live in the trees in our backyards and farms. Through hunting, trading, and recreational activities, the probability that humans encounter bats rises ... we need to reduce our massive and growing footprint across the planet.
>
> (Shah 2020)

Obviously due to the globalisation process with better communication and transport facilities, such viral diseases also spread with high speed to different continents and, therefore, systematic precaution for sanitation and adequate treatment facilities at different levels (individual, community, NGOs, and governance) are very much called for. Needless to say that in Latin America in 1991 cholera resurged and pneumonic plague resurged in India (Surat) in 1994. According to the World Health Organization (WHO), climate change often enhances diseases like dengue (Delhi is a case in point), malaria, diarrhoea, encephalitis (Bihar and eastern UP), Lyme disease and food-borne pathogens like salmonella, and, therefore, WHO has estimated additional 2,50,000 deaths in the whole world annually during 2030 and 2050.

Seventh, as per a study by McKinsey Global Institute, by 2030, the temperature in India could reach such a level that outdoor workers (about 200 millions or 20 crores) be forced to cut short their day working hours to beat the lethal heat waves, and the resultant loss in productivity could put 2.5% to 4.5% of India's GDP at risk annually. This would certainly put our economy in back gear.

Eighth, emissions from the cattle make the production of butter 3.5 times as damaging to the environment as plant-based spreads, as food production is responsible for up to 30% of greenhouse gases; recently scientists found for each kg of product, the 'mean average' CO_2 equivalent for plant-based spreads was 3.3 kg compared to 12.1 kg for dairy-based products – this was an increase of more than 3.5 times, based

Climate Change and Society 271

Table 7.1 Ten Hottest Years Globally (in a Sequence)

Sl. No.	Year	Remarks
1	2020	+1.25°C (hottest)
2	2016	
3	2019	
4	2015	
5	2017	
6	2018	
7	2014	
8	2010	
9	2013	
10	2005	

Source: Climate Central/NASA.

on an analysis of 212 plant-based spreads and margarines and 21 dairy butters (*The Independent in TOI*, 15 March 2020).

Finally, undoubtedly, global warming is visible to all since the late 1970s (the average world temperature has been at least 14°C, but that was never so in the past 200 years in the world history) – from the common persons to the specialist climate scientists: The 1970s were hotter than the 1960s, the 1980s were hotter than the 1970s, 1990s were hotter than the 1980s, 2000s were hotter than the 1990s, and the 2010s were hotter than the 2000s. January 2020 was the warmest month in 141 years of recordkeeping and 2010–2020 was the hottest decade. Ten hottest years globally were as follows in a sequence (Table 7.1).

At present, we have two major global ecological crises: first, the climate change; and second, the extinction of species of flora and fauna. Since the Industrial Revolution in Western Europe, there has been a substantial increase in earth's surface temperature and if no proactive mitigation steps are taken in time, we may experience up to 4°C increase in temperature by the end of the twenty-first century. There have been many extreme weather events (in both mean and spread) as, in 2015, two-thirds of India faced droughts and at global level glacier melting, shrinkage of lakes, rise in sea level, floods, droughts, cyclones, global warming, acid rains, longer and colder winter, and so on were more pronounced (Table 7.2).

Globally the year 2020 had 15 climate disasters out of which two occurred in India. Atlantic Hurricane (USA and Central America) caused maximum loss of $41 billion, floods in China caused a loss of $32 billion, West Coast fires caused a loss of $20 billion, cyclone Amphan in Bay of Bengal (India and Bangladesh) caused a loss of $13 billion, and floods in India (during June–November 2020) caused a loss of $10 billion as well as killed 2,067 people (more than all casualties in the remaining 14 climate disasters put together). Out of ten costliest climate disasters in 2020, five events occurred in Asia (Table 7.3).

But these figures of losses are on the lower side due to insured losses only, actual losses were much more, especially in the developing countries.

These and other global climate change events/disasters have caused massive losses to humans, animals, plants, and properties. In 1995, the United Nations Leipzig

272   Climate Change and Society

Table 7.2 Major Global Events of Climate Change

Sl. No	Major Global Climate Change Events	Country/ Continent	Period of Time	Climate Effects
1	Shrinkage of Lake Chad	Chad, Africa	1960–2002	Persistent drought has shrunk Lake Chad (once world's sixth largest lake) to 1/20th of its size in 1960 – now wetland in place of open water
2	Shrinkage of Lake Toshka	Egypt	1984–2001	From Lake Nasser reservoir (on the Nile river) water passed to Toshka Depression in the Western Desert, but flow to Toshka ceased in 2001 – so many lakes almost lost
3	Flood in Mississippi river	USA	28 January 2011–3 May 2011	Due to snowiest winters and violent early spring rainstorms, Mississippi and its tributaries overflowed their banks inundating lakhs of homes, crops, and woodland with muddy water
4	Flood in Indus river	Pakistan	August 2010	More than a million acres of land were flooded destroying crops, devastating towns (Sukkar, Dadu, and Mehar), 1,800 persons were killed, and one crore persons lost their shelters
5	Yellow river's course change	China	2001–2009	Yellow river was the cradle of Chinese civilisation but frequent devastating floods have changed its course – now it is known as 'China's sorrow'
6	Shrinkage of Lake Mead, Nevada/ Arizona	USA	2000–2010	Lake Mead supplies water to California, Arizona, Nevada, Las Vegas, and Mexico; since 2000 the water level is dropping due to lower snowfall – by July 2010, it was at 38% of its capacity; between 2001 and 2004 it dropped 18 m
7	Global warming	World over	(a) 1880–2009 (b) 2014–2018	(a) Earth's surface temperature increased by 0.7°C since 1880; two-thirds of warming since 1975 at 0.15– 0.20°C per decade (b) Five warmest years on earth since 1880
8	Helheim Glacier melt	Greenland	2001–2005	Helheim Glacier is crumbling into icebergs, glacier's flow to the sea has sped up
9	Inja Glacier melt	Himalayas	2000–2015	Major retreat and collapse of the lower tongue of the glacier and formation of new melt ponds; one-third of all Himalayan glaciers will melt by 2100
10	Ice melt, Mount Kilimanjaro	Tanzania (Africa)	1993–2000	Kilimanjaro is the tallest free-standing mountain, is made up of three volcanic cones, there is a major decline in its ice cap during 1993–2000
11	Flood in Kedarnath	Uttarakhand, India	June 2013	Cloudburst led to the death of 10,000 persons + and huge property losses

Source: Based on NASA data.

Climate Change and Society 273

Table 7.3 Ten Costliest Disasters in the World in 2020

Sl. No.	Disasters (Country/Region)	Financial Cost ($)	Human Casualties	Displacement of People
1.	Atlantic hurricane (USA and Central America)	41 billion	400	2,00,000
2.	Floods (China)	32 billion	219	3.7 million
3.	West Coast fires (USA)	20 billion	30	0.50 million
4.	Cyclone Amphan (India and Bangladesh)	13 billion	128	4.9 million
5.	Floods (India)	10 billion	2,067	14 million
6.	Floods (Japan)	8.5 billion	66	3.6 million
7.	Locust Swarms (East Africa)	8.5 billion	300	1.5 million
8.	Windstorms Ciara and Alex (Europe)	5.9 billion	16	NA
9.	Bushfires (Australia)	5.0 billion	34	65,000
10.	Floods (Pakistan)	1.5 billion	410	68,000

Source: Christian Aid, UK cited in Vishwamohan (2021); Clsnews.com, NYTimes, newindianexpress.com; earthobservatory.nasa.com.

Conference on Plant Genetic Resources pointed out that 75% of the world's biodiversity disappeared in agriculture due to Green Revolution and industrial farming. On the other hand, another UN agency, the Food and Agriculture Organisation (FAO), has estimated that 70–90% of global deforestation is caused by industrial agriculture, which has promoted monoculture into forests to grow commodities for exports, not for foods. Further, according to a grain.org report, the transnational food industry contributes to 44–57% of all anthropogenic greenhouse gas emissions. Furthermore, fossil fuel consumption is also largely responsible for the increase in emissions. It is a bitter truth that 68% of India's energy comes from thermal plants – mostly coal, and to some extent gas and oil. Thermal plants are largely responsible for carbon emissions besides transport vehicles, use of fuel wood, burning of paddy stems and wastes, etc. These thermal plants are owned by state governments, central government, and private companies (some being joint ventures). Maharashtra (28,294 MW) leads in thermal power capacity, followed by Gujarat (23,160 MW), Chhattisgarh (13,234 MW), UP (12,228 MW), Tamil Nadu (11,513 MW), M.P. (11,411 MW), and Rajasthan (10,226 MW).

UN's IPCC Findings

Both the range and depth of the climate change problem was recognised globally due to the collection of adequate scientific evidence by the IPCC, and, therefore, in 2007, US Vice President, Al-Gore (environmentalist) and IPCC were jointly awarded Nobel Prize for Peace 'for their efforts to build up and disseminate greater knowledge about man-made climate change and to lay the foundations for the measures'. In this regard, the UN's Inter-Governmental Panel on Climate Change (IPCC) (which was set up in 1988) published many comprehensive reports (in 1990, 1995, 2001, 2007, and 2014).

274 Climate Change and Society

Its Synthesis Report of AR5 (Fifth Assessment Report) (2014) found the following major trends:

(a) Anthropogenic emissions of GHGs are the highest in history; climate changes have widespread impacts on both human and natural systems.

(b) Oceanic uptake of carbon dioxide (CO_2) resulted in acidification of oceans; warming of 0.85°C increased during 1880–2012 and sea level rose by 0.19 m during 1901–2010 due to warming and ice melting. The sea ice extent in the Arctic has shrunk in every successive decade since 1979, with 1.07 × 106 km² of ice loss per decade.

(c) Due to continued emission of GHGs, there is likelihood of severe, pervasive, and irreversible impacts on humans and ecosystems.

(d) Limiting total human-induced warming to less than 2°C relative to the period 1861–1880 (with a probability of more than 66%) would require cumulative CO_2 emissions from all anthropogenic sources since 1870 to remain below 2,900 $GtCO_2$. – about 1,900 $GtCO_2$ had already been emitted by 2011.

(e) Risks are unevenly distributed and are generally greater for disadvantaged people and communities in all countries at all levels of development.

(f) Adaptation and mitigation are complementary strategies for reducing and managing the risks of climate change.

(g) Without additional mitigation efforts beyond those in place today, warming by the end of the twenty-first century will lead to high to very high risk of severe, widespread, and irreversible impacts globally.

(h) Multiple mitigation pathways would require substantial emission reductions over the next few decades and near zero emissions of CO_2 and other GHGs by the end of the twenty-first century; to implement these would pose technological, economic, social, and institutional challenges.

(i) During 2001–2010 relative to 1986–2005, the rise in sea level ranges from 0.26 to 0.55 m for RCP 2.6, and from 0.45 to 0.82 m for RCP 8.5; by 2065 average sea rise will be 24–30 cm, and by the end of the twenty-first century sea level will rise by 40–63 cm in more than 95% of the ocean area, relative to the reference period of 1986–2005.

(j) Emission scenarios leading to GHG concentrations in 2010 of about 450 PPM CO_2 or lower are likely to maintain warming below 2°C in the twenty-first century above the pre-industrial levels. These scenarios are characterised by 40–70% global anthropogenic GHG emissions reductions by 2050 compared to 2100, and emission levels near zero or below in 2100.

(k) By 2100 global sea level rise will be 10 cm lower with global warming of 1.5°C compared with 2°C; further coral reefs will decline by 70–90% with global warming of 1.5°C, while almost all (99%+) would be lost with 2°C.

(l) Limiting global warming to 1.5°C will require 'rapid and far-reaching' transitions in core sectors – land, energy, industry, buildings, transport, and cities.

(m) Global net human-caused emissions of carbon dioxide will need to fall by about 45% from 2010 levels by 2030, reaching 'net zero' around 2050, thus any remaining emissions would need to be balanced by removing CO_2 from the air.

However, new pieces of evidence point out that though 'glaciers are melting but no need for panic'. New researches by V. K. Raina, Indian Space Research Organisation

(ISRO), and Richard Armstrong do not support IPCC prediction in 2007 that all Himalayan glaciers might disappear by 2035; their following findings are notable (S. S. A. Aiyar 2021): (a) glaciers have been melting since the end of the last Ice Age but did not accelerate in recent decades despite a rise in temperature (V. K. Raina); (b) glacial melt contributes hardly 2% to the flow of Ganga at Prayagraj – rather rain is the main source of flow of Ganga, hence even the melting of all glaciers in a few centuries in future would impact the agriculture very modestly (V. K. Raina); (c) Richard Armstrong (National Snow and Ice Data Centre, USA) and his team of 11 scientists point out four factors contributing to river flows in Himalayan river basins – (i) snow on land melting; (ii) snow on glacier melting; (iii) glacial ice melting; and (iv) rainfall. To them even at above 2,000 m altitude, the contribution of glacial melt to river flows in the Ganga basin is less than 1%, of snow on glaciers 4%; of snow on land 43%; and of rainfall 52%. Contributions of these four in Indus basin are 2%, 6%, 67%, and 23%; and in Brahmaputra basin are 1%, 7%, 26%, and 66%, respectively; (d) Assessment Report 5 (published in 2014) monitored 2,018 Himalayan glaciers during 2001–2011 and found that only 248 glaciers were retreating, 1,752 glaciers were stable, and 18 were advancing. Thus, various studies confirm that many glaciers, not all, are retreating, hence melting is not leading to a crisis. To V. K. Raina, it is a myth that glacial melt is critical for river flow in the lean pre-monsoon season. To Richard Armstrong, snow will fall, melt, and feed the rivers even if all glaciers disappear in future (Aiyar 2021). Actually, all the rainfall in 8.60 lakh sq km of the Ganga basin flows to Ganga, so rainfall in this huge basin is more significant than the Gangotri glacier (the origin of Ganga).

Extreme Weather Events in India and the World

Various apparatuses of the world system have not come to the rescue of the poor in developing countries. We have failed in two global processes, as Sunita Narain (2018) rightly says: (a) 'the failure of climate change negotiations to mitigate emissions at a scale and pace that would not impact the world's poorest'; (b) 'the failure of global trade negotiations that have managed to turn the tables so that the poorest and most marginalised of the world are seen as the perpetrator to global fair trade practices'. Further, both rural and urban areas are suffering from climate change, and when the problem accentuates in rural areas, rural people migrate en masse to urban areas adding to the woes of the latter by putting an additional burden on urban infrastructures and facilities. Over the years following flood situations due to climate change in some cities of India are noteworthy :

(a) Flood situation in Mumbai in 2005 (26–27 July) due to continuous rains (944 mm in 24 hours) and choking of nullahs, 1,100 persons died in Maharashtra and 1,50,000 people were stranded in railway stations; again on 29–30 August 2017, Mumbai floods caused 35 deaths. Now Mumbai is the most traffic-congested city in the world, and Delhi ranks fourth in this regard. Mumbai takes 65% more travel time and Delhi takes 58% more travel time and this gets worsened during the floods.

(b) Chennai, too, got heavy rainfall due to the north-east monsoon (November–December) in 2015, hence it drowned; Comptroller and Auditor General (CAG)

276   Climate Change and Society

termed it as 'man-made disaster' (unregulated urbanisation) – Chennai received 1,049 mm of rain in November – it killed 500 persons in South India as a whole, and 18 lakh people were affected. On the one hand, plastic waste choking the drainage system, unplanned urbanisation by encroaching nullahs, and removal of wetlands were man-made causes, on the other hand, extreme rainfall event in a short time was also the cause of the Chennai floods in 2015.

(c) Chandigarh had deficit rainfall till 21 August 2017 and later it got 115 mm of rain in just 12 hours – it got 15% of its annual monsoon rainfall in just a few hours – hence it drowned.

(d) In 2017, during monsoon season, Bengaluru hardly had rainfall initially but later it got 150 mm of rainfall in just a day – about 30% of its annual monsoon rainfall – hence it drowned.

(e) There was a massive flash flood in Leh in 2010 (in present Laddakh Union Territory (UT))

(f) In the 2017 monsoon season, Mount Abu got over half of its annual monsoon rainfall in just two days, hence it drowned.

(g) Hyderabad, too, got heavy rainfall in the 2017 monsoon, resulting in devastation.

(h) In the fag end of September 2019, Patna experienced very excess rainfall resulting in water-logging and drowning in many lowland colonies like Kankarbag, Rajendra Nagar, Pataliputra Colony, S. K. Puri, Gola Road. Plastic waste choking nullahs, non-functioning of sump houses, clogging of Moin-ul-Haq Stadium – Rampur drain due to the ongoing construction works of Science City, dysfunctional existing drainage system at most places, and lack of adequate drainage system in new colonies caused long period water-logging in these localities. Western Patna has neither pucca roads nor drains, hence more water-logging occurred there too. On 13 November 2019, about a dozen illegal structures on Badshahi nullah (in South Patna at Sampatchak) were removed by the administration; only then nine major drains were cleared of encroachments.

(i) In 2013, there was a massive flooding due to cloudbursts in the Kedarnath temple city, killing more than 10,000 people (mostly pilgrims).

(j) On 19 August 2020, torrential rainfall in Gurugram (Haryana) – 118 mm in just 6 hours, leading to massive traffic jam, drowning underpasses, etc. On the other hand, water table has fallen drastically in Gurugram and its 99 villages are in "red zone" (a level beyond 30 metres below) over the last fifty years- water table depth 115 metres below ground level in Dundahera, at 72 metres in Chakkarpur and at 60 metres in Nathupur, a survey under Atal Bhujal Yojana, Jalshakti Ministry, Government of India, revealed in 2022.

The 2019 monsoon rains caused the deaths of about 1,900 persons across India and affected 25 lakh (2.5 million) people in 22 States as per the Ministry of Home Affairs, Government of India. This was the highest rainfall in India since 1994. Its details may be seen in Table 7.4.

Climate change has been found the major reason for such a record rainfall (2019) in the last 25 years, wherein 1,900 people died, 738 were injured, 20,000 animals were lost, 3.14 lakh (0.314 million) houses were damaged (1.09 lakh fully and 2.05 lakh partially), and crops in 14.14 lakh ha were also damaged. In total, more than 25 lakh (2.5 million) persons were affected (877 relief camps gave shelter to them in 277 districts

Table 7.4 Deaths and Damages Caused by 2019 Floods in India

Sl. No	States	No. of Districts Affected	Affected Persons	Deaths	Injured Persons	Loss of Animals	Damages of Houses (Full + Partial)	Damage of Crops
1	Maharashtra	22	7.19 lakh	382	369			
2	West Bengal	22	43,433	227	37 (+4 missing)			
3	Bihar	27	1.26 lakh	161	NA			
4	MP	38	32,996	182	38 (+7 missing)			
5	Kerala	13	4.46 lakh	181	72 (+15 missing)			
6	Gujarat	22	17,783	169	17			
7	Karnataka	13	2.48 lakh	106	14 (+6 missing)			
8	Assam	32	5.35 lakh	97	NA			
9	India (Total)	277 in 22 states	25 lakh +	1,900	738 (+46 missing)	20,000	1.09 lakh + 2.05 lakh = 3.14 lakh	14.14 lakh ha

Source: MHA, cited in *The Morning India*, 5 October 2019.

278 Climate Change and Society

in 22 States of India. It is notable that of all the disasters recorded in the world during 1998–2017, floods accounted for 44%; second, climate change is causing more havoc than geophysical events like earthquakes and tsunamis – floods and storms together account for 72% of all disasters; more than 2.6 crore (26 million) people globally are pushed into poverty annually as a result of disasters and during 1998–2017 India's economic loss due to disasters was $80 billion as per United Nations Disaster Risk Reduction (*The Times of India*, 2 October 2019). Further losses from sea level rise and storm surge along could lead to the global GDP back by 9%, and 36 crore (360 million) people will be at risk in coastal cities; as per World Resources Institute, in South Asia losses due to flooding would be $215 billion by 2030 and European coastal cities' loss will be $1.2 billion by 2030 (Venkatesh 2018).

Unfortunately, the frequency and intensity of extreme weather events, linked with climate change, have increased over the years. In Odisha and West Bengal States of India, an extremely severe cyclonic storm 'Fani' occurred on 26 April–5 May 2019, and affected 12 lakh people, killed 89 persons, and damaged properties to the tune of $8.1 billion. Similarly cyclonic storm 'Nilam' affected South India and Sri Lanka on 28 October–1 November 2012 and therein 75 persons were killed and it damaged properties to the tune of $56.7 million. Earlier the extremely severe cyclonic storm 'Phailin' occurred on 4 October–14 October 2012 affecting Odisha, Andhra Pradesh, Nepal, Thailand, and Myanmar, and killed 45 persons and damaged properties worth $4.26 billion. On 26 December 2004, earlier a very severe earthquake and tsunami had occurred in the Indian Ocean with an epicentre off the west coast of northern Sumatra, Indonesia (location at Banda Aceh) – with a magnitude of 9.3 MW, reaching a Mercalli intensity up to IX in certain areas. It killed 2,27,898 persons and damaged properties worth Rs1,500 crores ($15 billions) in 14 countries (including Tamil Nadu, India). Extremely severe 'Hudhud' cyclonic storm later occurred on 8–14 October 2014 in India and Nepal, and killed 124 persons and damaged properties worth $3.58 billion. Very severe 'Bulbul' cyclonic storm later occurred in West Bengal, Odisha, and Bangladesh on 11 November 2019; it killed 12 persons in West Bengal and Odisha, and 24 persons in Bangladesh. It affected 4.65 lakh people in West Bengal and damaged 60,000 houses. Earlier on 25 October–4 November 1999, super-cyclonic storm 'Bob 06' occurred in Odisha (India), Bangladesh, Thailand, and Myanmar. It killed 9,887 persons in Odisha as per Government of India – 8,000 persons killed in Jagatsinghpur in Odisha alone and caused damage of $4.44 billion – it damaged 16 lakh houses. It affected other countries too (Thailand, Bangladesh, and Myanmar), hence one estimate mentions 30,000 deaths (Wikipedia, 13 November 2019). Some experts are of the view that in the coastal district of Jagatsinghpur in Odisha earlier 1,70,000 mangroves were cut down for establishing an industrial project, hence the damage caused by this super-cyclonic storm was at the most because these mangroves were acting earlier as a safety net to the people and properties there. Floods and high waves along with tropical storms had caused 3,00,000 deaths in Bangladesh since 1970. In 2022, there was a long heat wave during March–May in North and Northwest India and later rainfall deficit there while there was a devastating flood in Odisha, Maharashtra, Gujarat, and north-east India due to climate change.

During 1980–2010, as per the Government of India report, 431 major natural disasters occurred in India. During the financial year 2018–2019, 2,405 persons died due to cyclonic storms, flash floods, landslides, cloudbursts, etc. (*The Times of India*, 9 July 2019). Extreme weather events led to about 16,000 deaths and economic losses of $142 billion in G20 nations, on average, annually during 1998–2017. India had 3,661 deaths, Russia 2,944 deaths, France 1,121 deaths, Italy 1,005 deaths, and Germany 475 deaths. Similarly, the USA had an annual average loss of $48,659 million, China $36,601 million, India $12,823 million, Mexico $2,955 million, and Australia $2,394 million.

In India, during 2020, extreme weather events claimed 1,400 lives, with maximum deaths due to thunderstorms (569), followed by rain and floods (452), cold wave (104), lightning (163), and cyclone (111). Statewise casualities due to extreme weather events in 2020 are as follows (Table 7.5).

Thus, it is clear that thunder storms and lightning took 815 lives in different States of India (as per IMD). Second, in India, September was the warmest month, followed by August (second warmest) and October (third warmest) in 2020 since 1901 (as per monthly mean temperatures, compared to their normal). Third, the annual mean land surface and air temperature for India in 2020 was 0.29°C above the 1981–2010 period average, but substantially lower than the warming in India during 2019 and 2016 (above 0.71°C). Further, the mean temperature exceeded the normal during September (by 0.72°C, warmest since 1901), August (by 0.58°C, second warmest), October (by 0.94°C, third warmest), July (by 0.56°C, fifth warmest), and December (by 0.39°C, seventh warmest), as per IMD. Fourth, as per IMD, 12 out of 15 warmest years were during 2006–2020 period – 2019 being the seventh warmest year; the five warmest years in India were – 2016 (+0.71°C), 2009 (+0.55°C), 2017 (0.541°C), 2010 (+0.539°C), and 2015 (+0.42°C). Fifth, annual rainfall in India as a whole was 109% of its Long Period Average (LPA), (based on the data of 1961–2010), in the 2020 year. Sixth, thunder storm and lightning claimed 815 lives in the whole of India – maximum from Bihar, UP, and Jharkhand. Finally, in 2020, cyclones took 111 lives in India as a whole; during 2020 five cyclones formed over the North Indian Ocean – of these Nisarga and Gati formed over the Arabian Sea and other three cyclones (Amphan, Nivar, and Burevi) formed over Bay of Bengal; the most devastating super-cyclone storm Amphan formed in the pre-monsoon season and crossed over Sundarbans on 20 May 2020 and took 90 lives mainly in West Bengal (as per IMD cited in Vishwamohan 2021).

Further following facts related to climate are also notable regarding India:

Table 7.5 Statewise Deaths in India due to Extreme Weather Events in 2020

Sl. No.	State	Rains and Floods	Thunderstorm	Lightning	Cold Wave	Total
1.	Bihar	54	280	45	NA	379
2.	Uttar Pradesh	48	167	53	88	356
3.	Jharkhand	–	122	–	16	138
4.	Assam	129	–	–	–	129
5.	Maharashtra	50	–	23	–	73
6.	Madhya Pradesh	–	–	72	–	72
7.	Kerala	72	–	–	–	72
8.	Telangana	61	–	–	–	61
9.	Himachal Pradesh	38	–	–	–	38
		452	569	163	104	1400

Source: Vishwamohan (2021).

280 Climate Change and Society

(i)   60% of its total land area is vulnerable to earthquakes
(ii)   More than 40 million ha is prone to floods
(iii) About 8% of its total area is prone to cyclones
(iv)   About 60% of its total area is prone to drought
(v)   About 55% of its total area lies in seismic zones III–V (vulnerable to severe earthquakes)
(vi)   Sub-Himalayan regions and Western Ghats are vulnerable to landslides
(vii) 8,000 km long coastline is prone to severe cyclone
(viii) India accounts for one-fifth of the world's deaths due to floods – more than 30 million (3 crore) persons are displaced in India due to floods every year! (GOI 2004).

In fact, the whole of South Asia is highly vulnerable to climate change, especially sea level rise, extreme weather events, and heat waves.

Urbanisation and Climate Change

In India, 100 cities have been chosen under the Smart City Mission with a total cost of Rs 25,000 crores, but climate adaptation has not been duly taken into account. Out of 47 cities with a population of one million each, ten cleanest cities in India (in 2020) are (1) Indore (continuously for the fourth year), (2) Surat, (3) Navi Mumbai, (4) Vijaywada, (5) Ahmedabad, (6) Rajkot, (7) Bhopal, (8) Chandigarh, (9) Visakhapatnam, and (10) Vadodara. On the other hand, ten dirtiest cities are (1) Patna, (2) East Delhi, (3) Chennai, (4) Kota, (5) North Delhi, (6) Madurai, (7) Meerut, (8) Coimbatore, (9) Amritsar, and (10) Faridabad. Similarly, though Goal 11 of the Sustainable Development Goals is committed to sustainable cities and communities making cities and human settlements inclusive, safe, resilient, and sustainable, yet urban administrators, mayors, and state governments in India have not done adequate planning in this regard.

In fact, the process of industrialisation is usually accompanied by the process of urbanisation (migration from rural areas to urban areas because of both 'push' and 'pull' factors). Usually, most of the developed countries have more than 75% urbanisation – Japan is the most urbanised nation (92%), followed by Brazil (87%), the UK (84%), the USA (82%), France 81%), Germany (77%), and Russia (75%). But now China (60%), too, is urbanising fast (urban population more than tripled in the last 40 years and increased fivefold during 1950–2019) but India's urbanisation is 34% only. This may be seen in Table 7.6.

Thus, it transpires that during 1950–2019, China's urban population increased maximum in the world by 400%, followed by Brazil (141%) and India (100%) but the UK experienced only 6% increase and Germany only 13% increase because these developed countries were already adequately urbanised in 1950 (79% and 68%, respectively). Now Japan is the most urbanised country 92% (with 73% growth during 1950–2019). It is also notable that India's urbanisation was more than China's in 1950 (17% and 12%, respectively) but by 2019 China surpassed India – 60% and 34%, respectively. That is, China has almost become a non-agrarian society while India is still an agrarian country with 66% rural population. However, urban areas are more polluted than rural areas on the whole in the entire world but cities and towns

Climate Change and Society 281

Table 7.6 Percentage of Urbanisation in Different Countries of the World (1950–2019)

Sl. No	Countries	1950 (%)	1980 (%)	2019 (%)	Increase during 1950–2019 (%)
1	Japan	53	76	92	73
2	Brazil	36	65	87	141
3	UK	79	78	84	6
4	USA	64	74	82	28
5	France	55	73	81	47
6	Germany	68	73	77	13
7	Russia	44	70	75	70
8	China	12	19	60	400
9	India	17	23	34	100

Source: UN, cited in *The Times of India*, 15 August 2019.

of developing countries are more polluted, ill-planned, and less clean than those of developed countries. It is estimated that 15 crore (150 million) people live in cities with perennial water scarcity, and by 2050 this number will increase to 100 crores (one billion)! (Venkatesh 2018). Urban areas have many problems due to climate change, e.g. water-logging and flooding, the spread of epidemics due to high population density (novel corona pandemic – COVID-19 is highly concentrated in cities), dengue, chikungunya, etc. Urbanisation has badly affected forests and green cover as well as water sources (ponds, lakes, rivers, etc.) by constructing buildings, roads, infrastructures, and public utilities over there. Climate change has worsened their situation in various ways.

As per the Ease of Living Index 2020, Bengaluru ranks first, Pune second, Ahmedabad third, Chennai fourth, and Surat fifth; Delhi ranks 13th (jump from 65th in 2018), Ghaziabad 30th and Faridabad 40th among 49 cities with a population of over one million; among 62 cities with a population of less than one million Shimla is most liveable, followed by Bhubaneswar, and Silvassa – Gurgaon ranks eighth in this category. Among municipal corporations of above one million population Indore 1, Surat 2, Bhopal 3, Pimpri 4, and among municipalities of less than one million New Delhi Municipal Council (NDMC) 1, Tirupati 2, Gandhinagar 3, and Karnal 4. Among mega-cities Chennai, Coimbatore, and Navi Mumbai rank 1, 2, and 3 in quality of life', while Bengaluru, Delhi, and Pune rank 1, 2, and 3, respectively in 'economic ability' list (Dash 2021).

How are emissions and climate change related? We will discuss this in the next section.

GHG Emission and Climate Change

G20 countries emit together about 80% of global GHG emissions, hence its report (2019) says that limiting global temperature rise to 1.5°C avoids over 70% of climate-related impacts in the water, health, and agriculture sectors. Further, Brazil leads by 82.5% renewable energy while Saudi Arabia, South Korea, and South Africa lag behind with shares of 0–5% in their total energy mix. India is investing most in renewable energy, and so is the case of Germany. The report also noted that CO_2 emission

282 Climate Change and Society

in G20 countries rose in all sectors in 2018 compared to 2017 (i) in building 4.1%, (b) in power 1.6%, and (c) in transport 1.2% (*The Times of India*, 12 November 2019). Hence, India should take a leading role in G77 countries to convince G8 countries to reduce emissions drastically.

Greenhouse gas (GHG) emissions contribute heavily to pollution and climate change. Globally agriculture contributes 15% of all GHG emissions and half of this comes from meat production; 30% of global land not covered with ice is used for growing crops for animals' consumption; as a University of Oxford study on British diets in 2014 found that meat-rich diets (i.e. eating more than 100 gm of meat daily per head) emitted about 7.2 kg of CO_2 daily compared to 2.9 kg of CO_2 emitted by vegan diets (Narain 2018). Thus, vegetarian food is to be preferred to non-veg but Sunita Narain also differentiated between 'luxury' meat and 'survival' meat – luxury meat is produced by using large areas, destroys rainforests for grazing, and uses more chemicals. Further, red meat contains antibiotics and US citizens eat 1.5 times more than the protein requirement (on average 122 kg annually per head)! It is also estimated that globally cities account for 80% of global GDP though occupying only about 0.51% of the total area; 150 cities in the world with 14% of the global population contribute to 40% of the global economy but at a huge ecological cost. All cities consume about 80% of the world's total energy and contribute to 70% of global GHG emissions resulting in global warming. Further urbanisation is growing at the cost of declining natural/semi-natural spaces – from 70% to 50% of total land area as per IPCC Fifth Assessment Report 2014 (Venkatesh 2018). Hence, cities may be 8°C warmer by 2100 AD due to the trapping of more heat and vulnerable to heat waves through the 'heat island' effect which may push up costs of climate change by 2.6 times by 2100 AD. Hence, researchers from Stanford University estimate that there is more than 50% likelihood of a 20% drop in productivity, and South–Southeast Asia and Sub-Saharan Africa will be hit the hardest (Venkatesh 2018). Due to no proper planning in transport, waste disposal and housing, industrialisation, urbanisation, and climate change together intensify in low and low-middle income countries (including India). Let us see GHG emissions globally per head and annually (in total) in some selected countries in Tables 7.7 and Table 7.8.

In France, the UK, and Portugal, a higher share of electricity is produced from nuclear and renewable sources; e.g. in 2015, only 6% of France's electricity came from fossil fuels compared to Germany with 55% of electricity coming from fossil fuels though both have a similar standard of living. Further oil-producing countries have higher per capita CO_2 emission, but they are very small, hence their total emission is quite less than the global average. Some of the poorest countries in Sub-Saharan Africa like Chad, Niger, and Central African Republic average per head footprint is only 0.1 ton per year – thus 160 times lower than that in the USA, Australia, and Canada each. Interestingly in just 2.3 days, the average American or Australian person emits as much as the average Malian or Nigerian emits in a year. In addition, Asia is the largest emitter (19 billion tons – 53% of global emissions), but its population is 60% of the world population – thus per capita emission in Asia is slightly lower than the global average. But China emits about 11.6 billion tons annually, hence one-fourth of global emissions. North America – dominated by the USA – is the second largest regional emitter with 18% of global emissions, while Europe emits 17% of global emissions. Africa emits 3–4% of global emissions;

Climate Change and Society   283

Table 7.7 Per Capita GHG Emissions (Including Land Use Change and Forestry) in the World (2014)

S. No	Country	Per Capita GHG (tons CO_2 emission)
1.	Qatar	49
2.	Trinidad and Tobago	30
3.	Kuwait	25
4.	UAE	25
5.	Brunei	24
6.	Bahrain	23
7.	USA	19.8
8.	Saudi Arabia	19.0
9.	Australia (2016)	18.3
10.	Canada (2016)	18.6
11.	Russian Federation	14.1
12.	Japan	10.4
13.	Germany	10.1
14.	South Africa	9.7
15.	China	8.5
16.	Brazil	6.6
17.	UK (2017)	5.8
18.	Portugal	5.3
19.	France (2017)	5.5
20.	India	2.5
	World Average (2017)	4.8

Source: CAIT Climate Data Explorer (2018).

Table 7.8 Annual GHG Emissions (Including Land Use Change and Forestry) in the World (2014)

Sl. No	Country	Annual GHG Emissions (in Gt. CO_2 emission)
1	China	11.6
2	North America	6.5 (18% of global)
3	USA	6.3 (15% of global)
4	Europe	6.1
5	EU	3.5
6	India	3.2
7	Russian Federation	2.0
8	Brazil	1.4
9	Japan	1.3
10.	Germany	0.8
11.	Saudi Arabia	0.6
12.	South Africa	0.5

Source: CAIT Climate Data Explorer (2018).

South America too emits 3–4% of global emissions. China, the USA, and the EU (27 countries of Europe) account for more than half of global emissions (Google, 11 November 2019). At COP 27, in Sharm-el-Sheikh (Egypt), India and other developing countries rightly raised the issue of 'big historical polluters' (industrialised nations), not current twenty big polluters in cumulative terms; for instance, since

1750 share of the USA in total global pollution has been 25%, of EU (27 nations) 18%, of China 14%, of Russia 7%, and of India only 4%. Similarly, energy use per capita in the USA is 6,804 kg of oil equivalent while that in India is only 637 kg of oil equivalent (2021). Hence, India argued for phasing down of all fossil fuels, not only coal. However, India should adapt cleaner technology sooner in coal use in a time-bound manner. Due to high air pollution in Delhi – National Capital Region (NCR), Patna, Allahabad, Gwalior, etc. – there is a 30% increase in respiratory problems (Asthma, COPD).

In 2016, global GHG emission was 49.3 billion tons (CO_2 equivalent) and 35.8 billion tons of GHG is due to CO_2 – out of which 32.1 billion tons is from fossil fuels. Various sources of global CO_2 emission (2016) were maximum from the power industry (41%), followed by other industrial combustion (industrial manufacturing and fuel production) (21%), followed by transport (18%), followed by building (10%), and 10% from non-combustion (industrial process, agriculture, and waste (Table 7.9).

As per Global Energy Monitor (Vishwamohan 2021) report (19 March 2021), globally 432 proposed coal mines would emit 1,135 million tonnes of annual CO_2 equivalent on a 20 year horizon, increasing by 30% over current emissions. The highest amounts of methane emission from such proposed coalmines are China (572 million tonnes, Australia (233 million tonnes), Russia (125 million tonnes), India (45 million tonnes), South Africa (34 million tonnes), the USA (28 million tonnes), and Canada (17 million tonnes), followed by Turkey, Poland, and Uzbekistan.

Coal is the dirtiest fuel/source of energy in the world as the carbon emissions from coal are almost double the emission from natural gas, and much more than those from petroleum. As per a study by the Centre for Science and Environment, New Delhi (S. Chakraborty 2019), Indian coal-fired thermal power plants are the most inefficient and polluting in the world – 75% of these thermal plants don't comply with governmental regulations. In fact, renewable sources of energy (solar, wind, etc.) need to be promoted as their costs of production have decreased due to technological advancement but, unfortunately, India's annual coal demand increased by 9.1%, during April 2018–March 2019, to about one billion tonnes; further coal is one of top five imports of India – total imports rose from 16.69 crore tonnes in 2013–2014 to 23.52 crore tonnes in 2018–2019. We are targeting the generation of 175 GW of solar and wind energy (by 2022) – India is the fifth largest producer of solar energy in the world and the sixth largest producer of renewable energy but China at present ranks first and Brazil ranks third in production of renewable energy in the world. Further, China increased its installed capacity in solar energy by 105.5 GW during 2014–2017, while

Table 7.9 Sources of Global CO_2 Emission in 2016

S. No	Sources	% of Total CO_2 Emission
1	Power industry	41%
2	Other industrial combustion	21%
3	Transport	18%
4	Building	10%
5	Non-combustion	10%

Source: EDGAR v4.3.2 database, EC-JRC/PBL, 2017, cited in State of India's Environment 2018 (CSE, New Delhi).

India increased it by only 14.3 GW. The USA and Japan installed twice the amount of solar capacity in the same period compared to India (S. Chakravorty 2019).

Global Energy Consumption Pattern

For correct appreciation of prevailing environmental issues, especially climate change, the most significant point is to know the pattern of energy use and consumption locally, nationally, regionally, and globally. This may be seen in Table 7.10.

Thus, from Table 7.10, it transpires that China (3,164 Mtoe) is the highest energy consumer (since 2009) due to continuous high economic growth, rising industrial demand, and increasing transport fuel consumption due to more vehicle fleets. On the other hand, the energy consumption in the USA (second highest with 2,258 Mtoe) was partially due to weather conditions (hot summer, cold winter). But both China and the USA had energy consumption growth rates in 2018 over 2017 at 3.7% and 3.5%, respectively. At global level, energy consumption increased by 2.3% in 2018 over 2017. The third highest energy-consuming country is India (929 Mtoe), followed by Russia (800 Mtoe), Japan (424 Mtoe), and South Korea (307 Mtoe). On the contrary, energy consumption decreased in the European Union (–1%), especially in Germany (–3.5%) partly due to decreasing consumption in the power sector, a milder winter, reducing consumption, as well as improvement in energy efficiency, as per Global Energy Statistical Yearbook 2019. New Zealand is the lowest energy-consuming country (only 21 Mtoe), followed by Portugal (22 Mtoe) and Norway (29 Mtoe). Unfortunately 850 million people (85 crores) still lack electricity access in the world. Further, in 2018, there was 2% growth of carbon emissions from energy use, the fastest for seven years. But what is the disaggregate data by the type of fuel consumed?

Hence, we may analyse the energy consumption by fuel at global and national levels. This is vividly shown in Table 7.11 for 2018.

From the above it is crystal clear that at global level, crude oil shares the maximum (33.62%) in total energy consumption (it declined from 46.1% in 1973 when there

Table 7.10 Highest and Lowest Energy Consuming Countries in the World (2018)

12 Highest Energy-Consuming Countries			12 Lowest energy consuming countries	
Country		Energy Unit (Mtoe)	Country	energy unit (Mtoe)
1.	China	3,164	New Zealand	21
2.	USA	2,258	Portugal	22
3.	India	929	Norway	29
4.	Russia	800	Romania	32
5.	Japan	424	Kuwait	36
6.	South Korea	307	Colombia	39
7.	Germany	301	Chile	40
8.	Canada	301	Uzbekistan	40
9.	Brazil	290	Czech Republic	43
10.	Iran	265	Venezuela	47
11.	Indonesia	251	Sweden	50
12.	France	243	Belgium	53

Source: Global Energy Statistical Yearbook 2019.

Note: Mtoe, million tonnes oil equivalent.

286   Climate Change and Society

Table 7.11 Percentage Shares of Types of Fuel in Total Energy Consumption in India and the World (2018)

Sl. No.	Fuel	World	India
1.	Crude oil	33.62%	29.55%
2.	Coal	27.13%	55.82%
3.	Natural gas	23.8%	6.17%
4.	Hydro electricity	6.8%	3.9%
5.	Nuclear power	4.4%	1.09%
6.	Renewable energy (solar, wind, geothermal)	4.0%	3.4%
	Total energy	100	100

was oil crisis at its worst at global level), followed by coal (27.13%), and further followed by natural gas (23.8%), thus total fossil energy (crude oil + coal + natural gas) share at the world level comes to 84.55%. On the other hand, in India, coal consumption is at the top of total energy consumption (sharing 55.82%), followed by crude oil (29.55%), and further followed by natural gas (6.17%), thus total fossil energy consumption in India is 91.54% (55.82+29.55+6.17)! Obviously, more coal consumption in India is due to the prevalence of thermal power plants (in total electricity generation it shares two-thirds). Though the share of natural gas is quite substantial at the world level – slightly less than one-fourth (23.8%) due to the huge stock of natural gas in Russia, its share is only 6.17% in India's total energy consumption, followed by hydroelectricity (only 3.91%), renewable energy (only 3.4%), and nuclear power the least (just 1.09%). At world level, the share of renewable energy is only 4% – slightly more than India (3.4%). Obviously, more use of fossil fuel for energy consumption means more carbon emissions, hence it causes environmental pollution more (especially in the air). Further, the quality of coal in India is inferior, causing more emissions.

If we look at the average per capita electrical energy consumption at the global level, the situation is in favour of developed countries, as may be seen in Table 7.12.

From the above, it clearly transpires that, on the one hand, the average per capita annual electrical energy consumption is the highest at 50,613 kWh annually in Iceland, followed by Liechtenstein (35,848), followed by Norway (24,006), followed by Kuwait (19,062), followed by Bahrain (18,130), followed by UAE (16,195), followed by Qatar (15,055), and followed by Canada (14,930), while the same in the poorest countries is quite negligible – 16 kWh in Chad, 17 in Guinea-Bissau, 27 in Somalia, 33 in Sierra Leone, and 36 in Burundi and Central African Republic! Even India (1,181 kWh) has far less than half the world average (2,674 kWh annually) and almost one-fourth of that in China. This clearly shows the high consumption rate in the developed countries of the West as well as some affluent gulf countries.

Undoubtedly various types of climate change have serious shocks (unpredictable weather events damaging sustainability) to individual households, communities, and the nation, and they respond to it in different ways. There may be no recovery or partial recovery or substantial recovery, and only rarely full recovery leading to restoration. This may be seen in Table 7.13.

The ability to recover from multiple shocks of climate change depends on various factors including the economic position of the household. The upper class, middle

Table 7.12 Average Per Capita Electrical Energy Consumption in Selected Countries

Sl. No (1)	Country (2)	Data Year (3)	Average Per Capita Electrical Energy Consumption (kWh per Head Annually) (4)
1.	Iceland	2014	50,613
2.	Liechtenstein	2012	35,848
3.	Norway	2014	24,006
4.	Kuwait	2014	19,062
5.	Bahrain	2014	18,130
6.	UAE	2014	16,195
7.	Qatar	2014	15,055
8.	Canada	2014	14,930
9.	Finland	2014	14,732
10.	Sweden	2014	12,853
11.	USA	2015	12,071
12.	Taiwan	2015	10,632
13.	Australia	2014	9,742
14.	South Korea	2014	9,720
15.	New Zealand	2014	8,939
16.	Singapore	2014	8,160
17.	Russia	2014	7,481
18.	Japan	2014	7,371
19.	Israel	2014	7,319
20.	Switzerland	2014	7,091
21.	China	2017	4,475
22.	India	2018	1,181
23.	Eritrea	2014	051
23.	Comoros	2014	051
24.	Haiti	2014	038
24.	Rwanda	2014	038
25.	Burundi	2014	036
25.	Central African Republic	2014	036
26.	Sierra Leone	2014	033
27.	Somalia	2014	027
28.	Guinea-Bissau	2014	017
29.	Chad	2014	016
	World average	2014	2,674

class, and lower class have substantial, partial, or no recovery, respectively, because the rich usually have a good cushion (savings) both in cash and kind, while the middle class has some cushion (savings), and the lower class has no cushion (savings) at all.

Second, the ability to recover from climate change also depends on the type or category of the community – its category in the eyes of the government of the day, its extent of forming as a political pressure group, its proximity to the State government or district administration, and finally, its level of collective consciousness. For instance, some governments favour while other governments disfavour *dalits* and/or minorities.

Third, the ability of recovery is very fast and comprehensive if there is democracy in a country, and it is further cemented if the party in power is socialist/social democratic as it often has a pro-poor and inclusive policy for all, hence the lower class benefits maximum then.

288 Climate Change and Society

Table 7.13 Multiple Shocks from Climate Change and Recoverability

Shocks from Climate Change	Recoverability
(i) Loss of life, injuries, or illnesses (not able to work)	(a) No recovery (b) Partial recovery (c) Substantial recovery (d) Full recovery/restoration
(ii) Loss of livelihood, jobs (without loss of life, injuries, or illnesses)	(a) No recovery (b) Partial recovery (c) Substantial recovery (d) Full recovery/restoration
(iii) Loss of environment/community (due to disruption/dislocation)	(a) No recovery (b) Partial recovery (c) Substantial recovery (d) Full recovery/restoration
(iv) Increase in prices of items of basic needs (food, clothes, medicines, shelter, education of children)	(a) No recovery (b) Partial recovery (c) Substantial recovery (d) Full recovery/restoration
(v) Decline in demand/prices of sale of items by the affected households	(a) No recovery (b) Partial recovery (c) Substantial recovery (d) Full recovery/restoration
(vi) Shortage of inputs – seeds, fertilisers, or increase in their prices (e.g. crops to be sown)	(a) No recovery (b) Partial recovery (c) Substantial recovery (d) Full recovery/restoration

Fourth, the ability of recovery also depends on the 'moral economy', i.e. the cooperation and assistance provided by the neighbours, kith, and kin or even local community-based organisations or other genuine civil society organisations (CSOs).

Finally, in a democracy, the media is the 'fourth estate', hence it publicises and pursues the cause of the affected people. However, as the daily life experience in India shows, media-in-itself is not a sufficient condition for full/substantial recovery because, often, the mainstream media follows suit what the Public Relations Department of the government or 'paid news' by the corporate, political party, or syndicate or what the ruling party states in press conferences! Only the alternative media objectively pursues early and comprehensive recovery of the affected poor people/communities, especially in remote areas.

In practice, as experience and empirical data suggest, full recovery or restoration is rare and exceptional in most countries of the world. On the other hand, it is also true that the politician-bureaucrat-contractor nexus operates in some parts of India wherein, in the name of drought or floods, reliefs or relief works are misused as a 'third crop' in a fraudulent manner, and the rainwater is not duly conserved for irrigation, domestic, and industrial purposes as P. Sainath (1996) found in many Indian states (Box 7.5).

Box 7.5: Everyone Loves a Good Drought!

Famous journalist P. Sainath empirically observed various shocks of climate change, especially drought and floods, in different parts of India. In some states/districts drought, despite being a natural disaster, was welcomed for getting government relief in different ways in different regions. To put it in his own words:

> Drought relief ... is rural India's biggest growth industry ... Relief can go to regions that get lots of rainfall. Even where it goes to scarcity areas, those, most in need, seldom benefit from it ... some of them call it *teesari fasal* (third crop) ... A great deal of drought 'relief' goes into contracts handed over to private parties. These are to lay roads, dig wells, send out water tankers, build bridges, repair tanks – the works.

Some experts agree that, barring erratic time and spread, most of the Indian districts get about 800 mm of rainfall annually, even the Kalahandi got 978 mm rainfall as the lowest during 1975–1995, otherwise 1250 mm. Palamu (Medininagar now in Jharkhand) district, then got 1200 mm of rainfall in a normal year; it got only 630 mm of rainfall in the worst year. Sarguja's annual rainfall is also 1200 mm while California, with only 300 mm annual rainfall, grows grapes. On the other hand, Ramanathapuram district in Tamil Nadu, too, gets an annual rainfall of 600 mm, but there the situation has never been so grave as in Sarguja, Kalahandi, and Palamu! P. Sainath makes another important point that drought and Drought Prone Area Programme (DPAP) were linked to 'fund flows in a big way' – most people, including political representatives, wanted their blocks included in DPAP! For instance, about 73% of sugar cane (water-intensive) produced in Maharashtra was grown in DPAP blocks! Annual rainfall in Lonavala, near Pune, has been 1650 mm annually, but it was included in DPAP! However, sometimes, it is possible that there is a 'long spell of drought' and a 'short spell of heavy rainfall' in a district or region. Even then, fraud in the name of 'drought relief' is a reality in some parts of India.

Source: P. Sainath (1996).

UNFCCC, Paris Agreement, and NDCs

UNFCCC (UN Framework Convention on Climate Change) that supports the global response to climate change, turned 25 in 2019; it mobilises both governments as well as non-state actors to act upon climate change; it has 197 Parties (members); the processes for submitting Nationally Determined Contributions (NDCs) every five years and for review, these have been duly defined. It collaborates with various UN bodies to pool resources and convening power to effectively implement the convention, Kyoto Protocol (1997), and other agendas like the 2030 Agenda for Sustainable Development. It also strengthens mitigation and adaptation action and mobilises support. Further, various conferences (Conference of Parties, COP) are organised with the help of scientists and other global climate stakeholders.

Second, UN Climate Change also supports all parties in their efforts to monitor and assess emission levels, reduce emissions, and design ambitious targets for keeping the temperature rise below 1.5°C from the pre-industrial level.

Third, it interacts with governments and partners in finding ways to increase developing countries' capacity to adapt; it supports these countries in preparing their national adaptation plans, especially through Open National Adaptation Plan Initiative; it also supports LDCs (Least Developed Countries) Expert Group and Adaptation Committee including financing for adaptation action.

Fourth, since inaction is costlier than timely action, it supports negotiations on a wide range of climate finance issues and assists developing countries in assessing the priority of their adaptation, mitigation, capacity-building, and financial needs.

Fifth, it also ensures access to innovative technologies, for both transfer of technology and finance; hence it prepares a work plan by identifying policy options, practices, and technologies with high mitigation potential through a climate technology database.

Sixth, capacity-building efforts are also its focus, through better coordination, systemic planning, and continuous evaluation. At COP25, the second capacity-building hub of the Paris Committee on capacity-building was held.

Seventh, Clean Development Mechanism (CDM) enables governments, organisations, businesses, and individuals to buy carbon credits from projects that reduce CO_2 emissions, thus lowering carbon footprint. CDM issued 5,09,95,101 certified emission reduction credits to 187 projects and 36 programmes in 55 countries in 2019. Thus, carbon markets enable the financing of clean development as well as a reduction in emissions. UN Climate Change also organised various capacity-building events and supported various countries with carbon pricing.

UN Climate Change's COP25 organised 25,000 meetings, 260 side events, 160 two-day exhibits, and 28,000 participants participated there (with 32 Heads of States/governments) and 190 national statements were made, and there 138 official documents were placed.

Needless to say that as per India's Nationally Determined Contributions (NDCs), by 2030, there will be a reduction in the emission intensity of the GDP by over a third (35%) and a total of 40% of installed capacity for electricity will be generated from non-fossil fuel sources. India has also promised an additional carbon sink (a means to absorb carbon dioxide from the atmosphere) of 2.5–3 billion tons of carbon dioxide equivalent through additional forest and tree cover by 2030 (trees fix carbon as part of photosynthesis, and soil, too, holds organic carbon from plants and animals).

The UN Conference on Climate Change was held in Paris (France) from 30 November to 12 December 2015; there 197 parties had consensus on the reduction of climate change. They agreed to reduce their carbon emission 'as soon as possible' to keep global warming below 2°C pre-industrial level, and efforts are being made to bring it down to 1.5°C. Thus, no detailed timeline and nation-specific goals were fixed, though in Kyoto Protocol it was fixed. On 5 October 2016, 133 parties ratified the convention and Paris Agreement came into force on 4 November 2016 to combat climate change and to adapt as well as to give support to help developing countries to do so. Therein the notion of voluntary Nationally Determined Contribution (NDC) from every country emerged – this was a 'bottom-up approach' accepted at an international forum of the UN. Various developed and developing countries prepared their NDCs. However, many developing countries consider a reduction in their emission norms, in absence of better and cheaper technology for renewable energy as well as a lack of fund transfer from developed countries to them, as a barrier to their prosperity. On the other hand, highly industrialised countries like the USA, Russian Federation, and Canada do not agree to Kyoto Protocol and Paris Agreement. Though, the USA had initially agreed to the Kyoto Protocol (1997) for 'common but differentiated responsibility', but in 2001 it withdrew from it and on 1 June 2017, the then US President Donald Trump announced to withdraw from Paris Agreement – a regressive step in connivance with energy and industrial lobbies there. However, the next US President Joe Biden joined it later. The core elements of Paris Agreement (2015) are the following:

(i) A collective commitment to limit average global temperature increase well below 2°C above pre-industrial level but to pursue below 1.5°C – no detailed timeline and nation-specific targets were fixed.
(ii) Every member nation to prepare and pursue NDCs voluntarily as early as possible – a 'bottom-up' approach.
(iii) The goal is to balance the amount of carbon that is added by the global economy to and taken from the atmosphere within the twenty-first century.
(iv) Developed countries collectively promised to raise a fund of $100 billion annually by 2020 to assist developing countries to reduce their emissions and adapt to climate change.
(v) Every nation must update and strengthen its promised contribution to climate change mitigation every five years, beginning in 2018.

The Government of India prepared Nationally Determined Contribution (NDC) on Climate Change on 2 October 2015. This is an official document as to what extent and how India intends to address the challenges of climate change, especially regarding urbanisation, transport, agriculture, health, water, and sea coasts. Needless to say that in 2007 at the Bali Convention on Climate Change, it was, by and large, agreed by most of the nations that there should be a 'paradigm shift' in curbing carbon emission from 'top-down' international decision-making to 'bottom-up' agreement (later also agreed in Paris in December, 2015). That is, instead of a global decision to reduce carbon emission everywhere in one stroke, every nation has the freedom to decide its roadmap to mitigate and adapt to climate change. It is believed by many scientists and eco-democrats that the 'bottom-up' approach of mitigation and adaptation will have larger co-benefits like lower air pollution, prudent use of energy, less potential of extreme weather events, etc. India has invoked Mahatma Gandhi who had once rightly remarked: 'Earth has everything to fulfill everybody's need but not anybody's greed'. Need versus greed paradigm is very relevant today, as it has a moral voice in addition to socio-economic and ecological upper hand. But some scholars (like N. K. Dubash, Radhika Khosla) have opined that, in reality, India's 'nature-friendly lifestyle' proposition did not hold good – almost 60% people were defecating in the open a few years back (but on 2 October 2019 India became open defecation free (ODF)). Delhi is the world's most polluted city (air pollution – mix of ozone, sulphur dioxide, nitrogen dioxide, carbon monoxide, and fine particulates – six times more than the permissible PM 2.5 limit), 15 out of 20 most polluted cities in the world with worst PM 2.5 counts are in India (including Gwalior, Delhi, Allahabad, Kanpur, Gaya,

Muzaffarpur, Jaipur, Raipur, and Patna), Mumbai's 60% population lives in unhygienic slums, many rural Indians use fuelwood for cooking, 75% of India's energy supply is provided by non-renewable sources, about 30 crores of the population live in poverty (implying unequal distribution of natural resources) and so on. India pledged at Paris Agreement (2015) to achieve by 2030 three things: to reduce carbon emissions intensity by 33–35% from 2005 baseline; second, it also promised to share non-fossil fuel-based electricity to 40% of total capacity through transfer of technology and low-cost finance; third, creation of an additional carbon sink of 2.5–3 billion tons of CO_2 equivalent (including methane, GHGs, nitrous oxide) through forest cover. However, India's NDC does not commit to any sector-specific mitigation obligation and its actual implementation of such pledges, in the ultimate analysis, is guided by UN's Paris Agreement (2015).

However, it is a well-known fact that per capita annual emission in India (2.5 tons) is less than the global annual average of 4.8 tons and far lower than that in developed countries like the USA (about 20 tons per capita) or even China (6 tons per head) in 2018. In fact, per capita emission in India equals only up to 8% of US per capita emission, that is, the intensity of energy consumption due to mechanisation of household and industrial–agricultural tasks in developed countries is much more than that in developing countries like India. On the other hand, due to the huge population of 137 crores, India's absolute carbon emission is quite high – 2 billion tons (5.2% of global total emission) – and its energy consumption is 5.9% of the global total. Hence, activist ecologists like Nagraj Adve and Ashish Kothari criticise that India's NDC wrongly justifies the projected rise in its emission by emphasising its development imperative because this obscures the fact that well-off will stamp their ecological footprint. India's NDC does not state about inequality between the rich and the poor within India itself; e.g. in India 1,75,000 households have assets of one million dollars or more whose per capita emission matches with the rich in the USA and Europe. Thus, the ecological footprint of the richest 1% of Indians is over 17 times that of the poorest 40%. So Nagraj Adve and Ashish Kothari rightly remark,

> Readventing risk and improving the capacity of people to adapt to climate change is linked to effective poverty eradication, improving food security through sustainable farming, promoting greater biodiversity, improving public health, and strengthening community resilience. These linkages have simply not been made explicit.

In addition, it is also observed by the critics that India's NDC uses the term 'non-fossil fuels' in place of 'renewable', targeting 63 GW (10 GW at present) by 2032, and it calls nuclear power as 'safe, environmentally benign and economically viable source' but incidents of Chernobyl in 1986 (in Ukraine, in the former USSR) and Fukushima Daiichi power plants (in 2011 in Japan) have proved otherwise. Further, in India, reactor construction has a history of cost over-runs and importing of foreign reactors would be very costly. Moreover, a huge displacement of local people for various electricity generation projects has been over-looked in India's NDC. According to them, India's NDC wrongly calls coal energy as 'clean energy' because coal emissions are 50% higher than those for oil, and 80% higher than those for natural gas. In fact, India is the world's third largest coal producer and has the fifth largest coal reserves in the world; even then in 2011, India's coal import reached 11% of the total coal demand. This fact has huge negative implications for climate change.

As far as the proposal for afforestation in new areas is concerned, this, too, is not very realistic because, on one hand, we have not been able to substantially check deforestation and, on the other hand, the processes of urbanisation, industrialisation, power generation, irrigation, etc. are taking away new agricultural or forest lands. According to a study by the Centre for Science and Environment (New Delhi), six lakh hectares of forest lands were diverted during 1992–2012 for the so-called development projects in India. However, as per Climate Change Performance Index (CCPI) 2020, India ranks ninth among 57 countries plus the European Union, and has reduced its emission intensity by 21% against the NDC target of 35% by 2030– by reducing fossil fuel, better waste management, and more renewable energy. It has hugely promoted the International Solar Alliance globally as well.

Needless to say that while the commitment of the developed countries to the Kyoto Protocol (1997) was binding on them, the Paris Agreement (2015) was not binding on the developed and developing countries. On Earth Day, 22 April 2016, 175 world leaders signed the Paris Agreement at UNHQs in New York; later 186 countries (including India) ratified it. Later, in 2019 (on 23 September), the UN Secretary-General Antonio Guterres convened a Climate Summit by bringing government, private sector, and civil society together to enhance and accelerate climate action on six core sectors – heavy industry, nature-based solutions, cities, energy, resilience, and climate finance: 'it is a race we must win'. In mid-2021, the European Union (with 27 countries after Brexit), proposed a carbon border tax or adjustment on imports in order to force emerging economies to adopt cleaner (non-fossil fuel-based) energy to manufacture goods. But BASIC nations (Brazil, South Africa, India, and China) opposed it on 7–8 April 2021 at a Ministerial Conference at New Delhi itself as being discriminatory and against the principles of equity and CBDR-RC (Common but differentiated responsibility and respective capabilities as per Earth Summit (1992) and Paris Agreement (2015); it amounts to a trade barrier.

Theorising Climate Change

Since climate change is a global collective issue, it needs a global collective response but all the nations, too, have to come up with their individual timely and time-tested response. In the original agreement of UNFCCC in 1992, at Rio, it was clearly mentioned that nations would 'protect the climate system for the benefit of present and future generations of mankind, on the basis of equity and in accordance with their common but differentiated responsibilities and respective capacities'. This was the notion of 'Climate justice'.

Thus, *climate justice perspective* has six assumptions therein (Dryzek et al. 2013):

(a) The climate system benefits and supports human life (thus the role of the natural world in the notion of justice was accepted).
(b) There is a responsibility of present governments to benefit both present and future generations of humans.
(c) Such responsibility extends to people beyond national borders.
(d) Equity is a basic ethic of international climate agreements.

294 Climate Change and Society

(e) The responsibility to address climate change would be, in part, different for different countries (wealthy nations caused most of the emissions due to industrialisation) despite the core ideal of equity.

(f) The actions required of nations would be based, on the one hand, not only on these different responsibilities but also on their different capacities to act, on the other hand. That is, the rich nations have more capacities (technological, economic, etc.).

Second, the *development rights approach's* conception of climate justice is based on 'the right of all people to reach a dignified level of sustainable human development free of the privations of poverty' (Eco-equity, 2008, cited in Dryzen et al. 2013, 81). Earlier Shue (1993) distinguished between 'luxury' and 'subsistence' greenhouse gas emissions. Hence, development rights approach accepts subsistence emissions for basic needs without penalty; on the other hand, it charges for and limits luxury emissions.

Third, *capabilities-based approach* emphasises political liberties, freedom of association, economic facilities, social opportunities, transparency guarantees, protective security, economic and social rights, health, bodily integrity, education, social affiliation and respect, and countries' right over their environment (Sen 1999, Nussbaum 2006). The UN used it for designing Human Development Index and Millennium Development Goals – focusing on subsistence, health, education, gender equality, and sustainable environment. Thus, climate justice requires a positive response to threats to basic needs and rights.

Fourth, Simon Caney (2005), on the other hand, talks of *right to environment*, not to suffer from drought, floods, crop failure, heat stroke, infections, diseases, damage to houses and infrastructures, and enforced relocation.

Fifth, Steve Vanderheiden (2008) synthesises approaches of Caney and Eco-equity – here climate justice is based on both environmental (stable climate), and development rights. To Holland (2008), all capabilities depend on a sustainable environment, hence the capabilities approach must attend to ecological systems. Therefore, he argues for a 'meta-capability' of 'Sustainable Ecological Capacity', at which level ecological systems could sustain themselves as well as other basic capabilities for human beings in the system at an 'environmental justice threshold'.

Further, *Green Growth* discourse, a variant of synthesis of development rights and stable environment, considers climate change as an opportunity, not a cost or crisis, hence argues for the investment in the environment for 'recoupling' environmental protection with growth – hence after 2008, it has been accepted in World Bank and Organisation for Economic Cooperation and Development. Needless to say that, by 2012, World Summit (Rio + 20) duly linked Green Growth with sustainable development, and the UN supported it. However, the pro-environment CSOs vehemently opposed that branding 'Green Economy, The New Enemy'. Hence, the final report of World Summit 2012 *The Future We Want* cautiously mentioned Green Growth as one of a set of means to achieve sustainable development. Consequently, Vazquez-Brust, Sarkis, and Smith (2014, 6) argued that 'in the west, Green Growth discourses have become colonised by underlying ideological structures, which utilise selected elements to further more longstanding agendas, such as the continued liberalisation of markets'. Hence, their *critical* Green Growth perspective, drawing from the South Korean (East Asian) Green progress, moves away from quantity to quality, from the consumption of physical to non-physical outputs, from technological to wider socially embedded innovation (organisational innovation, social networks, and R&D intensive specialisation). Hence, they stand for un-harmful economic activity that can support natural capital, with quality-orient action, low carbon, energy-efficient economic growth, 'new clean technology', etc. Here, *critical* Green Growth goes beyond the 'business as usual' perspective as it assigns a stronger role for the State than in European or North American Economy perspectives. Therefore, they argue for the need of 'institutionalisation of deliberative democracy processes' – and ecological values transcending national interest – in Development capitalism models. Hence, they distinguish their model from the *Green Leviathan* model by outlining four core features of their model:

(a) Flexible and diverse policy mix
(b) Value-driven, multi-stakeholder, multi-level governance
(c) Public trust and collaboration
(d) Appropriate measurements of progress

One more dimension of climate justice is that not only historical wrongs done by the industrialised countries but also their current higher degree of energy consumption, too, justifies that they should substantially take the responsibility for the prevailing climate change. But this view has been criticised by some western scholars (Singer 2004) on the ground that the present emission should be the guiding ethic for all countries because China, India, Brazil, and some other developing countries are emitting huge amounts currently.

Sixth, the *per capita equity approach* (Singer 2004) also adds that there should be a trade system in which countries, needing more than the permissible limit, may buy from other countries emitting less. But this approach ignores the fact that different people, in different places, may require more or less than others to live a decent life, and on the other hand, in that way the cumulative emission globally will ultimately not be reduced. Hence, development rights' concept of climate justice emphasises on '*the right of all people to reach a dignified level of sustainable human development free of the privations of poverty*' (Eco-equity 2008, cited in Dryzek et al. 2013, 81).

Seventh, development rights are to be supplemented by *climate duties*. Bell (2011) talks of four climate duties:

(a) Not to infringe other's human rights
(b) Not to emit greenhouse gases in excess of one's permissible quota
(c) Those exceeding their permissible quota must compensate others for their loss
(d) To promote fair and effective institutions for the protection of human rights, including development

However, we don't agree with (c) above because it is egalitarian and it will not curtail the culture of consumerism.

Finally, climate justice activists also emphasise on the *public and community participation* – to have a say in all the decisions that affect local people's lives. For instance, The Bali Principles of Climate Justice demanded that climate justice affirmed 'the right of indigenous peoples and local communities to participate effectively at every level of decision-making, including needs' assessment, planning, implementation, enforcement and evaluation, the strict enforcement of principles of prior informed consent, and the right to say 'No' (Dryzek et al. 2013). Thus, for a just and fair adaptation, the top-down, expert-driven theory and practice needs to be duly replaced by the bottom-up, eco-democratic participation in local contexts. Undoubtedly, the problem of climate change cannot be explained through an economic exchange or the so-called free and perfectly competitive market because future generations do not exist today. Hence, it has to be conceived as an ethical, political, and spiritual problem that must consider the present generations' collective responsibility and sense of justice towards future generations.

Policy Implications

The following suggestions for policy-making as per India's needs may be adapted:

(1) There should be a detailed and comprehensive sustainability plan for climate-resilient agriculture. Contract farming is not pro-farmers and results in the abuse of soil and desertification. The water-intensive crops like rice and sugarcane should not be grown in those areas where the water table has gone down (Punjab, Haryana, Western, UP, etc.) and where water is already scarce conventionally (Gujarat, Maharashtra, Chhattisgarh, MP, Rajasthan, etc.). In such areas, no subsidy for inputs is provided for the growth of paddy and sugarcane. The green revolution belt has resulted in desertification, salinisation of soil, and resistance to pesticides. By changing the crop cycle and using bio-organic manures, bio-pesticides, etc., the situation may be turned into an eco-friendly 'evergreen revolution' (M. S. Swaminathan's terms). Further drip and sprinkler methods of irrigation should replace the flooding of field methods soon. There should be 'procurement at the nearest place of distribution' – thus cutting transportation, storage, and manpower costs. In addition, the minimum support price should be made inclusive of farmers' total time spent in various field activities including supervision as well as the variability of weather (leading to crop losses); hence, it should be double the costs of production at least. The tendency to grow food crops for feeding cattle should be curtailed and 'luxury' meat production and consumption should be discouraged; vegetarianism may be encouraged for the sustainable use of natural resources. Doubling farmers' income by 2022 in India was a good policy decision but adequate means to achieve that was not taken in time. Modern agriculture is facing many challenges like resource scarcity, decrease in availability of land, water scarcity, toxic pesticides, monoculture, and soil degradation, hence there is a need for 'paradigm shift' (sustainability transition) to agro-ecology. According to UN FAO, 70% of food globally comes from small farmers but they are most vulnerable and even hungry. They also face a danger from the grand process of urbanisation that will swallow small farms (and farmers) by turning them into urban areas – 75% by 2050 AD. As Mindi Schneider, one of 70 international food scientists who appreciated UN FAO for taking an initiative on agro-ecology rightly remarked in 2014: 'Agro-ecology is more than just a science, it's also a social movement for justice that recognises and respects the right of communities of farmers to decide what they grow and how they grow it' (ecologyandfarming .com, 2 October 2014). In fact, agro-ecology's core components are as follows:

(a) Small farmer's participation in decision-making about agriculture and allied activities
(b) Use of critical traditional knowledge systems along with public interest scientific knowledge
(c) Enhancing plant genetic diversity
(d) Ensuring soil fertility and conservation
(e) Sustainable productivity through crop cycle, livestock system based on legumes, etc.
(f) Distributive justice and social inclusion (by enhancing livelihoods)
(g) Water conservation and its prudent use
(h) Eco-friendly (organic) agricultural practices, avoiding dependence on mechanisation and chemical fertilisers/pesticide, etc.
(i) Inter-connectedness of all agro-ecosystems' components and 'the complex dynamics of ecological processes' – taking a holistic view of human, ecological, and agricultural aspects
(j) Promoting agro-forestry and social forestry

Scholars like Wezel et al. (2011) consider agro-ecology as a (a) science, (b) movement, and (c) practice, at the same time; but sometimes there is some tension between these three or even within one (science vs research)! Similarly, Social, Technological and Environmental Pathways to Sustainability (STEPS) Centre (UK) speaks of three core dimensions of innovation in agro-ecology: technical, social, and political directions for change, distribution (social equity), and diversity (bio-diversity and cultural diversity). In its new manifesto, it poses fundamental questions about three Ds: direction, distribution, and diversity (Box 7.6).

Box 7.6: Questions Regarding Innovation in Agro-ecology

In a manifesto on innovation, sustainability, and development, STEPS Centre calls for a radical change in the whole innovation process (agenda setting, monitoring, evaluation, funding). Hence, for that, three sets of questions related to direction, distribution, and diversity should be addressed (3Ds):

(a) Technical, social and political directions for change: What is innovation for? Which kinds of innovation, along which pathways? And towards what goals?
(b) Distribution: Who is innovation for? Whose innovation counts? Who gains and who loses?
(c) Diversity: What – and how many – kinds of innovation do we need to address any particular challenge?

Source: STEPS Centre (2010).

The direction of change to be promoted by innovative agro-ecology is towards a 'strong ecologisation' of agriculture (Duru et al. 2014), and here ecosystem services are provided by biodiversity. Further, in order to eradicate hunger and poverty, distributive justice and equity is also the core component of agro-ecology movement that is oriented towards food sovereignty, equitable resources' distribution, and rights-based perspective. In addition, it accommodates the diversity of farms and agro-ecological practices, the diversity of innovations (to be promoted), and the diversity of actors to coordinate for a holistic and trans-disciplinary perspective through a synergy between direction, distribution, and diversity. Thus, innovative agro-ecology is a system of transforming food systems to achieve the goal of environmental, economic, and social sustainability, that is, questioning not only the 'green revolution' but also the whole agro-food regime. Hence, innovative agro-ecology attempts changes at five levels (Gliessman 2015) (Box 7.7).

Box 7.7: Levels of Agro-food System Changes through Agro-ecological Approach

Level 1: Increase the efficiency of industrial and conventional practices in order to reduce the use and consumption of costly, scarce, or environmentally damaging inputs.

Level 2: Substitute alternative practices for industrial/conventional practices.

Level 3: Redesign the agro-ecosystem so that it functions based on a new set of ecological processes.

Level 4: Re-establish a more direct connection between those who grow food and those who consume it.

Level 5: On the foundation created by the sustainable farm scale agro-systems achieved at Level 3, and the new relationships of sustainability at Level 4, build a new global food system, based on equity, participation, democracy, and justice, that is sustainable and also helps restore and protect earth's life support systems.

Source: Gliessman (2015).

Thus, 'open' innovative agro-ecology provides an alternative pathway to all, especially to the people of developing countries, by adequately addressing issues like degradation of natural resources, food insecurity, hunger and poverty, climate change, and associated socio-ethical issues (like questioning colonisation of seeds and other inputs). It also provides the 'counter-hegemonic process of internalisation and socialisation of agro-ecological knowledges' (Mc Cune et al. 2017). However, there has emerged a tension between agro-ecology as a science and agro-ecology as a movement and practices. Second tension exists between agro-ecology as a social movement and 'institutionality' (Giraldo and Rosset 2018). At this point, it may be argued that 'institutionalisation' may be taken as its success and scaling up, but 'it might also strip the agro-ecological movement of its freedom of manoeuvring and action as well as of its label as an "alternative movement"' (Bilali 2018). However, a critical and comprehensive dialogue between the local farmers, 'public interest' scientists, and the movement activists would overcome such tensions in due course of time.

(2) There should be comprehensive and advanced eco-sensitive urban planning for infrastructures with adequate and functional drains, water supply, sump houses

for taking out excess water from waterlogged areas, sewerage treatment plants, garbage segregation, lifting and recycling (processing) of wastes to convert into energy, all public offices/places to use alternative and renewable energy (solar and wind), afforestation on all vacant lands by removing encroachments, renovation of all water bodies, etc. Unfortunately at the time of Independence, Delhi had about 800 ponds, but on 90% of these water bodies new concrete structures, roads, etc. have been built up, thus destroying the healthy ecosystem and sustainability of urban life. So is the case of other big cities like Mumbai, Chennai, Hyderabad, Bengaluru, Kolkata, Jaipur, Patna, Lucknow, Allahabad, Kanpur – thus water does not properly flow into the rivers, causing traffic jams. Further public transport with green fuel (CNG or electricity) should be strengthened and the use of individual vehicles should be discouraged by charging extra for its ownership as well as parking, at least in metros. Individual motor vehicles usually occupy 80–85% of parking areas in cities. Public transport (bus) contributes only 3–20% of CO_2 emissions, but two wheelers, cars, and jeeps account for 60–90% of CO_2 emissions (Ahmed 2018). Again Bharat Stage VI is the need of the hour to ensure clean fuel and less emission. Hence, all diesel vehicles of 15 years old have been rightly banned in Delhi and should be banned in all metro cities immediately. However, electric vehicles are really useful in drastically reducing pollution. Undoubtedly, the cycle is the zero-emission mode, hence it may be promoted by making cycle tracks (as in most of the European countries), and reducing GST drastically on cycles so that poor people, too, may afford it. During 2017–2019, in 722 towns/cities of 15 countries, 'climate emergency' was imposed – plus 7,000 universities and colleges had also done so. Australia was the first country to impose 'climate emergency' in Darebin City on 5 December 2016 – and in 26 towns/cities of Australia 'climate emergency' was imposed by 2019. Ireland was the second country to impose 'climate emergency'. The UK also imposed 'climate emergency' in the entire country in May 2019. In New York City (US) 'climate emergency' was imposed on 26 June 2019. France decided in 2019: (a) to make 1,000 km long cycling track in Paris by 2020, (b) to ban cars in specific places, (c) to ban tourist bus entry in the city, (d) to plant urban forests in four crowded places, (e) to close coal power stations within three years, and (f) to ensure carbon-neutral France by 2050, and it also planned to impose climate emergency (Bhaskar, 11 July 2019). More than half the travel trips in most of the Indian cities are up to 5 km, hence walking (in Mumbai) and cycling (in Delhi) are preferred by the local people but proper and safe walking/cycling tracks are to be ensured. In many Indian cities, footpaths have been encroached on by adjacent shopkeepers or petty 'pheriwalas', 'thelawalas', etc., and hence these should be vacated.

(3) Since aeroplanes, refrigerators, air conditioners, and other machines as well as electronic gadgets emit huge CO_2, it is necessary to reduce their use drastically. Interestingly Greta Thunberg, a girl of 17 years from Sweden, started a massive campaign against climate change in 2019 and her mother Melena Ernman also started a campaign to make people feel ashamed of flying by aeroplanes as civil aviation is responsible for contributing 2–3% of total carbon dioxide in the world – rather one should travel by railways. In 2019, more than 14,500 persons in Sweden decided not to travel by air and this number reached 1,00,000 by 2020– hence there was a decline of 4.5% in domestic air travel in Sweden in

2019. Indians should follow it in spirit. However, the International Air Transport Association feels that carbon is the real enemy, hence electric planes through new technology will be a better mode without carbon emission (*Time*, cited in Bhaskar, 11 August 2019). It is also notable that though Mr Al-Gore, ex-Vice President of the USA, wrote much about climate change (global warming, etc.), yet he travelled most by air, hence he was highly criticised for a gap between his word and action. Similarly, the use of refrigerators and air conditioners, too, needs to be discouraged and their alternatives should be found in indigenous methods – though new refrigerators and air conditioners with less energy consumption (5 stars) are available in the market yet their emission problem has not been properly tackled so far.

(4) Though there is a National Action Plan on climate change and all the States and UTs have also prepared State Action Plans on climate change, yet either these have not been properly implemented before, during, and after extreme weather events by them or such plans are not really useful. Adaptation to climate change is very necessary. Adaptation process is of two types (Subbiah 2004) – (a) ex-ante (based on expectation, e.g. in anticipation of rainfall variability); and (b) ex-post (based on event realisation). Further, while ex-ante adaptation process aims at minimisation of the loss, with a long-term perspective, ex-post adaptation measures aim at sustenance in the immediate present. Second, while ex-ante process is backed by an adequate period of time, this is not the case in ex-post process which is simply a curative measure of a critical situation. Finally, ex-ante process is a preventive measure, hence to be preferred (for minimising loss) to ex-post process. The difference, on various counts, between the two may be seen in Table 7.14.

(5) As promised in NDCs, additional increase in carbon sinks may be achieved by restoring impaired and open forests, afforesting wastelands, agro-forestry, green corridors, plantations along railways, canals, roads, railway sidings, rivers, or urban green spaces. In this regard, it is notable that monocultures of plantations will not be useful as when these are harvested, carbon is released as wood is burnt. Hence, we should curtail deforestation at the maximum, and the area, allocated to the restoration of impaired and open forests and wastelands in the Forest Survey of India report, should be focused fully on natural forests and agro-forestry; social forestry (food forests) managed and protected by the local communities will have additional benefits undoubtedly (S. Byravan, 2019). As we have discussed earlier,

Table 7.14 People's Strategies to Cope with Climate Variations

Sl. No.	Ex-ante (Based on Expectation)	Ex-post (Based on Event Realisation)
1.	Diversify crops, livestock	Reduce or intensify inputs
2.	Occupational diversity	Change crops
3.	Invest or disinvest in irrigation, fertiliser, etc.	Depend on irrigation sources
4.	Accumulate assets	Buy or sell assets
5.	Purchase crop or weather insurance	Receive or provide transfers
6.	Make share-cropping contract	Seek non-agricultural employment
7.	Arrange to share with family, community	Migration
8.	Diversify income sources	

Source: A.R. Subbiah (2004).

tribals and other forest dwellers, who, usually conserve forests, suffer from various diseases and deficiencies, and highly depend on various kinds of forest produce, should have direct access to these forests in a meaningful way.

(6) As our healthy tradition says, '*sada jivan, uchcha vichar*' (simple living and high thinking) and '*aparigraha*' (non-collection of things), a concept given by Mahatma Gandhiji, is key for simple, minimalist, and sustainable lifestyle to be adapted in letter and spirit. Government of India rightly talks of Lifestyle for Environment (Life). In Denmark, the popular term 'hygge' (pronounced as 'hooga') means coziness, contentment, and well-being by enjoying simple things in life. Similarly, in Sweden, the popular term 'Lagom' (pronounced as 'lah-gum') means living a balanced, slower, and simpler life, with 'not too little', 'not too much' (Kim Edou, 2018) – this is in a way questioning of modern complex, highly ambitious, and consumerist lifestyle. Hindi poet, Rahim, prefers 'Santosh dhan' (satisfaction as wealth): 'Godhan Gajdhan, Bajidhan aur ratan dhan khan/ Jab awe Santosh dhan sab dhan dhoori saman' (the satisfaction as wealth is the ultimate wealth, other types of wealth – cows, elephants, horses, and even gems – are like dusts). Our neighbouring nation Bhutan prefers the concept of happiness and, instead of gross national product or gross domestic product, talks of 'gross national happiness' (GNH), hence it has rightly developed GNH Index, where material prosperity is not valued much, rather spiritual – social – religious happiness is valued the most. Now, this happiness index is appreciated globally and the World Happiness Report is published by the UN. As per World Happiness Report 2020, based on a survey of 156 countries, Finland, Denmark, Switzerland, Iceland, and Norway rank top five countries, while Afghanistan, South Sudan, Zimbabwe, and Rwanda rank at the bottom, and India ranks 144th (in 2019 its rank was 140th), Pakistan is at 66, China at 94, Nepal at 92, Maldives at 87, and Bangladesh at 107th position. Therefore, there is a need for a public policy to inspire and encourage people to only need-based consumption and the least wastage (of foods, clothes, medicines, materials, water, etc.) or what Mahatma Gandhi called, 'wantlessness' – reducing our wants and needs to the minimum essentials of life. This may be done by sustainably adopting 7 Rs: (a) *Reduce* (less use/consumption), (b) *Reuse* (use it again in some or other way), (c) *Recover* (recover from the used things to convert wastes/garbages into a resource), (d) *Recycle* (recycle after multiple uses), (e) *Remediate* (treating/cleaning soil or water due to spillage of oil, etc.), (f) *Replace* (substitute), and (g) *Restore* (to return to original state). Thus, the culture of consumerism ('more', 'bigger', 'faster') is to be eliminated by one and all in the long term. Only a few people require 'iodised salt' but now it is compulsorily supplied for all in branded packets, but at a huge (economic and health) cost! Similarly, the quality of water in some parts of India is bad due to iron, lead, fluoride, and other contaminations, but the so-called mineral water/packaged water (costing Rs 20 per litre) or purified RO (reverse osmosis) water is sold at most of places – thus a huge costly water market has been created by the vested business interests. Thus, the so-called free market has created consumerism and mass production; hence, we need to question not only consumerism and industrialism (as 'green-green' does) but should also question the capitalism (as 'red-green' does) which is the root cause of such problems to a large extent.

(7) There is an urgent need to restore and conserve rainwater in the whole India as we conserve only 17–20% of total rainwater, and the rest 80% goes to the ocean/sea through various rivers (and/or causes flooding/water-logging also), whereas a small country like Israel (with far less rain) conserves 80% of its rainwater! Further, even sewerage water is to be treated and reused for gardens, irrigating crops, etc.; this will solve the water scarcity problem in many cities like Bengaluru, Jaipur, Jodhpur, Ahmedabad, Mumbai, etc. Further, smart water metres should be installed by families to reduce water consumption, as many families have shown the way in Chennai and Bengaluru. As per NITI Aayog, in 21 cities in India, there may be a water crisis in near future – hence we should prepare actionable plans without any loss of time. We have to recognise the real value of water – hence its prudent use should be the norm of society. According to Rajendra Singh ('the waterman'), (a) 'only traditional techniques can guarantee clean water'; (b) 'linking crop pattern with rain is important to stop soil erosion'; (c) 'to stop soil erosion and siltation, we need to make the flow of the river slower, but not build a dam to obstruct its flow'; (d) 54% of India faces 'high' to 'extremely high' water stress (severe drought); (e) presently in India, about 72% underground water aquifer is overdraft; (f) NASA found that large-scale irrigation caused 108 cubic km of groundwater loss in Haryana, Punjab, Rajasthan, and Delhi during 2003–2008 (*The Times of India*, 14 November 2019).

(8) We have to be more cautious of the dangers of plastic and plastic products. The Government of India has banned the single use of plastic since 1 July 2022 as it has not only choked various drains and nullahs but the cattle eat it and die prematurely due to its dangerous chemical properties. The plastic items thrown in rivers or ponds are eaten by the fish and those people, who eat fish, finally also consume the part of plastic that is highly harmful. In India, about 60% of plastic products are recycled because the poor ragpickers 'recover' them from the garbage – in the USA only 10% of plastic products are recycled. Some plastic products may be recycled many times and recycled plastic products are cheaper and eco-friendly while virgin plastic is prepared from petrol (fossil fuel), hence that is costlier and not eco-friendly; plastic products should also be banned. Thus, a drive for 'no plastic use' be acted upon honestly by the Centre, States, and local bodies.

(9) Finally, India should always emphasise 'Common but differentiated responsibility' for mitigation and adaptation of climate change at global level because of 'historical wrongs' committed by many developed/industrialised countries acting as colonial powers in the last 300 years, hence they should assist developing countries with finance and eco-friendly technology transfer. Yet we should also make all voluntary genuine efforts at national, state, and local levels to reduce carbon emissions and practice adaptation through technological innovations in view of substantial absolute carbon emissions in India. In this regard, the concept of 'climate justice is quite relevant here key dimensions of climate justice were identified by the UNFCCC in 1992 at Rio that called upon the member nations to 'protect the climate system for the benefit of present and future generations of mankind, on the basis of equity and in accordance with their common but differentiated responsibilities and respective capacities' (cited in Dryzek et al. 2013, 76).

Some scientists like Johan Rockstrom et al. (2009) have suggested 'planetary boundaries' that define safe operating space for humanity with respect to the earth system. They identified planetary boundaries for nine processes:

(i)   Climate change
(ii)   Rate of biodiversity loss
(iii) Interference with the nitrogen and phosphorus cycles
(iv)   Stratospheric ozone depletion
(v)   Ocean acidification
(vi)   Global fresh water use
(vii) Change in land use
(viii) Chemical pollution
(ix)   Atmospheric aerosol loading

They rightly specified the pre-industrial value of the indicator, its current status, and the proposed boundary. For climate change, the boundary is 350 PPM of CO_2 in the atmosphere but humanity surpassed 400 PPM in 2012–2013 itself! Hence, such an alarming situation demands firm eco-friendly action with immediate effect. The Government of India has constituted Green India Mission in this regard (Box 7.8).

Box 7.8: Green India Mission: Policy and Reality

Under National Action Plan on climate change, Green India Mission is one of the eight Missions of Government of India that takes adaptation and mitigation measures:

(a)   Enhance carbon sinks
(b)   Adaptation of vulnerable species/ecosystems to the changing climate
(c)   Adaptation of forest-dependent communities

It is a centrally sponsored scheme – central share of 90% for north-eastern states and special category (hilly) states (Himachal Pradesh, Jammu & Kashmir, and Uttarakhand); 60% for other states.

Its objectives are:

(a)   Increase forest cover on 5 million ha of forest/non-forest lands
(b)   Improve the quality of forest cover on another 5 million ha
(c)   Improve ecosystem services including biodiversity, hydrological services, provisioning of fuel, fodder, timber, and non-timber forest produce (NT FP)
(d)   Forest-based livelihood income of 3 million households
(e)   Enhance annual CO_2 sequestration by 50–60 million tonnes by 2020

Concerns: As per a Parliamentary Committee report, Green India Mission is underfunded. It is for 10 years with an outlay of Rs 60,000 crores; but during 2017–2018, only Rs 47.8 crores were allocated for it.

304   Climate Change and Society

Second, as per Nationally Determined Contribution to UN FCCC, India has a target to sequester 2.523 billion tonnes of carbon by 2020–2030; India's present forest cover is 75 million ha, hence 30 million hectares of additional land would be required for forests. But Green India Mission is silent on this.

Third, during 2015–2016 and 2016–2017, GIM missed its targets by 34% – only 44,749 ha of land got green cover against the target of 67,956 ha.

Finally, unsuitable trees like eucalyptus were planted leading to more environmental problems like causing drought, declining the water table, and preventing biodiversity in the region.

Source: Government of India.

But what are practical actions in this regard? Let us discuss this in the next section.

Practical Actions for Climate Mitigation and Adaptation

We are of the firm view that India should opt for the following practical options and actions for both mitigation and adaptation:

(a) Actual renewable sources of energy like wind, solar, hydro, geothermal, bio-energy, etc. should be given top most priority – solar energy is cheaper than wind energy – to fossil fuels like coal, oil, natural gas, fuel wood.; further electricity vehicles be made more popular by providing more and better charging stations.

(b) Nuclear power is not environmentally safe (rather highly hazardous) in the long run, though it may appear cheaper, for the time being, hence it should be avoided. A large number of local people protested (during 2011–2015) against Kudankulam Nuclear Power Plant (Tirunelveli, Tamil Nadu) – mostly women – and one protestor was killed and many were injured; many were arrested and charged with sedition! Protestors feared that the plant's effluent would release toxins and impact the quality of fish, thus depriving them of their livelihood. The People's Movement Against Nuclear Energy was their organisation.

(c) There should be a promotion for public and private investments to raise energy efficiency levels consistently (e.g. Light Emitting Diode (LED) bulbs/tubes in place of conventional bulbs), and the Bureau of Energy Efficiency at national level and State Renewable Energy Development Authority should be more proactive.

(d) There should also be public, cooperative, and private investments to expand capacity in clean renewable energy; public–private partnership (PPP), and public–cooperative partnership (PCP) modes should be encouraged in letter and spirit.

(e) All States should prepare fool-proof and comprehensive State Action Plans on climate change; some States/UTs have been slack in this regard for a long time. There should be transparent mechanisms for their implementation at every stage.

(f) As per Article 12 of the Kyoto Protocol, there is a carbon market for the global reduction of carbon/GHG emissions through the sale and purchase of carbon credits – this is called 'Clean Development Mechanism' (CDM). During 2003–2014 out of total 7,589 CDM projects, 1,541 were from India (second highest in the world). Certified emission reductions issued to Indian projects is 19.10 crores (13.27%) – mostly in sectors of energy efficiency, fuel switching, industrial processes, municipal solid waste, renewable energy, and forestry. But, in the second commitment period, the number of CDM projects declined – only 307 projects from India were submitted out of total 3,227 projects. Hence, the Indian public and private sectors should be proactive, and new, sustainable, and effective market mechanisms should be created to cover the maximum countries of the world.

(g) A National Adaptation Fund with a corpus of Rs 100 crores has been set up by the Government of India (Ministry of Environment and Forests and Climate Change) to support adaptation actions to combat the major challenges of climate change in agriculture, water, forestry, etc., but it has not materialised effectively so far on the ground. Further, this corpus fund is quite meagre.

(h) New short-term and flood-resistant/drought-resistant varieties of seeds of various food crops and oil seeds are to be developed on a priority basis by various agricultural universities and research institutes; for instance, Khejri tree in Rajasthan and 'bamur' tree in Kalahandi (Odisha) are local and cheap species. Further, 'bamur' is appropriate for 'ridge farming' – growing trees along the ridges of fields or even tank bunds. Bamur (*Acacia nilotica*) leaves are good for manure, fodder, etc. and its wood is fit for making bullock carts and agricultural implements, door frames, etc. Exotic varieties of the eucalyptus plant, which is not useful to the farmers and sucks groundwater hugely, and silliness soil, should be discouraged.

(i) The urban areas of many parts of India are also experiencing the decline of water table – north Bihar (Darbhanga, Madhubani, etc.) with abundant groundwater earlier, faced a decline (of three to four feet) of groundwater in 2019 due to climate change as well as anthropogenic activities wherein there is no adequate pond water from rain for recharging groundwater. On the other hand, the surface, ground, and piped water supplied to the people is contaminated heavily and unfortunately contaminating the aquifer also in some parts of the country. In November 2019, Indian Standard Bureau found in the sample tests in 21 cities of India that supply water in Mumbai is the best (ranking first), followed by Hyderabad, Bhubaneswar, and Ranchi, but Delhi failed on 19 parameters out of 48 parameters. Raipur ranks 5, Amravati 6, Shimla 7, Chandigarh 8, Trivandrum 9, Patna 10, Bhopal 11, Guwahati 12, Bengaluru 13, Gandhinagar 14, Lucknow 15, Jammu 16, and Jaipur 17. Hence, not only the 100 smart cities taken up by the Government of India but others, too, should be given top priority for conserving rainwater, ('the purest form of water'), and also to ensure treatment of the contaminated water everywhere, more so in the poor settlements, and finally, supplying of the clean tap water regularly to all the people in all the settlements.

Conclusion

All of us (governments, NGOs, and citizens) should be fully conscious of all these environmental and climate change problems and actively act upon 'here and now'. In conclusion, we may observe that India should always emphasise on 'Common but differentiated responsibility' for mitigation and adaptation of climate change at global level because of 'historical wrongs' committed by many developed countries as colonial powers and, therefore, "historical polluters" in the last 300 years, hence they should assist the developing countries through the transfer of adequate fund, and innovative and cheaper technologies. Yet we should also make all voluntary genuine efforts at national, state, and local levels to reduce carbon emissions and practise adaptation through sustainable technological innovations in view of the substantial (absolute) emissions in India. India's eight missions should be made more proactive and should show visible and concrete outcomes in a time-bound manner. Capacity-building by multilateral organisations should be ensured. Thus, multiple initiatives are required to address the challenges of climate change, including the transformation of industrialism, consumerism, and ultimately capitalism! Hence, the 'paradigm shift' urgently needs new systems of deliberative (dialogic) eco-democracy, climate-resilient green agriculture (agro-ecology), and proactive good governance. We may conclude in the words of Luther Burbank:

> Nature's laws affirm instead of prohibit. If you violate her laws, you are your own prosecuting attorney, judge, jury and hangman'... (there is an urgent need for the) rejection of carbon heavy economic revival strategies if we are to avoid a vicious cycle of deadly viruses and extreme climatic conditions striking us repeatedly.
>
> (Cited in S. Chaulia 2021)

Now the question arises about other global environmental issues like various types of pollution, acid rain, ozone depletion, deforestation, loss of biodiversity, etc. We will discuss these in the next chapter.

Points for Discussion

(1) What are the major consequences of climate change for India and the world as a whole? Explain with concrete evidences.
(2) How does and to what extent the emission of greenhouse gases affect climate change? To what extent humans are responsible for it?
(3) What are policy implications of climate change? Discuss it in the light of India's NDCs.
(4) What are the major provisions of the Kyoto Protocol? To what extent have these been implemented globally?
(5) What are the main points of the Paris Agreement? Discuss the status of its implementation as well as deliberations in Conference Of Parties (COP) 27 at Sharm-El-Sheikh (Egypt) in November 2022.
(6) What are various theoretical perspectives on climate change? Discuss it critically.

Chapter 8

Global Environmental Issues

Introduction

Various global environmental issues are notable since 1970s. Among these, global warming and climate change, pollution, acid rain, ozone depletion, deforestation, loss of biodiversity, etc. are significant. In Chapter 7, we have dealt with climate change in detail; the rest of global environmental issues are dealt with here.

Here our first focus is on environmental pollution, especially water, air, noise, soil, and radiation pollution. Our approach is multidimensional and emphasis is on the convergence of resources, on the one hand, and that of various sectors/departments, on the other. After a brief introduction about the problem, we will analyse various causes of such pollution in India in the second part of the chapter and afterwards we will analyse the social consequences of environmental pollution in detail. Then we will deal with acid rain, ozone depletion, deforestation, and loss of biodiversity. In the last part, we will present our recommendations for policy-making in India.

For some decades, environmental challenges, especially in the form of pollution, have amplified at global and national levels. At the global level, five types of pollution are quite notable: air, water, noise, soil, and radiation. Since the industrial revolution in Europe in the eighteenth century, the emission of greenhouse gases (GHGs) has multiplied and this has serious environmental implications. For instance, 250 biggest listed companies in the world account for one-third of all man-made greenhouse gas emissions globally. Coal India, Gazprom, and ExxonMobil are three top most listed companies for the highest CO_2 emission, but unfortunately, people are using their products without questioning their credentials. To put it more rigorously, these 250 listed companies were expected to cut 3% emission per year to limit the temperature as per the goals set by the Paris Climate Agreement (2015). But actually, only 30% of these 250 companies have set goals to cut such emissions (*The Times of India*, 1 November 2017). In 2014, Delhi was marked by the World Health Organization (WHO) as the most polluted city in the whole world (annual mean PM 2.5 being 153 µg/m^3), and later in 2016 it slightly improved to 11th rank in the world (annual mean PM 2.5 being 122 µg/m^3). However, some experts and environmental organisations like Greenpeace International question it by pointing out that more data station points (10 in 2016 against 6 in 2014) for monitoring have diluted the earlier result! Further, the WHO report of 2016 is based on data (annual average) from 2008 to 2013. Hence, it alleges that if current data are taken into account, Indian cities would perhaps rank

DOI: 10.4324/9781003336211-8

308 Global Environmental Issues

worse. During November to January every year, Delhi becomes highly polluted due to burning of *parali* (stems of paddy) in Punjab, Haryana, and Western UP.

Here our objective is to show that environmental pollution in India has emerged as a very challenging problem with severe social consequences on health and hygiene, loss in economy, etc., but due to lack of sustained movement, it has not caught the full attention of the political class. However, if civil society organisations (CSOs) come together with mass mobilisation, it will have a massive effect on the polity.

Needless to say that 2017 was worse in terms of air pollution in India: on average as October 2017 was more polluted than October 2016 – average air quality index (AQI) for the former was 281 against 275 in October 2016. Further, October 2017 had 15 'Very poor' days compared to October 2016 with 5 'very poor' days. However, the highest AQI in October 2016 was 445 (on October 31) against the highest AQI in October 2017 being 403 (on October 20). AQI in Delhi on 1 November 2019 was the highest (504). AQI takes into account eight pollutants: (a) PM 2.5, (b) PM 10, (c) nitrogen oxide, (d) sulphur dioxide, (e) ozone, (f) carbon monooxide, (g) ammonia, and (h) lead. Anyway, 11 most polluted cities in the world as per the WHO report 2016 are as follows (Table 8.1).

Out of these 11 most polluted cities in the world (in 2016), five cities were in India: Gwalior, Allahabad, Patna, Raipur, and Delhi. Regarding PM 10 level, the highest polluted city in the world was then Orvishta (Nigeria) with an annual mean of 594 µg/m³ followed by Peshawar (Pakistan) of 540 µg/m³; Gwalior with 329 µg/m³ ranked tenth in the 10 highest PM 10 level polluted cities (and was the only Indian city among top ten highest PM 10 level). In this regard, Delhi ranked 25th position with 229 µg/m³ (annual average). In addition, WHO's 2016 report clearly mentions that urban air pollution levels were the lowest in high-income countries (Europe, Americas, and Western Pacific region); on the other hand, the highest urban air pollution levels were experienced in the low- and middle-income countries, especially eastern Mediterranean and South-East Asia regions, exceeding five to ten times the WHO safe limits. Further, the pollution situation of Indian cities worsened by 2019 (average annual data for 2010–2019) – Delhi tops the list in the entire world with 110 µg/m³ PM 2.5, followed by Kolkata (84 µg/m³) against the WHO norm of 5 µg/m³ (Table 8.2).

Table 8.1 Top 11 Most Polluted (PM 2.5 µg/m³) Cities in the World (2016)

Sl. No	Most Polluted (PM 2.5) Cities in the World	PM 2.5
1	Zabol (Iran)	217
2	Gwalior (India)	176
3	Allahabad (India)	170
4	Riyadh (Saudi Arabia)	156
5	Al Jubail (Saudi Arabia)	152
6	Patna	149
7	Raipur	144
8	Bamenda (Cameroon)	132
9	Xingtai (China)	128
10	Baoding (China)	126
11	Delhi (India)	122

Source: WHO report 2016 quoted in *Indian Express*, 13 May, 2016.

Table 8.2 Top Ten Most Polluted Cities in the World (2019)

Sl. No.	Cities	PM2.5 (Annual Average µg/m³)
1	Delhi (India)	110
2	Kolkata (India)	84
3	Kano (Nigeria)	83.6
4	Lima (Peru)	73.2
5	Dhaka (Bangladesh)	71.4
6	Jakarta (Indonesia)	67.3
7	Lagos (Nigeria)	66.9
8	Karachi (Pakistan)	63.6
9	Beijing (China)	55.0
10	Accra (Ghana)	51.9

Source: Health Effects Institute (USA) cited in *The Times of India*, 18 August 2022.

From this report it transpired that out of ten most polluted cities in the world, six are in Asia, three in Nigeria, and one in Latin America. Further, out of the world's 50 cities with the most severe increase in PM 2.5, 41 are in India, and 9 are in Indonesia, while, on the other hand, all the 20 cities with the greatest decrease in PM 2.5 (during 2010–2019) are in China.

Unfortunately, nowadays in central Delhi (Lodi Road and India Gate) one smallest, finest, and deadliest pollutant has emerged – PM 1, which is 70 times finer than the thickness of a human hair. At Lodi Road (New Delhi) SAFAR (Ministry of Earth Sciences) recorded in July 2017 the average volume of PM 1 during summer, winter, and monsoon at 46, 49, and 20 µg/m³, respectively. In 2006, in Delhi, particulate matter (PM 10) was 153 µg/m³ (annual average) that increased to 214 in 2008 and to 261 in 2010 (due to major construction works for Commonwealth Games), which declined to 222 in 2011 but again peaked to 260 in 2016. In November 2017, the air pollution in Delhi increased eight times plus to a good level (up to 50 µg/m³) in most of Delhi, as is clear from Table 8.3.

On 1 November 2019 (3:30 am) AQI of Delhi reached 'severe plus' (emergency) category of 504, while that in National Capital Region (NCR), Ghaziabad, and Greater Noida was 496, as per Central Pollution Control Board (CPCB). Second, the share of stubble (parali) burning in Delhi's pollution was 46% on 1 November 2019, as per the Ministry of Earth Sciences' Air Quality Monitor SAFAR; Dwarka and J. L. Nehru Monitoring Stations recorded an AQI of 499. Hence, emergency measures were started: odd–even car rationing system from 4 November (2019) in Delhi, closure of schools, banning entry of trucks in Delhi, stopping the construction activities, as well as starting artificial rain in some areas (*The Times of India*, 2 November 2019). Thus, Delhi has the stigma of one of the most polluted cities in the world, if not the most.

It goes directly into our bloodstream, causing serious heart and respiratory problems. WHO has not defined a safe limit for PM 1, though the same for PM 2.5 is 60 µg/m³, and for PM 10 it is 100 µg/m³. As per the WHO estimate, about 3.3 million people die annually in the world from air pollution by these particulate matters.

310 Global Environmental Issues

Table 8.3 Air Pollution Level in Various Parts of Delhi (1 November 2017)

Sl. No	Area	Level of Pollution	Air Quality Range	Index
1	Rohini	458	1. 0–50	Good
2	Shadipur	456	2. 51–100	Satisfactory
3	DTU	424	3. 101–200	Normal
4	Anand Vihar	409	4. 201–300	Poor
5	Lodi Road	401	5. 301–400	Very poor
6	R. K. Puram	413	6. 400–500	Severe
7	Siri Fort	402	7. Above 500	Severe plus (emergency)
8	North Campus (Delhi University)	358		
9	Mandir Marg	347		
10	Burari Mod	340		

Source: Dainik Jagran, 2 November 2017, cited in Sharma 2017.

Magnitude and Causes of Air Pollution

Various studies have pointed out different causes of air pollution in India, especially in Delhi. Some of the findings of major studies are presented in Table 8.4.

Further thermal power plants (Badarpur), too, contribute substantially to air pollution in Delhi. However, if we look at the national scene, we find that thermal power plants (mainly coal) are a significant source of carbon emissions besides vehicles, firewood, and burning of solid waste as well as construction works. In total, 68% of India's total electricity power comes from thermal power plants (1,94,200 MW), owned by States or Centre or private firms/Joint Ventures: Maharashtra (28,294 MW) has the largest thermal capacity, followed by Gujarat (23,160 MW), Chhattisgarh (13,234 MW), Uttar Pradesh (UP) (12,228 MW), Tamil Nadu (TN) (11,513 MW), Madhya Pradesh (MP) (11,411 MW), and Rajasthan (10,226 MW).

In Delhi, air pollution is basically caused by five factors (*The Times of India*, 18 October 2017/*Dainik Jagran*, 6 November 2017):

a) Emission from coal-fired thermal power plants – 13 thermal power plants within 300 km radius of Delhi emit secondary particles contributing to 30% of PM 2.5 in winter and 15% in summer.
b) There are more than one crore (10 million) vehicles registered in Delhi – since 2000 AD, vehicles in Delhi increased by 97%, particulate matter by 75% and NO_x by 30%; in winter, vehicles contribute 20–25% to particulate matter and 6–9% in summer; in Delhi from vehicles, 217.7 tonnes of carbon monoxide, 84 tonnes of NO_x, 66.7 tonnes of hydrocarbon, and 0.72 tonnes of sulphur dioxide are emitted daily and 9.7 tonnes of particulate matter daily. More than half of vehicles in Delhi do not have pollution certificates; there are 50 lakh vehicles on roads in Delhi daily, but no adequate staff to check these vehicles.
c) Burning of paddy stems (*'Parali'*) – Punjab 19.6 million tonnes, UP 21.9 million tonnes, and Haryana 9.1 million tonnes annually – contributes 140 µg/m³; PM

Global Environmental Issues 311

Table 8.4 Major Causes of Air Pollution in Delhi

Year	Study Sources	Main Findings
2003	Ministry of Environment and Forests (White Paper), GOI	a) During 1970–1971 to 2000–2001 Contribution of vehicles to Delhi's particulate matter increased from 23% to 72%
2004	*Atmospheric Environment* journal	a) Over 80% of emissions of major pollutants including particulates, NO_x, and CO came from vehicles
2008	Central Pollution Control Board and National Environmental Engineering Research Institute (NEERI)	a) Road dust as the biggest contributor 52.5% to particulate matter in Delhi's air, industries 22%, vehicles' particulate emissions 6.6% b) For NO_x, industries contribute 79% and vehicles 18% c) For CO, vehicles contribute 59% and hydrocarbons 50%
2010	SAFAR (Ministry of Earth Sciences, GOI)	a) Most of PM 10 from road dust, vehicular emission not the major source of air pollution b) Vehicle and industry sources as dominant sources of PM 2.5, followed by construction works
2013	*Atmospheric Environment* (by Dr Sarath Guttikunda) journal	a) Vehicles' emission contributed 90% to NO_x, 54% to total suspended particulate matter and 33% to SO_2
2013	Dr Pramila Goyal (IIT Delhi)	a) Heavy commercial vehicles as major source of particulate matter (92%) followed by two wheelers
2013	UN Environment Program (by IIT Delhi and Dr Sarath Guttikunda)	a) Higher pollution level attributed to exhaust emissions from trucks (passing after 9 pm)
2015	IIT Delhi and IIT Bombay (by Dr Sarath Guttikunda and Rahul Goel)	a) Trucks contribute more than 60% of total vehicle pollutants PM 2.5, CO and SO_2 in the Greater Delhi region
2015 December	IIT Kanpur (for Delhi Government)	a) In winter, 46% particulate emissions from trucks, 33% from two wheelers, 10% from four wheelers, 5% from buses, and 4% from LCV b) Four major contributors to PM 2.5 are dust, vehicles, domestic fuel burning, and industrial pollutants

Source: Adapted from *Indian Express*, 6 January 2016.

10, and 120 µg/m³ PM 2.5; on 22 October 2017, at least 2,334 instances of paddy stems burning were recorded by NASA in north India.

d) Ready-mix concrete mixture, used for construction activities, emits a large amount of fly-ash, accounting for 10% of PM 10 emissions (14.37 tonnes daily) and 6% of PM 2.5 emissions (3.5 tonnes daily).

e) Dust/smoke from municipal wastes also adds to particulate matter, especially in three landfills at Bhalaswa, Ghazipur, and Okhla (Delhi).

f) In Delhi, in addition, 36% of the urban poor were using unclean cooking fuel before COVID-19, but during this pandemic, 49% of the urban poor used unclean cooking fuel (during the lockdown in 2020, as per an environmental NGO, Chintan) (Gandhiok 2021).

Therefore, air quality on 6 and 7 November 2017 deteriorated drastically in Delhi and NCR: PM 2.5 concentration reaching up to 1,000 µg/m³ at Delhi Technological University (against the average of 529), 971 at Sirifort (against the average of 333), 976 at Ghaziabad (against the average of 432), 894 at Punjabi Bagh (against the average of 355), 865 at Noida (against the average of 353), 862 at Anand Vihar (against the average of 309), and 819 at Mandir Marg (against the average of 350), thus 16–20 times the normal quality of air. A study by S. N. Tripathi (of IIT Kanpur) during March 2018 in Delhi (at IIT Delhi and IITM, Delhi) found that PM 2.5 particles were composed of organic compounds (44%), chloride (14–17%), ammonia compounds (9–11%), nitrates (9%), sulphates (8–10%), and black carbon (11–16%). PM 1 particles comprised 47% organics, 13% sulphates and ammoniums, 11% nitrates and chlorides, and 5% black carbon. The study also noted that the average black carbon concentration was the highest at IIT Delhi as mostly affected by combustion-related emissions like traffic and burning of solid fuels (Agarwal 2021).

New studies have found that even kitchens may also be as polluted as outdoor, PM 2.5 emission reaching up to 310 µg/m³ during summer (Pallavi Pant, *The Times of India*, 4 April 2017). Actually, frying and grilling generate ultrafine particles that are quite harmful. Due to the use of fossil fuel burning for cooking (3.22 lakh households use it), air pollution, especially PM 2.5 level, increases 12% in Delhi (*The Hindu*, 30 December 2015). In rural areas, many families use fossil fuel for purposes of cooking, light, mosquito repellent, fertilisers, etc. The Government of India gave power connections to 4.10 crore households (without power until then) in rural areas by December 2018. In 2014, electricity consumption in Russia (the coldest area) was 6,603 kWh per capita against 6,353 kWh in Europe, 4,229 kWh in South Africa, 3,927 kWh in China, 2,601 kWh in Brazil, and just 806 kWh in India. Needless to say that per capita consumption of power indicates the degree of industrialisation and the elite lifestyle (consumerism) of the people at a particular time and place. However, too much use of electricity and other forms of energy by the people for luxury in most of the developed countries is not a good sign for sustainable development as it contributes substantially to GHG emissions.

Magnitude and Causes of Water Pollution

Every year 'World Water Day' is celebrated on 22 March. But except for observing some rituals, we hardly take any concrete measures for water conservation. While India conserves only 17–20% of its total rainwater, Israel, on the other hand, conserves 80% of its rainwater – this shows the failure of both policy planning and practice in India. Actually, 80% of India's drinking water source is groundwater. There was a 65% dip in the water level of India's wells in the last decade. As per the *Lancet* journal, in 2015, water pollution led to 64 lakh deaths, air pollution to 18.10 lakh deaths, and total 90 lakh deaths in the world from all types of pollution. Total pollution deaths in India then were 25.15 lakh. Both surface and ground water are polluted to different degrees due to different factors in different regions of India. There are 63,968 habitations with contaminated water in India, and out of these, 28,000 habitations in India (especially in West Bengal, Bihar, Jharkhand, UP, Assam, Manipur, Rajasthan, and Chhattisgarh) are heavily affected with arsenic and fluoride contaminants in water. Other habitations (among total 63,968 ones) suffer from salinity, nitrate, and iron content. In fact, the Department of Drinking Water, Government of India, admitted in the parliament that during 2010–2015 more than 16,528 persons died due to diseases, caused by water contamination, across India (*The Times of India*, 1 April 2017). The Government of India on 1 August 2022 admitted in Rajya Sabha that arsenic was found in ground-water in parts of 209 districts in 25 states and Union Territories – uranium in parts of 152 districts in 18 states, lead in parts of 176 districts in 21 states, iron in parts of 491 districts in 29 states/UTs, cadmium in parts of 29 districts in 11 states, and chromium in parts of 62 districts in 16 states; further, there are 14,079 iron affected, 671 fluoride affected, 814 arsenic affected, 9,930 salinity affected, 517 nitrate affected, and 111 heavy metal affected habitations in India – at present in total 26,122 such contaminated habitations. Further, global research by ORB Media found that in New York and Washington, plastic fibre contamination in tap water was 94%, Lebanon (Beirut) 93%, India (New Delhi) 82%, Uganda (Kampala) 80%, and Ecuador (Quito) 79%. Actually, microscopic fragments enter from synthetic fibre clothing, tyre dust, and micro beads (*The Times of India*, 16 October 2017). Water situation in India is not satisfactory (Box 8.1), both surface and underground water.

In Delhi, the ammonia level in Yamuna river was recorded very high at 3.4 parts per million (ppm) on 14 January 2021 (at 6.30 am) – 0.5 ppm is the upper limit of ammonia as per the Bureau of Indian Standards (BIS), and a maximum limit up to 0.9 ppm of ammonia can be treated, beyond it toxic cancer – causing by-products like chloramines are produced. Every year in January, the ammonia level increases in Yamuna water: 2.2 ppm during 4–31 January 2019, 2.7 ppm during 4–12 January 2020, and 2.4 ppm during 15–17 January 2020. In 2020 itself, the ammonia level remained above the treatable level for 33 days. This directly affects the functioning of Chandrawal, Wazirabad, and partly also Okhla water treatment plants – thus about one-third of Delhi's daily water supply is affected (Singh 2021). As per Delhi Jal Board (DJB), the source of ammonia is 85 km upstream in dyeing and other industrial units in Panipat (Haryana) and inter-mixing of industrial waste in Sonipat (Haryana), hence stern action needs to be taken against these. Second, DJB plans to establish ozonation units that would solve 95% of ammonia up to 4 ppm. Third, building a concrete wall between drain 6 (carrying industrial waste) and drain 8 (carrying raw water towards Delhi) would also be helpful.

Box 8.1: Status of Water Situation in India

'A 21st Century Institutional Architecture for India's Water Reforms' report submitted to Union Government of India clearly mentions that water tables are falling in most parts of India, 60% of India's districts face groundwater over-exploitation and/or serious quality issues. In North Bihar (a region of abundant ground water earlier) too, the water table has gone down 3–4 ft. Ninety nine villages in Gurugram (Haryana) are in red zone (beyond

314 Global Environmental Issues

30 metres depth). Second, there is fluoride, arsenic, mercury, and uranium in groundwater. Third, water use efficiency in India's agriculture is among the lowest in the world (25–35%). Fourth, the drying up of India's peninsular rivers is mainly caused by over-extraction of groundwater. Fifth, cities produce 40,000 million litres of sewage daily but only 20% of the total sewage is treated. Consequently, there are various kinds of conflicts between towns and villages, and also between social classes in a particular habitation/locality. Finally, various rivers and other water bodies are highly polluted with plastic, leading to health and hygiene hazards, as well as choking drains/nullahs resulting in water-logging.

The status of water in Delhi is worse. Though there were 1,000+ water bodies recorded in revenue records in Delhi, yet 80% of these exist only on paper as these are encroached by parks, school, and hospital buildings constructed, and dumping sites and sewages created. About 20% of Delhi's population does not get safe drinkable pipe water regularly; it adds insult to injury. During the Chhath festival in 2021 in Delhi, women were seen taking 'Sacred' bath in the Yamuna river with full of froth!

In 2012, Municipal Corporation of Delhi (MCD) got water samples tested and the results were alarming: 100% of the samples in Narela, 87% of the samples in Sadar Paharganj, 70% of the samples in Karol Bagh, 61% of the samples in the south zone, 58% of the samples in civil lines zone, 50% of the samples in Shahdara, and 33.3% of the samples in the central zone were found unfit for drinking. Three main causes of water pollution in Delhi are as follows: Algae growth and related contamination in water pipes, mixing of sewage water with drinking water due to breakage of pipelines, and the outlived water pipe system. Therefore, the 'drinking water supply and sewage system' is converted into 'sewage supply and drinking water system'!

Similarly, there were 2,300 ponds in Ernakulam district in Kerala in the 1980s but declined to 800 ponds in the second decade of the twenty-first century.

Further, the Naini lake in Nainital (Uttarakhand) suffers from mounds of debris and dead fish.

In Dal lake (Kashmir) about 20 million litres of untreated sewage are drained daily, and further high tourist footfall leads to high waste dumping. Dal lake lost 24.5% of its area during 1850–2017 due to the unregulated change in land use.

In Bengaluru, too, 67 lakes surveyed have no drinkable water. In Bengaluru's Subramanyapura lake weeds have taken roots, and chemicals and sewage in the lake water have spoiled its quality – froth found during pre-monsoon rains.

In Chandigarh, too, Sukhna lake is choked due to silt in a nearby village.

Source: Based on *Hindustan Times*, 26 March and 22 March 2017.

Due to rapid urbanisation, public and private encroachments, etc., water bodies are shrinking across India, especially in cities (Table 8.5).

Further, in Jharkhand, Jadugora-based Uranium Corporation of India Ltd. (UCIL) has an operating mine since 1967 and extracting about 1,000 tonnes of uranium ore daily, accounting for 20% of raw material for India's nuclear power generation. Adrian Levi researched there and reported (*The Times of India*, 15 December 2015) that this mining has exposed workers, in particular, and surrounding villages, in general, to

Table 8.5 Loss in Water Bodies Due to Rapid Urbanisation in Major Indian Cities (2000–2016)

S. No	Cities	Loss in Water Bodies (%)
1	Surat	95
2	Raipur	80
3	Bengaluru	79
4	Kolhapur	75
5	Ghaziabad	75
6	Delhi	62
7	Guwahati	60
8	Udaipur	50
9	Chennai	50

Source: Adapted from *Hindustan Times*, 8 June 2017.

radiation, heavy metals, and other carcinogens, including arsenic. In fact, toxins leaching into underground aquifers and the Suvarnarekha river are contaminating the food chains – from fish to vegetables. It is also mentioned in the report that a study done by Deepak Ghosh found that in the water of Suvarnarekha river and adjacent wells, the level of radioactive alpha particles was 160% higher than the safe limit set by WHO. Further, the bursting of a pipe carrying toxic slurry on 24 December 2006 resulted in the deaths of many fish in Suvarnarekha's tributary.

In the Sonbhadra district of UP, 250 villages have excess fluoride in the groundwater and 150 villages have fluoride levels in groundwater beyond the safe limit. The Central Pollution Control Board, in 2013, categorised Sonbhadra district as 'critically polluted'. There 16 scheduled tribes reside and they cannot afford treatment for black/brown spots on teeth, gastro-intestinal discomfort, pain in muscles, thickening of bone density, and calcification of ligaments (crippling a patient). There is no treatment for skeletal fluorosis and prevention is the only option. The dumping of wastes by aluminium, chemical, mining, and cement factories in Sonbhadra has polluted both surface and groundwater. In addition, coal fly-ash from thermal power plants is also a source of fluoride. Large dams built in this region have exacerbated the problem of fluoride that exists in geological rocks, and from these, it leaches into groundwater. In 2014, National Green Tribunal (NGT) constituted a committee to monitor potential hazards of industries, and it was found that over 35 million tonnes of fly-ash are generated annually by ten thermal power plants located in that region. Actually, fly-ash has fluoride and leaches into groundwater and contaminates surface water too. The fly-ash pond of Obra Thermal Power plant overflows and contaminates river Renu, a tributary of Son river. Further, over 2,250 tonnes of spent pot lining (SPL) is generated daily by the aluminium industry in the region; the SPL contains fluoride and is highly hazardous waste. Unfortunately, defluoridation plants, established under corporate social responsibility (CSR) by the private sector, are usually not working (D. Sinha and Y. Mittal 2017).

In addition, even rain water is becoming highly acidic in different parts of India due to rain water in the atmosphere mixing with polluting gases like oxides of sulphur and nitrogen emitted from power plants, automobiles, and some industrial units.

316 Global Environmental Issues

Namami Gange (National Mission for Clean Ganga) was approved in May 2015 with a sanctioned amount of Rs 20,000 crore for 2015–2020 with 100% central funding for various cleaning and treating activities. Actually, treatment capacity of only 3,500 million litres daily (MLD) is available out of a total of 8,250 MLD of waste water generated in all the towns/cities in the Ganga river basin. Besides core schemes of sewage treatment plants (STPs), secondary schemes of bathing ghats, crematoria, and river surface cleaning are also included. Ganga's total length is more than 2,500 km covering five states of India: Uttarakhand, UP, Bihar, Jharkhand, and West Bengal. According to Sadhguru Jaggi Vasudev, out of 800 streams and tributaries of Ganga, 470 have become seasonal, flowing only for four months in a year, leading to 44% reduction of water in the river (*The Times of India*, 2 October 2017). Further, Kanpur alone, with more than 400 functional tanneries, generates 450 MLD of sewage but can treat only 160 to 170 MLD. Four major cities of Kanpur, Allahabad, Varanasi, and Patna have 1,091 industrial units (sugar, paper, distilleries, and tanneries) in the Ganga basin. Most of the untreated effluents pass into Ganga, causing health hazards. (*The Hindu*, 1 September 2018). Abhishek M. Singhvi raised three points about Namami Gange (*The Hindu*, 25 August 2018): First, out of 221 projects sanctioned for activities like sewage treatment at a total cost of Rs 22,238.73 crores, only 58 projects were completed till mid-August 2018; second, against the programme aimed at treating 2,278 million litres a day of sewage through sewage treatment plants (STPs), only 329.3 million litres a day were completed by mid-August 2018; third, as per CAG report, during 2016–2017, the level of pollutants in Ganga river across UP, Bihar, and Bengal was six to 334 times higher than the prescribed level. This shows the sordid state of affairs on the ground, and Namami Gange could not achieve its target. According to the Ministry of Jalshakti (Government of India), in total, 315 projects were sanctioned, out of which, by 31 July 2020, 132 projects were completed, 149 projects were in progress and 27 projects were under tendering; second, against the total sanctioned cost of Rs 28,854.11 crores, total expenditure was Rs 9,065.92 crores; third, against 151 sewage infrastructure projects with a sanctioned cost of Rs 23,120.63 crores, 54 projects were completed with an expenditure of Rs 6,476.24 crores, 72 projects were under progress and 25 were under tendering; fourth, against 76 projects of *ghats* and crematoria (with a sanctioned cost of Rs 1,089.31 crores) 51 were completed, 30 were under progress, and 1 was under tendering – by July 2020 expenditure in this was Rs 720.59 crores. By July 2022, a total of 374 projects were sanctioned, 210 were completed, and 132 were in progress; 162 sewage infrastructure projects were sanctioned, 92 were completed, and 54 were in progress; 95 ghats and crematoria projects were sanctioned, 70 were completed and 16 were in progress as per Jalshakti Ministry. Still, Ganga water is unfit for drinking and bathing in UP, Bihar, Jharkhand, and West Bengal stretches.

Yamuna River: The status of Yamuna river is also very bad. Yamuna's total length is 1,376 km. Though only 2% of Yamuna river's total length falls in Delhi, it receives about 70% of the pollution while passing through Delhi. There are 21 major waste water drains in National Capital Territory (NCT) of Delhi – of these 18 drains join the Yamuna, and three drains join the Agra/Gurugram canal. Under Yamuna Revitalisation Project 2017, at least 14 sewage treatment plants (STPs) were planned to be constructed by May 2019 in Phase I and these would treat 67% of river's pollution coming from two major drains at Najafgarh and Delhi gate (*Hindustan Times*, 9 August 2017). After the first stretch of about 180 km, most of Yamuna ceases to

Global Environmental Issues 317

exist after monsoon as at Hathnikund all the water from Yamuna is diverted to the western and eastern Yamuna canals. Only 160 cusec water is released in Yamuna for a minimum ecological flow. In fact, six barrages and 17 hydropower plants interrupting the flow, encroachments in the flood plains, loss of vegetation, extraction of sand, and pollution caused by agriculture and industry are the main reasons for the drying of river Yamuna. Further, during festivals like Durga Puja, Chhath, the pollution level in Yamuna rises substantially, e.g. on 11 October 2016 at Kudesia Ghat (Delhi) it shot up to 102 total suspended solids (TSS) against 56 on 7 October 2016, 38 Biochemical Oxygen Demand (BOD) against 25 on 7 October 2016, 120 chemical oxygen demand (COD) against 80 on 7 October 2016, and 640 total dissolved solids (TDS) against 609 on 7 October 2016; of course, these are much higher than the prescribed safe limits by Bureau of Indian Standards (BIS) (*Indian Express*, 24 October 2016).

Sukhna Lake: Sukhna Lake, near Chandigarh, had almost silted up in the 1970s. A team was led by P. R. Mishra, of Central Soil and Water Conservation Research and Training Institute, to check the soil erosion, they constructed four check dams and planted trees in the Shivalik hills. He motivated local farmers of the Sukhomajri village (Panchkula, Haryana) to refrain from extensive grazing, and the latter agreed in return for dam water. In five years, farmers' income increased from higher crop yields, forage grasses ('bhabbar' and 'mungri'), and the rearing of livestock. The villagers formed the Hill Resource Management Society (HRMS) in 1983 and regulated the use of water and forest resources. HRMS levied a fee on water users, sold bhabbar grass to the people and paper mills, and leased ponds for pisciculture for village development. It was a '*bottom-up*' approach to development. HRMS earned a profit of Rs 43,800 in 1986. Sukhomajri became India's first village to pay income tax! Production of bhabbar grass and food crops doubled, and milk production increased to 658 litres daily. In 1990, HRMS earned a profit of Rs 68,800 which increased to Rs 1.7 lakh in 1996. But due to conflict with Dhamal village, three dams fell into disuse by 1997. HRMS' earning halved in 1998, after the Forest Department claimed 55% stake in the revenue earned. Household income rose to Rs 40,000 a year, the groundwater level was up to 42 mbgl, milk yield to 995 litres daily and crop yield to 4.2 million kg in 1999–2000. In the 2000s, revenue from forest and water dried up, bore wells increased and ground water level dipped to 55 mbgl but average household income increased (mainly due to milk production) to Rs 60,000 a year. Haryana Urban Development Authority (HUDA) acquired 32 hectares of the village land in 2012, fourth dam was in disuse, so the crop yield was reduced to 600 kg (2016) and farmers now prefer dairy and work outside in brick kilns (Sengupta, 2017).

There have also been many oil spills in the rivers, seas/oceans all over the world. There are severe effects of oil spills, especially pollution caused by oil spills on the marine environment in the world (Box 8.2).

Box 8.2: Effects of Oil Spill/Pollution on Marine Environment

(1) Highly complex ecosystems exist within the marine environment, and substantial fluctuations in abundance and diversity occur as normal functioning.

318   Global Environmental Issues

(2) Key mechanisms for environmental damage from oil spill are smothering and toxicity; severity depends on the type of oil spilt and how quickly it dissipates relative to the location of resources sensitive to oil pollution.

(3) Sea birds are specifically at risk – penguins in particular – and penguins respond well to cleaning but others may not survive for long when released back into the wild after cleaning.

(4) Short-term impact of oil spill can be severe; long-term damage is restricted to geographically discrete areas.

(5) Well-designed reinstatement/restoration measures may enhance natural recovery processes.

(6) During Gulf War I and II, due to the burning of oil wells, there was severe damage to marine life, as the most vulnerable organisms are found on the sea surface or shorelines.

(7) Mangroves are salt-tolerant trees/shrubs and are habitats of crabs, oysters, etc. and nursery areas for fish and shrimp; these are highly vulnerable to oil contamination.

Source: ITOPF; London (UK) www.itopf.org.

More specifically, local people, especially women, fight for community rights in case of oil spill/pollution in rivers and oceans. For instance, the women of Cuninico (Peru) successfully fought for food, water, health, education, and economy (Box 8.3).

Box 8.3: How Cuninico Women Fought for Community Rights (Peru)

On 22 June 2014, there was an oil spill from the State-owned Petroperu, Norperuano Pipeline, to the tune of 2,358 barrels of crude oil at km 42 point on section I of the pipeline and the damage from the creek to the indigenous Cuninico community in Urarinas district in Loreto Province of Peru. The contract with the company to clean up the oil spill prohibited the men from 'speaking out, or complaining about the situation, or speaking against Petroperu or government'!

Hence, the women, under the leadership of Flor de Maria Parana (47 years old), raised their voices in an organised way because of two types of consequences of the pollution problem: First, the socio-economic problems like lack of food and water as the fish, crops, water, etc. were highly contaminated; second, health and hygiene problems like respiratory diseases, skin infections, diarrhoea, and even spontaneous abortions. Hence, women as mothers, for the first time, realised their community's suffering. After a few months from the oil spill incident, Flor de Maria Parana was designated as Cuninico's 'indigenous mother', with the same decision-making power as the 'apu', the leader elected by the community. In 2016, in Chile, at a hearing of the Inter-American Commission on Human Rights, regarding this oil spill, Flor de Maria Parana strongly put forth her points about the contamination of water resulting in a lack of food and drinking water, and various diseases, especially among children. After such a crisis, three grassroots organisations were duly formed: Organisation of Indigenous Women of the Maranon (Ordemin), Organisation of Native Women of Maranon (Orgamunama), and Association of Indigenous Women (Admic) – all women from Cuninico. Hence, in 2017, the Commission strongly recommended to the Peru Government to attend to the Cuninico and three other communities affected by the oil spill. Women demanded 'remediation', a decent and healthy life, and good education for their children. They made their organisations a part of the working groups with the Presidency of the Council of Ministers. In 2016, some residents of Cuninico were detected with high levels of heavy metals like mercury and cadmium in their bodies. As per Alicia Abanto, an officer at Peru's Ombudsperson, oil spill results in not only environmental problems but also social problems, due to the heavy presence of outsiders, like increased alcohol abuse, leading to sexual assaults against women and minors.

Source: F. G. Delgado and V. Romo (2020).

Social Consequences of Pollution

Fine particles, especially from PM 1 and PM 2.5, affect the human body (as well as animals) in the following ways:

a) It affects lungs – worsening chronic obstructive pulmonary disorder (COPD) and reduction in lung function.
b) It affects blood – particles pass through walls of blood vessels, affecting the flow of blood and also causing thrombosis.
c) It affects the vascular system – atherosclerosis, reduction in the diametre of blood vessels and high blood pressure.
d) It affects the brain – an increase in strokes, brain ischaemia, cognitive disorders, neurodegenerative illness.
e) It affects the heart – changes in heart function, increase in heart rhythm problem, lungs of people living in Delhi/NCR are turning black while those of hill-residing people are pink and healthier.
f) It affects reproduction – fertility problems, miscarriage, foetal growth problems, premature birth, and low-weight birth.

In view of the above, many specialist doctors in Delhi suggested (in November 2017, 2018, 2019, 2020, and 2021) avoiding morning walk in Delhi/NCR to save lungs. They even advised against exercising/running a marathon in the open because normally one breath six litres of air per minute at rest while during exercising one may inhale up to 20 litres of air per minute, thus causing more toxicity (due to polluted air). Further, running a marathon in polluted air conditions can deposit about two table spoons of toxic ash in lungs. Air pollutants trigger allergies, cough, viral fever, lung infections, high blood pressure, asthma, diabetes, heart disease, etc. more among children than adults because they breathe faster than adults per unit of body weight and inhale more contaminants. Further, they spend more hours daily outdoors than

320 Global Environmental Issues

adults and also play on the open ground where particulate matter and air toxins are higher. Often children living and studying in schools, near traffic busy roads, have under-developed lungs.

Nowadays 'smog' (smoke+fog), and also particulate matter, a major consequence of vehicle emission and burning of 'Parali' (paddy stems) in Punjab, UP, and Haryana, has become a very dangerous pollutant for Delhi/NCR people. As per a WHO study, air pollution killed at least six lakh (600,000) people in India against a total of 30 lakh (30,00,000) such deaths in the world in 2012 (China had 8 lakh such deaths then), but this is an underestimate because figures of India do not include NO_x and ozone and pre-term births or low birth weight (*The Hindu*, 28 September 2016). In 2015, India's air pollution further deteriorated causing 11.98 lakh (1.198 million) deaths against 11.80 lakh (1.18 million) deaths from air pollution in China in that year (*The Hindu*, 17 November 2016). Further, as per WHO, *indoor air pollution* kills 43 lakh (4.3 million) persons annually in the world, and 13 lakh (1.3 million) such deaths occur in India against 15 lakh (1.5 million) such deaths in China (*The Hindu*, 30 December 2015). Open defecation also leads to air and water pollution; as per UNICEF, 1 rupee spent on sanitation saves Rs 4.30 medical cost averted, time saved, and mortality averted – saving, in total, Rs 50,000 per household every year (Sharma 2017). Thanks to the Swachh Bharat Mission (of GOI), on 2 October 2019, the entire nation was declared ODF (Open Defecation Free).

A study has shown that children in Delhi are worst affected by poor air quality with 2% poor lung capacity; and 19% bad lung capacity; further, due to dust and particles in Delhi, 92% of children using unpacked transport faired poor while 8% of children using packed transport were found poor in terms of lung capacity (*The Hindu*, 26 November 2015). Unfortunately, we are breathing air pollutants at more than the safe level in most cities, thus causing serious health hazards. It may be perused in Table 8.6.

In addition, on 2 October 2018 (at 4 pm), the overall air quality index (AQI) score in Bhiwadi (Rajasthan) was recorded at 304 – a very unhealthy level of pollution – while, on the other hand, on the same day, Tirupati (AP) recorded a healthy AQI score of 35 (*The Hindu*, 3 October 2018).

Table 8.6 Air Pollutants as People Breathe in Different Cities in India (2 October 2018)

S. No	Cities	SO_2	NO_2	CO	PM 2.5	PM 10	Air Quality Code
1	Ahmedabad	71	234	17	48	NA	Poor
2	Bengaluru	10	30	46	32	51	Good
3	Chennai	7	15	50	68	NA	Good
4	Delhi	13	42	23	279	NA	Poor
5	Hyderabad	4	62	30	119	96	Moderate
6	Kolkata	6	61	58	127	117	Moderate
7	Lucknow	4	57	30	235	NA	Poor
8	Mumbai	20	47	97	82	107	Moderate
9	Pune	34	13	58	87	89	Good
10	Vishakhapatnam	6	47	30	NA	122	Moderate

Source: *The Hindu*, 3 October 2018.

Global Environmental Issues   321

Table 8.7 Health Hazards Caused by Various Pollutants (India)

Sl. No.	Gases	Health Hazards
1	SO$_2$ (sulphur dioxide)	a) Short-term exposure can harm respiratory system, making breathing difficult b) It can affect visibility by reacting with other air particles to form haze and stain
2.	NO$_2$ (nitrogen dioxide)	a) It aggravates respiratory illness, causes haze to form by reacting with other air particles b) It causes acid rain and pollutes coastal water
3.	CO (carbon monoxide)	a) High concentration in air redness oxygen supply to heart and brain b) At very high level, it may cause dizziness, confusion, unconsciousness, and even death
4.	PM 2.5 and PM 10	Particulate matter pollution can cause irritation of the eyes, nose, and throat; coughing; chest tightness and shortness of breath; reduced lung function; irregular heart beat; asthma attacks; heart attacks; and premature death (of people with heart or lung disease)

Source: *The Hindu*, 3 October 2018.

Various pollutants cause health hazards in Table 8.7.

In addition, as per a study by the Energy Policy Institute, the University of Chicago (US), the residents of Patna are losing 7.7 years of life due to bad quality of air, though such reduction in life at Patna was only four years in 1998; further people in Siwan are losing 9.1 years of life due to bad air, and Sheohar, Bhojpur, Buxar, Gopalganj, Siwan, Muzaffarpur, and Vaishali have air quality worse than Patna (*The Morning India*, 6 November 2019). In Delhi/NCR, as per the same study, people are losing ten years of their life due to bad air quality. On 2 November 2019, AQI of Patna was 428, of Muzaffarpur 382, and of Gaya 294. Further, due to nitric oxide in polluted air, human lungs in Delhi/NCR have turned black, while, due to better air quality, human lungs in Himachal Pradesh are red. In Delhi/NCR people are inhaling bad air equivalent to 40 cigarettes per head daily! Air Pollution (PM 2.5) caused 54,000 deaths in Delhi, 25,000 deaths in Mumbai, 12,000 deaths in Bengaluru, 11,000 deaths in Hyderabad, 11,000 deaths in Chennai, 6,700 deaths in Lucknow in 2020 – and 1,60,000 deaths in five most populous cities in the world in 2020; further pollution-related cost percentage of GDP was 13% in Delhi (Rs 58,895 crores), 14% in Lucknow, 9% in Mumbai, 8% in Hyderabad, 6.8% in Bengaluru, and 6.8% in Chennai. Air pollution-related economic losses were Rs 80,000 crores in Lucknow, Rs 58,895 crores in Delhi, Rs 26,912 crores in Mumbai, Rs 12,365 crores in Bengaluru, Rs 11,637 crores in Hyderabad, and Rs 10,910 crores in Chennai in 2020, according to Greenpeace Southeast Asia (*TOI*, 19 February 2021).

Similarly, water pollution is also dangerous to human and animal lives; thousands of tanneries and associated factories in Kanpur and Agra have highly polluted Ganga and Yamuna rivers' water, respectively. In nine districts of UP (Ghaziabad, Aligarh, Gautam Budh Nagar, Bijnor, Rampur, Sambhal, Bulandshahar, Amroha, and Hapur), supplied water contained highly toxic elements like arsenic and lead, high level of dissolved solids, and also has a 'pungent smell'. In over 300 schools of these nine districts of UP, water quality is very poor and has a high level of toxic chemicals, especially because of pollution from local industries and slaughter houses discharging effluents into rivers/ponds and drainage system – ultimately going to groundwater (*The Times of India*, 10 August 2017). These have serious health hazards like affecting organs permanently, serious skin diseases, dental decay, respiratory problems, heart problems, high blood pressure, bone deformity, etc. As a part of Namami Gange (Ganga Rejuvenation) programme, the Quality Council of India surveyed 578 water bodies in five states (UP, Bihar, Jharkhand, West Bengal, and Uttarakhand with a maximum of 329 water bodies in UP). Of these, 96% are ponds, 2% lakes, and 2% tanks. Out of 578 water bodies assessed, only 56% were found functional, 28% dried up, and the remaining 16% are Eutrophic (full of algae and aquatic plants, depletion of fish species, and overall deterioration of water quality). In UP out of 329 water bodies assessed, only 42% are functional, 37% dried up, and 21% are Eutrophic (Ghazipur in UP has a maximum number of functional bodies). In fact, encroachment and use of water bodies for dumping solid waste are causes of their drying up. In total, 411 water bodies out of 578 water bodies are surrounded by settlements (Vishwa Mohan 2021a).

Noise Pollution

What is noise after all? There is a difference between sound and noise. An *excessive or unwanted sound* is noise as it disturbs a person, group, neighbour, community, or people at a work or residing place; when the noise increases a certain limit of ears' capacity, it becomes noise pollution. The problem of noise pollution is closely associated with two new processes of modernisation: urbanisation and industrialisation. The problem of noise pollution massively increased in the nineteenth and twentieth centuries, and further increasing in the twenty-first century. Developed industrialised countries like the USA realised it in the early twentieth century, as the US Congress reported to President as back as 1937: 'The large city and essentially its central business district is so characteristically a place of noise that a sudden wave of silence frequently proves to be oppressive to the urbanite for he is accustomed to distracting sounds of all kinds'. It adds more perceptively:

> Screeching brakes, screaming trolly cars, rumbling trucks, rasping auto horns, barking street vendors, shouting news boys, scolding traffic whistles, rumbling elevated trains, rapping pneumatic hammers, open cut-outs, and now advertising sound trucks and aircraft with radio amplifiers, when added together, constitute a general din'.
>
> (NBS 1971, 3)

The situation, thereafter, in cities in the developed countries has improved substantially but due to following a similar development path, cities in developing countries like India are experiencing worse than that – in fact, during general elections, festivals (especially Diwali – the festival of light), the victory of India in cricket matches (symbol of nationalism), etc., the noise pollution in India crosses all the limits! In January 2000, for the first time, the Ministry of Environment and Forest (Government of India) fixed the norms of permissible noise levels in residential, commercial, industrial areas, and silence zones (Table 8.8).

Global Environmental Issues   323

Table 8.8 Permissible Noise Levels in Different Areas/Zones (India)

Area Code	Category of Area	Limits in db (Decibel)	
		Day Time	Night Time
A	Industrial area	75	70
B	Commercial area	65	55
C	Residential area	55	45
D	Silence zone	50	40

Source: Ministry of Environment and Forest, GOI, 2000.

However, in practice, in all these areas/zones, usually the noise level is more than the permissible limit, especially in the metropolitan cities – Mumbai, Delhi, Bengaluru, and Chennai are the worst in this regard: with average noise level (in 2011) being 129 db in Chennai, 130.66 db in Bengaluru, 132 db in Delhi and Mumbai against the maximum permissible limit being 85 db as per WHO!

Sources of Noise Pollution: Major sources of noise pollution are as follows:

(a) Transport motor vehicles (especially diesel fuel operated), railways, airports, ports, highways (24-hour functional), motor boats, autorickshaws/bikes
(b) Industrial factories/power plants
(c) Recreation media/areas – TV, radio, music, theatres, music system
(d) Air conditioners/coolers/washing machines, mixers, vacuum cleaners
(e) Diesel generators
(f) Boring machines for irrigation/drinking water and pump sets
(g) Flour/oil/rice mills
(h) Construction works
(i) Fire crackers and rounds of firing from fire arms
(j) Bomb blast/explosion

Effects of Noise Pollution: These are as follows:

(a) *Effects on Health* – The foremost effect of noise pollution is pain in the ears, burst of the eardrum, and partial or total deafness finally. Various studies have confirmed that most of the traffic constables (up to 75%) on regular duty at the busiest crossings in Indian cities suffer from partial or total deafness. According to WHO, noise above 30 db may cause sleeping disorders, and above 80 db may cause hearing impairment. Further, there are cardiovascular (high blood pressure, stroke) or neurological effects too. For instance, a Russian Scientist, N. N. Shatlov, in a study of 589 factory workers, found that two types of noise have different effects. First, *continuous noise* results in 'arterial tension, downward trend in venous pressure, reduced peripheral resistance and bradycardia'. Second, *intermittent noise* causes 'hypertension, rising arterial pressure and frequent capillary spasms' (cited in NBS 1971). Further, a higher level of noise results in more accidents, leading to deaths/injuries.

324 Global Environmental Issues

(b) *Psychological Effects*: Borsky notes four factors that enhance or impede noise acceptability:

   (i) Feeling about the necessity or preventability of noise

   (ii) Feeling the importance of the noise source and the value of its primary functions

   (iii) Types of living activities affected

   (iv) Extent to which there are other things disliked in the residential environment

   In fact, the level of annoyance increases with the level of noise increase. Further, due to aircraft noise, many persons cannot sleep. In addition, headache, insomnia, and nervousness are highly associated with annoyance.

(c) *Social Effects*: Since noise-preventing devices are costlier, usually consumers are not willing to buy such products. But collective actions may control the effects of noise in the community. Regulations are made to reduce the incidence of deafness caused by noise and to minimise noise disturbances in the community. Occupational safety measures in factories on construction sites and acoustical treatment of buildings by municipal bodies. Local citizens may also claim damages for noise disturbance. Further, public transport is encouraged in lieu of individual vehicles.

(d) *Ecological Effects*: Due to noise of various animals (fish, dogs, cattle, etc.), birds and plants, too, suffer. Hence, there is less growth rate of greenery/plantation in noise-ridden cities. Birds and animals are too sensitive to noise, resulting in less breeding, less chirping, and less enjoyment. Animals change their natural behaviours or they relocate to avoid noisy areas. Marine animals/organisms are also badly affected by noise, especially from commercial vessel traffic, oil and gas exploration, seismic surveys, and military sonar. Artificial noises change the acoustic environment of marine and terrestrial habitats affecting animal's ability to hear, make it difficult to find food, locate mates and avoid predators, or impair its ability to navigate, communicate, reproduce, and participate in normal behaviours (Box 8.4).

Box 8.4: Changing Behaviour of Birds Due to Noise (World)

The following changes in birds' behaviour due to noise are notable:

(a) In a study in 2007, it was found that urban European robins (*Erithacus rubecula*), due to more noise in the daytime, changed their singing at night (quieter) – thus adjusted their timing of singing but in lieu of sleep, it affected their behaviour.

(b) In the Western US, the scrub jays avoid nesting in noisy areas (near gas wells) – thus noise 'acts as a form of sensory pollution, forcing animals to adapt their calls to be heard over it, or leave the area altogether'. But when the scrub jays left (and relocated), the forest of their previous habitat declined. In New Mexico, in normal situations, the birds collect and bury pine seeds in preparation for winter; they fail to collect all the seeds they bury, and these become trees in future. But near the gas wells, without scrub jays to plant the seeds, the pines are disappearing.

(c) A study in 2013 (at Boise State University) placed many speakers in the woods which played the sounds of a busy highway at regular intervals. There was a stop where migratory birds used to stop and take rest. Due to traffic noise played through speakers for four days, one-fourth of those birds did not stop there for rest. After the speakers were off, these bounced back.

(d) Noise pollution kills the sex life of animals too. A study in Melbourne (Australia) by Kirsten Parris et al. found the correlation of noise with an increase in the frequency of their calls. The mating call of male pobblebonk frogs could be heard up to 800 m away by interested females. At very noisy sites, this is reduced to 14 m. As female frogs of some species prefer lower-pitched calls, due to noise, if male frogs change their call to a higher frequency to be heard, female frogs don't like that.

Source: K. Parris and R. Mc Cauley (2016), Noise Pollution and the Environment, Australian Academy of Science, science.org.au.

Social Inequality in Environmental Noise Exposure

Some social groups have more noise exposure; e.g. the poor people live in low-quality (polluted) environments, and similarly, they work in adverse (unhealthy) situations that have a higher prevalence of noise. Consequently a higher level of noise leads to various kinds of effects, beginning with health, psychological, and social effects. Further, a less healthy life often contributes to increased vulnerability to noise-linked health effects. On the other hand, the upper class usually has better health and can afford better houses, noise-reducing devices like earplugs (preventive) as well as better curative measures. Further, in a multiracial society, the blacks or immigrants usually have a higher level of noise exposure. In US cities, due to more segregation, such social inequality in environmental noise exposure is more pronounced than in the European cities. A study from Hong Kong found a statistically significant association between lower income, lower educational attainment, and higher noise exposure at the street block level (Lam and Chan 2008).

Nuclear Radiation

Usually, radiation exists in nature as light and sound. The term 'radiation' originates from the phenomenon of 'waves radiating' (travelling outward in all directions) from a source. Radiation, however, is essentially a transmission of energy in the form of waves of particles through space or through a material medium. This includes the following four types:

(a) Electromagnetic radiation – radio waves, microwaves, infrared, visible light, ultraviolet, x-rays, and gamma radiation (χ)

(b) Particle radiation – alpha radiation (α), beta radiation (β), and neutron radiation (particles of non-zero rest energy)

326 Global Environmental Issues

(c) Acoustic radiation – ultrasound, sound, and seismic waves
(d) Gravitational radiation – in the form of gravitational waves or ripples in the curvature of spacetime

As radiation expands when it passes through space, and as its energy is conserved (in a vacuum), the intensity of all types of radiation from a *point source* follows an inverse-square law in relation to its distance from its source. In view of the surging global demand for more energy, nuclear energy appears to be a good alternative as it is carbon-neutral but it has severe risks. For example, nuclear accidents at Three Mile (USA) in 1979, at Chernobyl (Ukraine) in 1986, and at Fukushima (Japan) in 2011 have been disastrous, especially the Chernobyl accident had multiple effects even up to thousands of kilometres away from the meltdown site, and even after more than two decades following that accident.

Consequences of Radiation: Radiation has the following negative consequences for the health of *humans and other animals:*

(A) Ionising radiation may cause cancer or even genetic damage. Non-ionising radiation may also cause damage to living organisms like burns. As per WHO, radiofrequency electromagnetic fields (including microwave and millimetre waves) are carcinogenic to humans.

(B) Nuclear radiation also has *ecological consequences:* rats showed changes in sleep behaviour after drinking water poisoned with 'only' 400 Bq (Becquerel) per litre, and onions showed a significantly elevated rate of chromosomal aberrations at levels as low as 575 Bq per kg (Henrik Von Wehrden et al. 2011). Further, there are physiological and morphological changes in some cases, and their long-term consequences on the ecosystem as a whole. In addition, radiation levels in mosses, soil, and glaciers have remained greatly elevated in several locations across Europe. Moreover, lethal mutations (after Chernobyl accident) were also observed in various animal and plant species. There are also changes in genetic structure in various organisms including fish and frogs. Again, biological accumulation through the food chain may negatively affect some species, especially those at higher trophic levels and those depending on strongly affected food items. After the Chernobyl accident, humans were evacuated, leading to a major change in the land use – agricultural fields were abandoned, vegetation in place of former urban areas, and the number of many wildlife species increased in response, e.g. common crane, and eagle owl increased in number, but on the other hand, species attached to farming, like the white stork, stopped breeding! Further, Chernobyl accident had effects on ecosystem services to humans – especially due to caesium-137 (radioactive material), having a longer life, freshwater, and associated fish are not safe for human consumption, in both present, and future; agricultural productivity is low – crops and timber were contaminated across the Europe! Timber in Sweden, fish in Finland, crops in the entire Scandinavia, wild foods (mushrooms) in Poland, game meat in Germany, and berries in Finland (Von Wehrden 2011).

In fact, there are four major challenges for the conservation management of nuclear disasters (ibid):

(a) Direct conservation via decontamination and restoring polluted ecosystems (soils and water bodies) is often technically or financially not feasible.

(b) Nature conservation measures are often thought secondary to measures to preserve and restore human health in affected regions.
(c) Conservation managers often lack a solid scientific basis to understand and mitigate how nuclear disasters affect biodiversity and ecosystem services.
(d) Effects of nuclear disasters on biodiversity and ecosystem services are spatially and temporally heterogeneous, hence difficult to follow the appropriate conservation measures, e.g. pieces of evidence of both rise and decline in the number of wildlife in the affected regions were found.

(C) There are also *emotional consequences* of nuclear disasters – anxiety, depression, post-traumatic stress disorder, and psycho-somatic symptoms. In this regard, mothers of young children and cleaning staff are the top risk groups. Thus, there is also an increase in mental health problems in the long run.

Acid Rain or Acid Deposition

Acid rain has also been a global environmental issue since the late 1960s. The term 'acid rain' was used for the first time by Robert Angus Smith in 1852 during his research on rain water chemistry in various industrial cities in England and Scotland (UK). It occurs due to air pollution In fact, rain, by nature, is acidic, but it is acidified more by the air pollution from factories, transports, and power stations (especially thermal power stations as well as homes/offices). Industrialisation in the western countries, especially North-West Europe and North-East US, resulted in acid rain there in the 1970s and 1980s. In 1980, some North-East States of the USA and the Ontario Province of Canada sued the US Environmental Protection Agency to take action to control acid precursor emissions emanating from seven states. US Congress constituted National Acid Precipitation Assessment Programme (NAPAP) and directed it to carry out scientific, technological, and economic study of acid rain issue under the Acid Precipitation Act of 1980 (Kumar 2017). Acid rain also occurs in Asia and parts of Africa, South America, and Australia. There are two types of acid deposition:

(a) *Wet deposition:* acidic rain, fog, mist, and snow, if the acid chemicals in the air are blown into the areas of wet weather; acid can fall on the ground in the form of rain, snow fog, or mist. It actually affects various plants and animals.
(b) *Dry deposition:* In case of dry weather, acid chemicals may get into the dust or smoke and fall on the ground through dry deposition – these stick to the ground, buildings, homes, vehicles, trees, etc.; dry deposited gases and particles may be washed away from these surfaces by rains.

Cause of Acidification: Primary causes of acid rain are sulphur dioxide (SO_2) and nitrogen oxides (NO_x) (and to some extent ozone) when the latter are emitted into the atmosphere, and carried by the wind and air currents. Actually, SO_2 and NO_x react with water, oxygen, and other chemicals and thus form sulphuric and nitric acids. In fact, the burning of fossil fuels, in order to generate electricity or to start and operate vehicles, generators, and machines, manufacturing works, oil refineries and other industries are the main sources of producing SO_2 and NO_2. In addition, other sources are a smattering of iron and other metallic (Zn and Cu) ores, the manufacturing of sulphuric acids, and also the operation of acid concentrators in the petroleum industry. These are anthropogenic sources. On the other hand, the natural sources of sulphur pollutants are oceans and volcanic eruptions too (Kumar 2017). Compared to SO_2, the level of NO_X is small; however, its contribution to the production of acid rain is increasing.

Chemical reaction during acid rain formation involves the interaction of SO_2, NO_X, and O_3. When pollutants are emitted into the atmosphere by tall smoke stakes, the 'molecules of SO_2 and NO_X are caught up in the prevailing winds, where they interact in the presence of sunlight with vapours to form sulphuric acid and nitric acid mists' (Kumar 2017, 55). Needless to say that such acids remain 'in vapour state under the prevalent conditions of high temperature'. On the other hand, 'when the temperature falls, condensation takes the form of aerosol droplets' – these, due to the presence of unburnt carbon particles, are 'black, acidic, and carbonaceous'.

Coal is particularly rich in sulphur, and when it is burnt, its elements get oxidised ($S+O_2 \rightarrow SO_2$). Thus, oxidation of sulphur to SO_2 takes place directly in the flame; hence SO_2 is discharged to the atmosphere from the smoke stacks. The degree of acidity is measured by the pH value (potential hydrogen). No doubt, the pH of normal rain water is also acidic because water reacts, to a slight extent, with atmospheric carbon dioxide (CO_2) to produce carbonic acid. Acidic water with a pH of less than 6.5 is contaminated with pollutants, hence unsafe to drink. The pH level of common bottled water is 6.5–7.5, bottled water labelled as alkaline – 8–9, ocean water – 8, and acid rain – 5–5.5.

Effects of Acid Rain: Acid rain has many effects on both nature and humans (Kumar 2017):

(a) Various water bodies (lakes, rivers, ponds, streams) directly get acid rain, as both dry and wet depositions run off to forests, fields, roads, etc., and then reach the water bodies.

(b) The two pollutions – sulphur dioxide (SO_2) and nitrogen oxides (NO_X) – causing acid rain harm human health, as these gases' fine particles are blown by winds to long distances and due to inhaling enter into human lungs, causing asthma, bronchitis, and allergies.

(c) Acid rain may harm trees and plants, too, or even some areas of forests – as leaves and needles turn brown (due to acid rain) and fall down soon though these should be green.

(d) Marble and lime stone consist of calcium carbonate ($CaCO_3$), and these are used in buildings and monuments but these are gradually eroded away by acid rain's calcium carbonate and sulphuric acid – resulting in the dissolution of $CaCO_3$ to give aqueous ions.

Measures: Various damages, caused by acid rain, to water bodies may be removed by adding lime. Further, some chemicals like caustic soda, sodium carbonate, slacked lime, and lime stone may be used to raise the pH value of acidified water of water bodies. However, liming is quite costly and even then it cannot fully de-acidify water.

Therefore, individuals, communities, organisations, and government bodies/agencies should be made conscious of energy conservation and reducing emissions by

Global Environmental Issues   329

using public transport, using vehicles emitting low NOx and using energy-efficient electrical appliances; but the major role is to be played by industries. It is now duly recognised that globally the issue of acid rain is often over-shadowed by climate change; but earlier it was a reality in industrialised countries, later it was stopped with the efforts of scientists (biologists), policy-makers, and poeple's organisations (Box 8.5).

Box 8.5: The Bittersweet Story of How We Stopped Acid Rain in North

In the mid-1980s, in Canada's Killarney Provincial Park, the lake floor was visible at the depth of 50 ft. But the lake was altered by acid rain, as it was situated near the nickel and copper smelters. Even the algae were no more as the water was lifeless. In other lakes in North-west Ontario (Canada), the situation was similar. In Europe, Canada, and the USA, acid rain stripped forests bare and wiped lakes clear of life, and harmed human health and crops in China, caused by sulphur dioxide and nitrogen oxides emitted by fossil fuel combustion by vehicles, and smelters, and coal-burning utilities in industries. In 1963, Gene Likens collected a sample of rain water at the Hubbard Brooke Experimental Forest in New Hampshire's White Mountains – it was 100 times more acidic than presumed. Further, evidence also came from experiments at north-west Ontario's Experimental Lakes Area (ELA). Its soft-water lakes were far enough from sources of pollution that they had escaped the effects of acid rain, acting as a control. Scientist David Schindler and his team (during 1976–1983) lowered the pH of lake 223 from 6.8 (close to neutral) to 5.0 (slightly acidic), and even before reaching 5.0 (pH) it harmed the fish and at pH 5.6 most of the lake trouts' preferred food had died as acidic water dissolved their protective coats. According to Schindler, 'Lake trouts stopped reproducing not because they were toxified by the acid, but because they were starving to death'. Carol Kelly reached ELA in 1978, and Schindler ordered the crew 'to take the lake down to a given pH, and then add enough acid to hold it there'. It was discovered that alkali-producing microbes were capable of buffering some of the acidity, helping the lake chemistry to recover. Thus, acid could be neutralised by bacteria living in every lake. Then Carol Kelly travelled elsewhere in Canada, the USA, and Norway to lakes acidified atmospherically and found that acid-neutralising bugs exist in the sediment in most lakes which could recover if pollution causing the acid rain were eliminated. This study's photographs of starving fish from lake 223 and efforts of environmental groups like Canadian Coalition on Acid Rain persuaded the policy-makers to make laws for quality standards. But the sceptics denied acid rain for long, under the influence of industries and rich elites. Finally, the US legislated Clean Air Act in 1990, though earlier the USA accused Canada of acidifying lakes in the boundary waters. The USA established acid rain programme then, and Canada also took similar action. Lake 223 is no longer acidic. Compared with the 1990 level, sulphate ions in the atmosphere have dropped substantially; but nitrates from agriculturally emitted ammonia released from fertilisers and livestock feed still contributes to nitric acid precipitation. Further, acid rain from both sulphur and nitrogen is an increasing problem in Asia. The

330   Global Environmental Issues

> acid rain problem was solved in North America because it became a non-partisan issue. A similar partnership of people and scientists is the need of the hour for tackling climate change.
>
> Source: bbc.com/Future (Lesley Evans Ogden, 7 August 2019).

Analysis of rain water samples from Nagpur, Mohanbari (Assam), Allahabad, Srinagar, and Kodai Kainal showed a pH level ranging from 4.77 to 5.32 in 2012, receiving 'acid rain' (rain water with pH below 5.65 is considered acidic). This may be seen in Table 8.9.

Ozone Depletion

Ozone is a colourful gas in the upper atmosphere and is one of the main causes of global warming. It absorbs most of UV (ultraviolet) radiation. Depletion of the ozone layer means that a belt of the naturally occurring gas 'ozone' is depleting. It exists about 15–30 km above the earth. It acts as a shield from the harmful ultraviolet radiation that is emitted by the sun. As all sunlight contains some UVB, even with normal stratospheric ozone levels, it is essential to protect our skin and eyes from the ultraviolet rays of the sun.

In the early 1970s, some scientists found that CFCs release chlorine into the stratosphere that ultimately destroys the ozone layer. It is also notable that CFCs are used all over the world as coolants in air conditioners and refrigerators as well as other appliances and equipment.

Ozone layer's depletion, above the Antarctic, due to huge pollution, came to light in the 1980s. Due to the low temperature of the Antarctic, CFCs convert faster into chlorine. In the South, especially during the summer, when the sunshine is quite longer during the day, such chlorine reacts with ultraviolet rays and damages the ozone on a larger scale even up to 60–65%. On the one hand, concentrations of the ozone are affected by the release of manufactured substances (that interact with and destroy ozone), and, on the other hand, other factors like increased circulation between the stratosphere and the troposphere, and the chemistry of the atmosphere in relation to the release of particulates and other substances from human activities and the burning of biomass also affect. Further various factors influencing ozone concentration in the stratosphere are directly linked to climate change (that is linked to human activities).

Table 8.9 Declining pH level in Rainwater at Different Places in India (1981–2012)

S. No	Stations	pH in 1981–1990	pH in 2001–2012
1	Srinagar (J&K)	7.15	6.11
2	Mohanbari (Assam)	5.95	4.77
3	Allahabad (UP)	6.67	5.32
4	Nagpur (Maharashtra)	5.87	4.87
5	Kodainkainal (TN)	6.01	5.19

Source: IMD and Indian Institute of Tropical Meteorology, cited in *The Times of India*, 4 March 2017.

Effects of Ozone Depletion

(A) **Effects on Human Health:** Except for one positive benefit to humans from sun-rays, there are more negative effects on humans (Solomon 2008). Long-term effect of UV-B radiation is skin cancer that has increased over the twentieth century – melanoma causes most of the deaths from skin cancer. The probability of surviving a melanoma decreases with age and is lower for boys than girls. Some inherited or acquired traits that increase oxidative stress are associated with melanoma and its precursor lepion dysplastic nevus (Pavel et al. 2004). UV radiation-induced immunosuppression substantially increases the progression of both melanoma and non-melanoma cancers. Further, climate and ambient temperature influence the behaviour of school children, and hence their sun exposure, more than ambient solar UV radiation, increases the risks of all forms of skin cancer.

Second, the effect of UV radiation is acute (latent) or long term after a short exposure to greater intensity or long term following chronic exposure to less intensity. The effect of UV radiation is generally from indirect exposure. UV radiation contributes to cataract formation. Ozone depletion would affect 5–20% of the US population, hence also leading to huge expenditure. Further, there is a close association between nuclear cataract and occupational sun exposure, especially for those between 20 and 29 years of age (Neale et al. 2003). Further, there is also a close association between pterygia (inflammatory, proliferative, and invasive growth that occur on the conjunctiva and cornea of the human eye) and environmental UV radiation (Al Bdour and Al Latayfeh 2004) for all age groups.

Third, UV radiation also affects the immune system – the degree of suppression and forms of cell-mediated immunity affected may vary due to quantity, quality, and timing of UV radiation; frequency of exposure; and extent and location of the body surface irradiated.

Fourth, on the positive side, UV radiation provides vitamin D – 90% of people get their vitamin D from UV radiation. Sunlight exposure prevents rickets during childhood; further, it plays a protective role regarding internal cancers (colon, breast, prostrate, and ovarian tumours). To some extent, sun exposure helps in auto-immune diseases (multiple sclerosis (MS), diabetes mellitus type 1, rheumatoid arthritis (RA), and inflammatory bowel diseases).

(B) **Effects on Ecosystem:** Plants exposed to ambient UV-B experience huge reduction in herbivore by insects if compared to plants cultivated under filters which particularly exclude UV-B component of solar radiation. Further, most terrestrial animals are protected from UV-B radiation by hair or feathers but Arctic collembolan (springtail) species respond to UV-B radiation in relation to the degree of pigmentation in their integument. Above-ground UV-B radiation can affect plant interactions with microbiota. Mycorrhizal fungi, associated with roots and important for plant mineral nutrition, decreased by 20% when plants were exposed to supplemental UV-B radiation above ground. Further, UV-B radiation elicits acclimatisation responses including increased activity of antioxidant enzymes, increased DNA repair capacity, and accumulation of phenolic compounds that serve as 'sun screens' or UVR filters. Overall, the effects of increased UV-B radiation on crop production are small and less than those associated with climate change. Regarding forestry, plants' responses to an increase in UV-B radiation include the following:

(i)   A reduction in height and biomass
(ii) Smaller and thicker leaves
(iii) Increase in reflective surface waxes
(iv) Increase in UV-B radiation absorbing compounds and antioxidant defence systems
(v)   Direct effects on photosynthesis and stomatal conductance (Searles et al. 2001)

(C) **Effect on Aquatic/Marine Ecosystems:** Phytoplankton forms the foundation of the marine food chain. Oceans in polar regions and clear lakes at high altitudes, where UV-B radiation penetrates far into the water, are more vulnerable. DNA damage is closely related to the penetration of UV-B into the water. The high sensitivity of picoplankton to ambient solar radiation may act as a primary driver of species composition and population structure and govern the dynamics of the microbial food web in clear oceanic waters (Llabres & Augusti 2006). Again, exposure to UV-B radiation changes the ability of picoplankton to take up nitrogen from the matrix (Sobrino et al. 2004). Phytoplankton (producer) can adapt to radiation if exposed for long to solar UV radiation, and they have protective mechanisms against high solar radiation. Scientists have found a direct reduction in phytoplankton production due to ozone depletion-related increase in UV-B radiation that causes damage to early stages shrimp, crabs, amphibians, and other marine animals. Consumers like zooplankton are also important in the food chain in salt water and freshwater systems, and can be affected by UV-B radiation. Fish can also be affected by UV-B radiation (though relatively less). However, the skin and eyes of fish may be damaged by UV radiation. The oceans provide at least 20% of their animal protein to about 18% of the world's population. The main causes of the decline in fish population are predation and poor food supply for larvae, over-fishing, increased water temperature, pollution, and disease.

(D) **Biogeochemistry and Atmospheric Processes:** The interactions of atmospheric chemistry, biogeochemistry, and UV-B radiation play a significant role in controlling life processes, climate, and their effects on the concentration of greenhouse gases. Nitrogen and phosphorus, necessary for terrestrial and marine ecosystems, may be affected by UA-B radiation. Soil organisms are vulnerable to damage from UV-B radiation but are rarely exposed to solar radiation under natural conditions. Further, solar UV-R has substantial effects on the chemical speciation of trace metals, e.g. iron, manganese, copper, chromium, and mercury, and may change their bio-availability to plants and animals. Solar UV-R has substantial effects on mercury, volatilisation, solubilisation, and methylation which are important for humans who may be exposed through the food chain, especially in polar regions like northern Canada. As far as atmospheric chemistry is concerned, the quality of the air depends on many factors: How fast chemicals are released and reactions of the substances UV-B radiation cause the photolysis of several atmospheric trace gases, e.g. sulphur dioxide, formaldehyde, and ozone. The direct effect of stratospheric ozone depletion is a potential reduction by about 30% in the amount of ozone transported into the troposphere (known as Stratosphere Troposphere Exchange (STE)) (Fusco and Logan 2003). Since 1990 there has been a significant increase in the surface ozone (up to 20%) in Antarctica and the oxidation potential of the Antarctic boundary layer is greater now than prior to 1980 (Jones and Wolf 2003; Solomon 2008).

(E) **Effects on Materials:** Synthetic organic polymers (plastics and rubber) and most natural biopolymers, e.g. wood, hair, and proteins absorb solar UV-R and, thereafter, undergo photodegradation. Most of the polymers are photolabile materials, slowly losing their desirable physical and mechanical properties on routine exposure to solar UV-R. Further, popular composites of plastic and wood are also susceptible to degradation unless protected with UV stabilisers. About one-third of plastic produced in North America and Europe is used in building products (pipes, cable coverings, window frames/doors, and organic protective coatings). Stratospheric ozone depletion accelerates light-induced degradation reactions in some materials (outdoor plastics) reducing their service life.

Now the question arises as to what efforts have been made to check it?

Montreal Protocol: There was the UN's Montreal Protocol agreed on 15 September 1987 for taking suitable measures. It came into force on 1 January 1989. All 197 member countries of the UN ratified Vienna Convention along with its Montreal Protocol. This has successfully led to the phasing out of production and consumption of major man-made 100 chemicals referred to as ozone-depleting substances (ODSs), e.g. CFCs, carbon tetrachloride (CTC), and halons. Developed countries agreed to phase out all ODSs by 2020 and developing countries will phase it out completely by 2030.

Hydrochlorofluorocarbons (HCFCs) are about 2,000 times more potent than CO_2 in terms of their global warming potential. Unfortunately, these gases are used in refrigerators, air conditioners, aerosols, and foam applications. There is a multilateral fund for the implementation of the Montreal Protocol, established in 1991; it provides technical and financial assistance to developing countries whose annual per-head consumption and production of ODSs is less than 0.3 kg to comply with the measures of the Protocol. Activities of the multilateral fund are implemented by four agencies:

(a) UNEP (United Nations Environment Programme)
(b) UNDP (United Nations Development Programme)
(c) UNIDO (United Nations Industrial Development Organisation)
(d) World Bank

Many developing countries have exceeded their reduction targets for phasing out ODSs, with the support of the multilateral fund. As non-ozone-depleting alternatives for phasing out CFCs and HCFCs, HFCs (hydrofluorocarbons) are used but some of these gases have high GWP (global warming potential) ranging from 12,000 to 14,000. Hence, parties to the Montreal Protocol agreed on 15 October 2016 in Kigali (Rwanda) to phase out even HFCs – 80–85% reduction by the late 2040s. The Kigali Amendment came into force on 1 January 2019. Parties have phased out 98% of ODSs globally compared to the 1990 level. Further, during 1990–2010, Montreal Protocol's measures reduced greenhouse gas emissions by the equivalent of 135 gigatons of CO_2, thus 11 gigatons a year. It has further been saving two million people each year by 2030 from skin cancer. The Kigali Amendment is further preventing emissions of up to 105 million tonnes of CO_2 equivalent to greenhouse gases. Thus, the Montreal Protocol is

334 Global Environmental Issues

helping to mitigate the problem of global warming and climate change substantially, on the one hand, and it also contributes to the realisation of Sustainable Development Goals, on the other. The Government of India's Ministry of Environment, Forests, and Climate Change is the nodal Ministry to implement Montreal Protocol and is on the right path.

Deforestation

Deforestation is primarily the conversion of forest land for other purposes or services as per human demand. According to FAO, deforestation is 'the conversion of forest to another land use or the long term reduction of tree canopy cover below the 10% threshold'.

India has a forest cover of 7,34,204 sq km – 21.71% (in 2022) of the total geographical area against the target of 33%. Further, there is 2.9% tree cover. However, all this forest cover is not dense; the intra-differentiation of forests in India may be seen in Table 8.10.

From Table 8.10, it transpires that only 2.54% of total surface land is very dense forest cover, and additionally 9.7% is moderately dense forest cover – thus, in total, only 12.3% of total surface land is dense forest. Further, the open forest is 8.7% and scrub forest is 1.2%. On the other hand, 77.7% of the total surface land is non-forest. Further, area-wise Madhya Pradesh (77,482 sq km) has the largest forest cover in India, followed by Arunachal Pradesh (66,688 sq km), Chhattisgarh (55,611 sq km), Odisha (51,619 sq km), and Maharashtra (50,778 sq km). In addition, in terms of forest cover as a percentage of total geographical area, the top five states are (a) Mizoram (85.4%), (b) Arunachal Pradesh (79.6%), (c) Meghalaya (76.3%), (d) Manipur (75.5%), and (e) Nagaland (75.3%). Further, the total mangrove cover in India is 4,975 sq km. The total bamboo-bearing area is 16 million ha. Wetlands within forest areas add richness to the biodiversity; there are 62,466 wetlands covering 3.8% of the area within the Reserve Forest Area (RFA) of India.

Finally, the total carbon stock in India's forest is 7,124.6 million tonnes (2019). The annual increase in the carbon stock is 21.3 million tonnes – 78.2 million tonnes CO_2 equivalent. In fact, globally there has been 3.16% reduction in forest cover during 1990–2015. The total global forest cover is about 30.6% at present compared to 31.6% in 1990 – annual loss of 18.7 million acres. In 2016, the University of Maryland found that 73.4 million acres of global tree cover were lost. The Government of India

Table 8.10 Differentiation in Forest Cover in India (2014)

Type of surface land	Area	%
(I) Forest:	83,471 km²	2.5
(a) Very dense forest	3,20,736 km²	9.7
(b) Moderately dense forest	2,87,820 km²	8.7
(c) Open forest	42,177 km²	1.2
(d) Scrub forest	25,53,059 km²	77.7
(II) Non-forest		

Source: A Down to Earth Annual State of India's Environment.

Global Environmental Issues   335

claims 24.56% of the geographical area as total forest cover plus tree cover, but many experts and NGOs do not agree to this, as many projects were cleared on forest lands.

Causes of Deforestation

(a) Rapid and subsidised industrialisation-leading to 'an overheated engine' (Myrdal 1968);
(b) Rapid urbanisation
(c) High population growth/pressure, especially in developing countries-more forest land used for agriculture
(d) Culture of consumerism – more (quantity), bigger (size matters), and visibility – how much is enough for a person, or a country?
(e) Shifting cultivation (Jhoom)
(f) Diversion of forest lands for development purposes for local people – housing, school/hospital buildings, roads, dams, tourism, etc.
(g) Forest fires – frequency due to more heat and less rains, and intensity increasing (e.g. in MP in India, also in California in the USA, and Australia)
(h) Nexus between forest officials-contractors-politicians-smugglers (e.g. Veerappan case in Tamil Nadu)

Loss of Forestlands Cover: According to the Centre for Science and Environment (New Delhi), there was a huge forest land diversion in India for various purposes during 1990–2012 – it is notable that the 'economic reforms' started in India in 1991, and hence the policies of liberalisation and privatisation had a great negative impact on Indian forests (see Table 8.11).

Thus, from above it is clear that during 1990–2012, in total 11,83,967 ha of forest land was diverted for different purposes – 14.8% for industries, 14% for power projects, 12% for irrigation, 13.5% for mining, 5.5% for social services, 5.1% for transport, and 3.9% for defence. More surprisingly, 31.12% of total forestlands diverted was in the form of regularisation of encroachments! It is relevant to quote Madhav Gadgil (1990) here:

Table 8.11 Diversion of Forestlands in India 1990–2012

Sl. No.	Purpose/Sector	Diversion of Forestlands	
		Area (ha)	%
1.	Defence	46,087.93	3.89
2.	Transport (rail and roads)	60,363.10	5.10
3.	Social services	65,165.24	5.50
4.	Irrigation	1,41,641.76	11.96
5.	Mining	1,59,660.34	13.49
6.	Power project (hydel, thermal, wind, and transmission lines)	1,67,118.00	14.12
7.	Others (including industries)	1,75,498.24	14.82
8.	Regularisation of encroachments	3,68,432.07	31.12
9.	Total forestlands diverted	11,83,966.68	100.0

Source: CSE (2014).

336 Global Environmental Issues

Table 8.12 Loss of Primary Forest and Tree Cover in India (2001–2018)

Year	Primary Forest Loss (ha)	Tree Cover Loss (ha)
2001	–	62,441
2002	11,718	53,035
2004	19,166	74,217
2008	20,702	86,045
2012	18,804	95,181
2015	20,997	1,16,374
2016	30,936	1,75,478
2017	29,563	1,89,677
2018	19,310	1,32,429
2002–2018	3,10,624	–
2001–2018	–	16,25,397
% loss (2001–2018)	3.0%	–
Loss/year (2001/2–2010)	15,376.23	66,993
Loss/year (2011–2018)	21,529.71	1,27,239

Source: rainforests.mongabay.com (Deforestation Statistics for India)–(2 September 2020).

> where forest cover nominally exists, the biomass is being degraded at an accelerating pace... (as) all decision-making power is concentrated in the hands of an elite comprising the urban-industrial sector, the bigger land holders in tracts of intensive agriculture, the bureaucracy, and the politicians. None of these depend directly, on the health of the forest resources in their immediate vicinity.
>
> (pp. 140–141)

Further, there has been a loss of primary forest cover and tree cover, during 2001–2018 to the extent of 3,10,624 ha and 16,25,397 ha, respectively. This may be seen in Table 8.12.

From the above, it is clear that during 2002–2018 total loss of primary forest was 3,10,624 ha while during 2001–2018 the total loss of tree cover was 16,25,397 ha. Second, during 2001–2018 the percentage of loss of primary forest was 3%. Third, the loss has been increasing in succeeding years, e.g. during 2001/2–2010, the average loss of primary forest per year was 15,376.23 ha and that of tree cover during the same period was 66,993 ha but during 2011–2018 both types of losses increased substantially – primary forest loss increased to 21,529.71 ha (28%) while tree cover loss increased to 1,27,239 ha (90%). This speaks volumes of the reality on ground.

Further, tree felling across India almost doubled between 2016 and 2019; 17,31,957 trees were felled in 2016–2017 and it increased to 30,36,642 in 2018–2019, thus 76,72,337 trees were cut during 2016–2019. Due to various factors, a sizeable number of tree-felling data may not be available. During 2016–2019, the following details of tree cutting in different states are eye-openers (Table 8.13).

Consequences of Deforestation: Following consequences are notable:

(a) Since forests act as a major carbon sink (as plants utilise carbon dioxide in the process of photosynthesis, and store it in the form of carbohydrates that reach the

Table 8.13 Tree Cutting in Different States During 2016–2019

States	No. of Trees Cut	States	No. of Trees Cut
Uttarakhand	1,05,461	West Bengal	1,76,685
Telangana	12,12,753	Haryana	1,72,194
Maharashtra	10,73,484	Gujarat	1,65,439
Madhya Pradesh	9,54,767	Himachal Pradesh	1,61,677
Chhattisgarh	6,65,132	Manipur	1,23,888
Odisha	6,58,465	Assam	1,18,895
Andhra Pradesh	4,95,269	Bihar	91,850
Jharkhand	4,34,584	Karnataka	40,776
Arunachal Pradesh	3,25,260	Andaman & Nicobar	12,467
Punjab	2,28,951	Sikkim	8,630
Rajasthan	2,28,580	India (total)	76,72,337
Uttar Pradesh	2,05,551		

Source: *The New Indian Express*, 10 February 2020.

soil as dead organic matter and contribute to the soil carbon sink), due to deforestation less CO_2 is absorbed by the plants, and atmospheric CO_2 concentration increases with time due to unavailable sink. More productive forests (tropical forests) store more carbons, but tropical forests are a highly threatened ecosystem at present. Different values of forest ecosystems are in Figure 8.1.

(b) Loss of livelihoods, especially for the tribals and other forest dwellers. In MP/Chhattisgarh the poorest local people gain about 30% of their living from forest produce – higher than the returns from agriculture. Various ecological services also provide a 'safety net' to local people.

(c) Change in land-use pattern (clearing of forests) affects both hydrometrological and global CO_2 concentrations leading to more warming; due to deforestation cloud formation shifts to higher elevations from lowland plains, hence climate change – mainly resulting in low rainfall, air pollution, and soil erosion.

(d) Deforestation has a direct effect on drinking water, fisheries, and aquatic habitats, floods, and droughts; increasing siltation reduces the life of dams, and agriculture

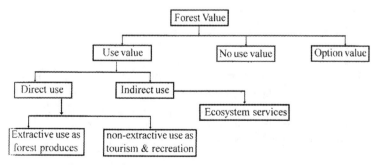

Figure 8.1 Values of Forest Ecosystems. Source: Rima Kumari et al. (2019), DOI: 10.5772/intechopen.85804.

is also affected due to poor quality irrigation and crop yield. Deforestation leads to the dwindling of the water table (in TN up to 30 m). Deforestation increases soil erosion and siltation rates, hence more flooding, e.g. in the Yangtze river basin (China), major river basins in East Asia and the Amazonian basin (South America).

(e) Since two-thirds of all known species and 65% of 10,000 endangered species (as per IUCN) are in tropical forests, their deforestation leads to biodiversity loss. As per WHO, a loss in biodiversity affects people depending on traditional medicines. Further, deforestation is limiting the habitats of many species, hence many animals are entering human settlements, hence human–animal conflicts (e.g. in West Bengal, Sikkim, Jharkhand, Odisha, etc.). In Himalayan Biodiversity Hotspot, every year 20 elephants and 50 persons are killed in human–animal conflicts (Rima Kumari et al. 2019).

(f) Deforestation results in economic loss – loss of tropical forests annually results in about a loss of $45 billion (Rimi Kumari et al. 2019).

(g) Deforestation leads to a decline in natural manures, hence farmers are forced to use chemical fertilisers that have various negative effects on the health of both soil and humans.

(h) Deforestation leads to soil erosion too, among other causes. The Government. of India's scheme of soil testing and issuing of Soil Health Cards is restricted to seeing the nitrogen and phosphate in the soil and recommending appropriate fertiliser for micronutrients to grow suitable crops but Sadhguru Jaggi Vasudev of Isha Foundation launched a global 'save soil movement' in 74 countries to ensure enough organic content in the soil by providing inspiration, incentives, and finally disincentives. While others see soil as a mere 'resource', he sees it, as 'a source of life', as soil organisms are to be kept alive (Vishwamohan 2022).

Therefore, all major causes of deforestation must be addressed at the earliest locally, nationally, and globally, by the government; NGOs and the local people through sustainable alternatives, without depriving the poorest and marginalised people. Sustainable forest management, including joint forest management (JFM), should be preferred to departmental regimentation wherein the forest officials take stern action against the poor people for taking away even small amounts of forest produce, but even sexually exploit tribal women and actively connive with the big culprits (nexus of contractors–politicians–smugglers). Though Compensatory Afforestation Fund Management and Planning Authority (CAMPA) was established by the Government of India in 2009, under the Chairmanship of the Union Minister of Environment, Forests and Climate Change for the monitoring, technical assistance, and evaluation of compensatory afforestation works, yet various environmental NGOs have pointed out that, on the one hand, there is a trend of easily diverting forest lands for the so-called 'development' works in the name of 'ease of doing business', and, on the other hand, there is no serious and adequate compensatory afforestation for the diverted forest lands. However, some committed forest officers have done appreciable work in forest conservation and have been recognised nationally and internationally (see Box 8.6)

Global Environmental Issues   339

Box 8.6: Rajaji Tiger Reserve Ranger Wins International Award in Uttarakhand, India

Mahendra Giri, a Range officer, posted at Rajaji Tiger Reserve's Motichur range (Uttarakhand), was awarded the prestigious International Ranger Award in 2021. He was the only ranger from Asia to win it. Such an award was given to ten professionals across the globe by the International Union for Conservation of Nature and the World Commission on Protected Areas based on their contribution towards forest conservation. This award carries $10,000 (Rs 7,26,150) to their organisations along with a uniform patch for these winners. Mahendra Giri successfully curbed human–wild life conflict in and around Rajaji Tiger Reserve as well as contributed to tiger translocation from Corbett Tiger Reserve to Rajaji Tiger Reserve. Ironically, his father Rameshwar Dayal Giri, a forest guard at Motichur range, was killed earlier by an elephant, while on duty (in 1996). Consequently, Mahendra Giri got a Government job in forest department on compassionate ground and proved his mettle.

Source: *The Times of India*, 28 March 2021.

Loss of Biodiversity

India has been a biodiversity-rich nation for quite a long time – one of the 12 richest biodiversity nations. India has 81,000 animal species and 45,000 plant species – 6.5% of the world's fauna and 7% of the world's flora, respectively. However, many (73) species of animals in India are critically endangered; important ones are as follows:

(1) Ganges Dolphin (North and East India, Brahmaputra)
(2) Red Panda (Sikkim, Arunachal Pradesh)
(3) Indian Rhinoceros (Assam)
(4) Lion-tailed macaque (Western Ghat)
(5) Nilgiri Tahr (Western Ghat)
(6) Gharial (crocodile species) – Chambal river, Brahmaputra, Irawati
(7) Great Flying Bustard (Gujarat, Maharashtra)
(8) Indian Pangolin (India, Pakistan, Sri Lanka)
(9) Dhole (wild dog) (Assam, Bengal, Gujarat, Nilgiri)
(10) Big Cat (Asiatic Tiger, Royal Bengal Tiger, White Leopard)
(11) Gaur (*Bos gaurus*), or Indian bison (South, South-East Asia)
(12) Water Buffalo (*Bubalus arnee*) (Assam)

Further, there are also many (two dozens) animal species threatened in India. As per *Science* (journal), between 22% and 47% of the world's plant species are endangered. There are 1,500 endangered plants in India of which the major ones are as follows:

(a) Milk wort (*Polygala irregularis*) (Gujarat)
(b) Bird's foot (*Lotus corniculatus*) (Gujarat)
(c) Assam catkin yew (*Amentotaxus assamica*) (Arunachal Pradesh)
(d) Moa, skeleton, fork fern, whisk fern (*Psilotum nudum*) (Karnataka)
(e) Ebony tree (*Diospyros celebica*) (Karnataka)
(f) Umbrella tree, Kudaivel (Tamil Nadu) (*Acacia planifrons*) (Tamil Nadu)
(g) Indian mallow (Atibalaa) (Thuthi – Tamil) (*Abutilon indicum*) (Tamil Nadu)
(h) Musli (*Chlorophytum tuberosuna*) (Tamil Nadu)
(i) Malabar lily (*Chlorophytum malabaricum*) (Tamil Nadu)
(j) *Nymphaea tetragona* (J & K)
(k) Spider wort (*Belosynapsis vivipara*) (Madhya Pradesh)
(l) *Colchicum luteum* (Himachal Pradesh)
(m) Malayuram/malavuram (*Pterospermum reticulatum*) (Kerala, Tamil Nadu)
(n) Jeemikand (*Ceropegia odorata*) (Gujarat, Rajasthan)

IUCN has assessed 1,20,372 plant and animal species of which 32,441 are threatened with extinction globally (till July 2020).

Causes of Biodiversity Loss in India

(a) Habitat fragmentation-reduced by 10%
(b) Over-exploitation of plant and animal species due to 'culture of consumerism'
(c) Pollution – of soil, water, and air
(d) Global warming and climate change
(e) Unplanned and unsustainable development – during 1951–1980, more than 5,02,000 ha of forests were diverted for river valley projects
(f) Ignorance about species and ecosystems
(g) Introduction of exotic species (e.g. *Eucalyptus*) at the cost of native species
(h) Neglect of traditional knowledge system (especially of tribals and other forest dwellers)
(i) Intensive modern agriculture's use of chemicals (fertilisers, pesticides, herbicides, insecticides, etc.)
(j) High population pressure on natural resources
(k) Socio-economic inequity – gap between protectors and consumers of natural resources
(l) Unequal ecological exchange at global level

Consequences of Loss of Biodiversity

It has five major impacts on humans:

(a) It increases the number of disease-carrying animals in local population – a range of biotic disease agents affect human, animal, and plant health.
(b) Changes in ecosystem services affect livelihoods, income, local migration, and also political conflicts – especially because the tribals and other forest dwellers directly depend on various plants and wildlife.
(c) Loss of biodiversity also badly affects aesthetics, recreation, and tourism sectors.

Global Environmental Issues   341

(d) Loss of biodiversity also leads to the loss of alternative aspirations/vision and positive sources of abundant knowledge from nature – uniformity usually results in one-dimensional/fractional life and ignores the inter-connectedness and inter-dependence of the whole ecosystem.
(e) Loss of biodiversity also leads to the forceful dominance and hegemony of humans over other (non-human) animal species, thus their sensitivity to other sentients (animals and plants) is lost.

A new study by Durham University, Birdlife International, and New Castle University found that the impact of 32,000 km of fortified international borders would be to affect 696 species of non-flying mammals – three political barriers pose the biggest conservation challenges along US–Mexico, India–Myanmar, and China–Russia; on India–Myanmar border, 128 species of non-flying mammals could be unable to cross if barriers are completed – border fencing problem, combined with political barriers, lead to a threat to many species of wildlife. In fact, biodiversity flourishes without borders.

Policy Implications

For some decades, the air pollution level has deteriorated in most of the urban areas in India in general and that in Delhi/NCR in particular. There should be a genuine political and administrative will and associated concrete genuine plan of action for the mitigation of different types of pollution. Hence, central and state governments and their subordinate agencies like Central/State Pollution Control Boards as well as the local bodies (in both urban and rural areas) should be proactive, going beyond the rhetoric and empty promises. An integrated approach with stringent measures for defaulters/polluters is very much required in this regard. The conventional view of 'polluter pays' should go to a paradigm shift to 'polluter is punished' (curative) as well as 'prevention is better than cure'. National Green Tribunal (New Delhi) took cognisance of the increase in polluted water stretches in India (from 302 to 351) during 2016–2018 and directed all state governments and UTs to restore minimum standards for all the rivers, making these rivers suitable for at least bathing purposes (Ghosh 2018). Hence, these implementing authorities should be more sincere and sensitive to the serious problem of various forms of pollution.

Second, the focus should be on long-term measures, not just on short-term fire-fighting measures like odd–even vehicle restriction, e.g. in Delhi, the first phase of 1–15 January 2016 odd–even drive had some success but the second phase of 15–30 April 2016 drive was a failure as to beat the drive, the rich people bought more new cars or second-hand cars. On 4 November 2019, again odd–even vehicle restriction was imposed in Delhi, but not of much use. In fact, adequate public transport facilities like local trains (in Mumbai), metro, CNG buses, etc. should be enhanced tremendously (in terms of more coaches and more trips) in lieu of private cars/scooters/bikes. Further, adequate cycle tracks, battery rickshaws, and pedestrian tracks should be built/developed in all the cities/towns. Diesel vehicles/DG sets should not be allowed in highly polluted cities in India (out of 20 most polluted cities in the world) like Delhi, Gwalior, Allahabad, Gurugram, Ghaziabad, Faridabad, Gaya, Jaipur, Bhiwadi, Noida, Lucknow, Jodhpur, Muzaffarpur, Mumbai, Varanasi, Agra, Moradabad, Jind,

342 Global Environmental Issues

and Patna. In phase I, all diesel vehicles older than 10 years should be ousted from there. Bharat Stage VI emission norms should be adopted at the earliest without going through Bharat Stage V, and the automobile industries should ensure the manufacturing of new vehicles as per Bharat Stage VI norms with immediate effect. Further, 'no parking zones' in crowded areas be declared and implemented strictly, and parking fees should be enhanced. In addition, the use of CNG should be mandatory in all public transports in all cities of one million or more population.

Third, brick kilns, hot mix plants, stone crushers, thermal power plants, etc. in and around such cities should be totally shut down, and road dust should be removed regularly through cleaning machines, especially during nights (as done in the West, and even in Surat). Further, there should also be water sprinkling to remove dust on the road side, and on empty sites, grass and other green plants should be grown so that dust may not fly as it contributes a lot to PM 2.5/10. In this regard, we should adopt the better road-cleaning practices of Surat Municipal Corporation in other cities/towns.

Fourth, the congestion pricing method in high pollution areas during peak hours should also be tried, e.g. the drivers are to be fined heavily for using busy roads at certain places and time in order to reduce car use at peak hours. This model is successful in London and Singapore. Hence, better traffic management practices should be adopted in Indian cities, especially Delhi/NCR.

Fifth, more use of clean energy devices like solar, wind, LED bulbs/tubes, LPG, CNG gas to save energy (saving is producing in one sense) and avoid pollution (from the use of fossil fuel, etc.). Use of coal/firewood in hotels and eateries, and pet coke and furnace oil in factories be stopped strictly. Needless to add that establishing a coal-based thermal power plant takes five years for operation, whereas setting up a renewable power (wind/solar) unit takes five or six months only; further, coal-based thermal plant takes large forest areas, uses a huge quantity of water, and creates most polluting industrial units, hence to be avoided. The Prime Minister of India was awarded, on 3 October 2018, UN Champion of Earth Award (along with France) for popularising solar and other renewable energy sources, hence India should lead by example in this regard.

Sixth, building materials should be strictly transported in covered vehicles, and such materials should be kept covered in the shops, stores, and work sites. Further, treated wastewater should be re-used in all constructions of more than 500 m carpet area.

Seventh, the public should be made aware of taking individual measures like the use of masks, air purifiers, indoor plants like areca palm, snake plants, English ivy, Boston fern, etc.; and carpooling, use of cycles, and so on. Walking habits should also be inculcated among people of all age groups.

Eighth, the use of non-biodegradable plastic as carry bags, etc. should be strictly banned in order to prevent choking of drains, eating of these by cows and other animals, exposure to hazardous chemicals used in plastic products, as well as to avoid affecting surface and groundwater. Maharashtra Government banned production, sale, and use of plastic bags of less than 50 microns and sizes of less than 8 × 12 in. UP, too, decided on such a plastic ban. The Government of Bihar also banned the use and sale of plastic bags of less than 50 microns (in urban areas effective from 25 October 2018 and in rural areas from 25 November 2018). The Government of India also banned single-use plastic in October 2019, and from 1 July 2022, it is fully banned in the entire country. But it still needs to be enforced effectively. Further, plastic waste may be successfully used in road-building. Mumbai generates 8,000 tonnes of waste every day and plastic contributes up to 5% of that. Further, there should be awareness generation for 7 Rs: reduce, reuse, recover, remediate (cleaning/treating water/soil), recycle, replace (substitute), and restore (return to original state). Maharashtra has exempted food-grade virgin plastic milk pouches and polyethylene terephthalate (PET) bottles, that is biodegradable, on the one hand, and for reuse and recycling of such items a 'buyback depository system' has been introduced wherein buyers are to pay a deposit that they can reclaim on returning the bottle/pouch (Bavadam, 2018).

Ninth, water pollution due to arsenic, fluoride, iron, etc. are causing serious health hazards, hence it should be addressed soon and effectively and treated water supply, especially in areas of contaminated water, be ensured to all the needy people.

Tenth, the drive for rainwater harvesting should be launched as a mass movement because India conserves only 17% to 20% of its rainwater while Israel conserves 80% of its rainwater. 'Rally for rivers', launched in late 2017, by Sadhguru Jaggi Vasudev (in South and North India), had thousands of genuine participants. He stated that, 50 years ago, Bengaluru had three rivers and 1,000 lakes but now these rivers have disappeared, and the lakes are reduced to 81 of which 40 are polluted with chemical wastes (*The Times of India*, 2 October 2017). Hence, mass mobilisation by genuine civil society organisations (CSOs) as catalyst agents should be launched at the earliest. Everybody should be activated for four Ps: *Pet (belly), Pados (neighbourhood), Prakriti (Nature), and Parivesh (Social milieu)*.

Eleventh, all water bodies should be restored (e.g. encroachment/construction on most of about 1,000 water bodies done in Delhi since the twentieth century), and duly cleaned as a mass drive till the sewage treatment plants are commissioned; it is also advisable to use bio-remediation techniques of sewage-eating microbes that will fasten the process of improving the quality of river water. The National Mission for Clean Ganga rightly approved six such projects at Patna and Allahabad. This is cheaper (costing Rs 7 lakh) and less time-consuming (6–8 months); about 54 drains are identified on Ganga for this purpose. All municipal bodies/industries enroute should be punished if untreated water and effluents are drained by them into rivers. Factories should be compelled to treat industrial wastes and effluents, and concerned states should include, with transparency, the water footprint in their yearly reports. Central and State Pollution Control Boards should monitor it regularly and transparently. Various groups/individuals grabbed huge lands of Yamuna and Hindon rivers by illegally making farms, but some of these were removed by the local authorities in mid-2022.

Twelfth, water conservation and management should not be left to the engineers, rather all stakeholders like natural scientists, social scientists, local people, civil society organisations, etc. should be involved in participatory and '*bottom-up approach*' for both groundwater and surface water management. Central Water Commission (CWC) and Central Groundwater Board (CGB) should complement each other and focus on demand-side, not supply side, by involving the local farmers, civil society organisations, retired technocrats, etc. Various Central and State agencies often prepared faulty models on their own (without involving local people) which failed subsequently and their repairing/rebuilding led to the huge expenditure without adequate rate of return to local people. Ideally these two bodies (CWC and CGB) should be merged for a better and integrated approach to water management. Drip/sprinkler irrigation should replace flood irrigation through canals and tubewells as the latter results in a huge wastage of water, salinisation of soil, soil loss, and damage to crops.

344   Global Environmental Issues

Thirteenth, the over-use of dangerous chemical fertilisers and pesticides in agriculture and horticulture is leading to contamination of both surface water and groundwater. About 100 dangerous pesticides, banned in developed and other countries, are still used in India, and only 18 pesticides are banned in India – resulting in water and food contamination and poisoning. Anupam Varma Committee recommended banning of only 12 more pesticides and a review of 27 chemicals in 2018, and suggested phasing out of six pesticides by 2020! 'India for safe food' leader Kavitha Kuruganti has criticised this Committee and the Ministry of Agriculture and Farmers' Welfare for not banning many pesticides like monocrotophos, endosulphan, glyphosate, isoproturon (herbicide), etc. and thus protecting the business interests of MNCs like Monsanto that profit from glyphosate-tolerant GM crops selling both GM seeds and associated chemicals (*The Times of India*, 16 October 2017). Further, from solid waste and liquid waste, organic fertilisers may be prepared, and thus 'from waste to wealth' model may be realised. Needless to say that only 4.3 crore (43 million) tonnes of municipal waste out of 24.7 crore (247 million) tonnes has been remediated by government agencies as per Ministry of Housing and Urban Affairs, Government of India (November 2022); five states with maximum legacy waste are – Maharashtra (4.2 crore tonnes), Delhi (2.4 crore tonnes), Karnataka (1.8 crore tonnes), Tamil Nadu (1.6 crore tonnes) and Rajasthan (1.1 crore tonnes). Unfortunately, India has 1873 dumpsites and 20,359 acres of land are covered under such waste, legacy is targeted to be remediated by 2026.

Fourteenth, there should be a pragmatic and real convergence of planning as well as pooling of resources by various Ministries, Departments, and Government/Semi-Government bodies like the Ministry of Jalshakti, the Ministry of Environment, Forests and Climate Change, the Department of Drinking Water (Ministry of Rural Development), the Ministry of Urban Development, the Central/State Pollution Control Boards, the Ministry of Agriculture and Farmers' Welfare, the Department of Minor Irrigation (at State level), the Ministry of Surface Transport and Shipping/Waterways, the Ministry of Railways, the Ministry of Civil Aviation, the Ministry of Earth Sciences, the Ministry of Health, etc., so that effective inter-connected measures may be taken with better and simple technology, with less expenditure and less time.

Fifteenth, more plantation is needed near factories and other pollution sources as well as all open spaces, because planting trees cut down air pollution by 27%, and is cheaper than technological interventions – hence there is a need to learn from nature and respect it (*The Morning India*, 8 November 2019). A good shady tree like peepal (Fecus religiosa), mahua (Madhuca longifolia), vat (Fecus benghalensis) reduces temperature by 6°C.

Sixteenth, the menace of '*parali*' (paddy stems) burning, especially in Punjab, Haryana, and Western UP, leading to air pollution in Delhi/NCR, is to be tackled sensitively and sincerely. But Punjab and Haryana Governments resorted to short-term measures in 2018 by providing various agro-equipment/machines like Happy Seeder, paddy straw chopper, cutter, mulcher, shrub cutter, zero tiller drill, combined harvester with straw management equipment, etc., with a subsidy of 50% to individual farmers and 80% to cooperative societies/farmers' associations. But this will increase the load of machines on farmers who already have more than required tractors – 4.5 lakh tractors against 1 lakh tractors actually required in Punjab. Similarly, highly subsidised polyhouse (up to 25 lakh rupees) has not been a success in Punjab as 80% of polyhouses are in disuse. On the other hand, the farmers should be given incentives to harvest 'parali' (in lieu of burning it) but the Punjab Government's request for invest-ment of Rs 2,000 crores was turned down by GOI. Further, Punjab Government is not using Mahatma Gandhi National Rural Employment Guarantee Act (MNREGA) fund of Rs 4,000 crores, hence this may be utilised for 'parali' – cutting by farmers (Devinder Sharma, 2018). Confederation of Indian industry launched a pilot project in Punjab in 19 villages over 16,000 acres of agricultural land (total target was one lakh acres) and made 75% of farmland stubble-burning free through alternative prac-tices – about 25,000 tonnes of paddy straw were recycled back into the soil, saving fine particulate matter (PM) being released into the air, farmers saved in field prepara-tion for Rabi sowing by 7–10 days, reduction in weedicide application by 50–70% and lower irrigation requirement by 40%, the yield level remained comparable with traditional fields and overall production costs decreased in general (Banerjee 2019). The Government of Punjab launched agriculture mechanisation scheme in 2018 pro-viding subsidy for Happy Seeders, zero tillers, mulchers, etc., but the subsidy scheme should be extended beyond 2020, and it may be integrated with bank and National Bank for Agriculture and Rural Development (NABARD) schemes to facilitate credit linkages. Further, subsidy should be extended to other machines too, giving farmers more options, as a CII pilot project found that a combination of rotavator and Roto seeder was effective in checking stubble-burning, and also to tractors of more than 45 HP capacity which is necessary for Happy Seeder. Finally, to avoid partial burning of straw, super straw management system (with quality control) should also be popular-ised and ex-situ straw management for composting, biogas, fodder, fertilisers, fuel, fibre boards, etc. is also called for ensuring stubble-burning free air (Banerjee 2019). The Supreme Court of India directed the State Government of Punjab, Haryana, UP, and Delhi on 6 November 2019 to pay Rs 100 per quintal to small and marginal farmers to stop burning of paddy stubbles. This needs to be addressed sooner and effectively.

Finally, other global environmental issues like acid rain, ozone depletion, noise pollution, deforestation, loss of biodiversity, and nuclear radiation, too, need to be addressed by the central and state governments and local agencies in letter and spirit.

Conclusion

Various global environmental issues like air, water, and noise pollution, acid rain, ozone depletion, deforestation, loss of biodiversity, etc. are pertinent in the contem-porary world. Delhi has emerged as the most polluted city in the world (2019) as its air quality index (AQI) is often in the severe or very severe (dangerous) category. Similarly, many Indian cities and towns have severe/very severe/dangerous air quality. Further, water situation is equally bad, both surface and underground water. Over the years, there have been many oil spills in different seas/oceans or rivers leading to social, economic, and environmental consequences. Similarly, due to more urbanisa-tion and industrialisation, increasing traffic during the day and night in different cities (Mumbai is now called the Maximum City!), new construction activities, burning of paddy stems, long spells of dry weather noise, etc., hence have resulted into various kinds of pollution and their ill effects on human and animal health as well as global warming. Nuclear radiation, acid rain/deposition, ozone depletion, deforestation, and

346 Global Environmental Issues

loss of biodiversity have huge global environmental effects. Global efforts by UN agencies and other bodies have controlled acid rain and ozone depletion to a large extent. There has been extinction or near extinction of thousands of plant and animal species in the world, more so in India. Many forests in the USA, Australia, Amazon, etc. have faced huge fires in the recent past. Therefore, sustained efforts by UN bodies, international and national environmental NGOs, national and sub-national governments as well as local people at large are urgently called for.

Now a pertinent question arises, commons for whom? For the local communities, for the rich people inside and outside (in the name of global commons) or the State? This question will be dealt with in the next chapter.

Points for Discussion

(a) What is the range and depth of air pollution in India, with special reference to Delhi?

(b) What are the major consequences of air and water pollution in India?

(c) What is the magnitude of water pollution in India? What are its causes?

(d) What are acid rain's effects? What measures are taken against it?

(e) What are the major effects of ozone depletion? What measures should be taken to stop it?

(f) What are the major causes and consequences of deforestation in India?

(g) What is more important in energy producing: price, safety, or unrestrained production?

Chapter 9

Commons for the Communities

A Critique of the 'Tragedy of Commons'

Introduction

Commons or common property resources (CPRs) or common natural assets/ resources are

> the resources accessible to the whole community of a village, and to which no indi-vidual has exclusive property rights. In the dry regions of India, CPRs gathering areas include village pastures, community forests, waste lands, common threshing grounds, waste dumping places, watershed drainages, village ponds, tanks, rivers/ rivulets, and river beds, etc.
>
> (N. S. Jodha 1986, 1169)

Further, Bina Agarwal (1995, 2) adds to this:

> a wide variety of essential items (that) are gathered by rural households from the village commons and forests, for personal use and sale: food, fuel, fodder, fibre, small timber, manure, bamboo, medicinal herbs, oils, materials for house building and handicrafts, resin, gum, honey, spices, and so on.

Thus, not only in dry regions but also in other regions, people use CPRs. The salient features of the CPRs are as follows:

(a) These are natural resources or assets – not a 'built environment' (a feature of urban parks, etc.).
(b) These are common or communal – not a personal/family property.
(c) These are primarily situated in rural areas.
(d) These provide various benefits to ordinary people for their survival or livelihoods or nutritions, health, etc.
(e) There are some (unwritten) norms about the use and sharing, maintenance, restriction, regulation, and control of such resources in everyday life.
(f) These contribute to improving village equity.
(g) These generate employment/livelihood, sometimes more than farm or public works.
(h) Women and children play a key role in accessing and gathering such resources.

DOI: 10.4324/9781003336211-9

348   Commons for the Communities

(i)   The poor people face conflicts, both from the poor fellows (due to competition over scarce resources) and from the rich (who encroach upon and try to control these).

Thus, these CPRs are sources of physical products; they provide employment and income opportunities, and also give social, and ecological benefits. Various empirical studies by N. S. Jodha (e.g. in 1986, 82 villages' survey in 7 States in dry tropical west and South India; in 1985, 6 villages' survey in arid Rajasthan; and again in 1985, survey in 3 arid districts in Rajasthan, and 2 district in Madhya Pradesh (MP)) have put forward three significant findings related to commons, income, and social equity – and these hold good to almost all agro-ecological zones of India (Beck and Nesmith 2001):

(i)   Commons add 12–25% to the total income of the poor households.
(ii)   The poorer the household, the more significant the commons' contribution.
(iii) Commons provide rural social equity as these are accessed by the poor more than by the rich.

Therefore, the concept of 'sustainable livelihoods' (SL) has become popular as it addresses coping and adaptive strategies. As United Nations Development Programme (UNDP) (1998, 1) rightly points out that sustainable livelihoods 'are derived from people's capacity to access options and resources, and use them to make a living in such a way as not to foreclose options for others to make a living either now or in the future'. Needless to say that the commons provide the poor people ample scope for such coping and adaptive strategies for realising sustainable livelihoods, and thus assisting in the reduction of poverty in some way or other. In a comparative study of CPRs in India and West Africa, the following conclusions have been found (Beck and Nesmith 2001):

(a)   CPRs are vital, especially in the lean or pre-harvest season or other times of stress.
(b)   Women are usually involved in accessing and using CPRs, but often not in management.
(c)   CPRs have a 'redistributive effect', especially assisting the poor in different ways.
(d)   Privatisation and commercialisation progressively lead to the exclusion of the poor people from these livelihood resources.
(e)   Indigenous institutions for the management of CPRs are under stress due to the pressures from the processes of modernisation and globalisation, and consequently various conflicts amongst their users (who have limited influence on such institutions).

Regarding dependence on CPRs, there is a class basis. For instance, the poor people (small farmers, landless labourers, artisans, etc.) increasingly depend on CPRs due to lack of alternatives. Jodha (1986) found that 84–100% of the rural poor in 80 villages (in 20 districts of seven dry tropical States) depended on CPRs for fuel, fodder, and food. On the other hand, only 20% of the rich (large farmers, shown as 'others'), except in very dry villages in Rajasthan, depended on CPRs. The intermediate category of farmers depended on CPRs less than the poor, but more than the rich. This may be seen in Table 9.1.

Commons for the Communities   349

Table 9.1 Extent of Households' Dependence on CPRs in Dry Regions of India

States[1]	Household Categories (%)	CPRs, Contribution Per Household				
		Fuel Supplies[2] (%)	Animal Grazing[3]	Employment Days (No.)	Annual Income (Rs)	Income as % of Total Income
Andhra	Poor	84	–	139	534	17
Pradesh	Others	13	–	35	62	1
(1, 2)						
Gujarat (2, 4)	Poor	66	82	196	774	18
	Others	8	14	80	185	1
Karnataka	Poor	–	83	185	649	20
(1, 2)	Others	–	29	34	170	3
MP (2, 4)	Poor	74	79	183	733	22
	Others	32	34	52	386	2
Maharashtra	Poor	75	69	128	557	14
(3, 6)	Others	12	27	43	177	1
Rajasthan (2,	Poor	71	84	165	770	23
4)	Others	23	38	61	413	2
Tamil Nadu	Poor	–	–	137	738	22
(1, 2)	Others	–	–	31	164	2

Source: N. S. Jodha (1986).

Note: 1 indicates parentheses is the number of districts and villages included for each state. 2 indicates fuel gathered from CPRs as a proportion of total fuel used during three seasons covering the whole year. 3 indicates animal unit grazing days on CPRs as a proportion of total animal unit grazing days.

Unfortunately since the early 1950s with the initiating of land reforms in most parts of the Indian nation, the area of CPRs declined by 31% to 55% in the 80 villages studied by Jodha (1986). Second, due to such reduced area of CPRs as well as the population growth, the average number of persons per 10 ha of CPRs ranged from 13 to 101 in 1951, but by 1982 the same measure increased to 47,238 (depending on the sample village). Thus, it resulted into over-exploitation and degradation. For example, the number of common property products ranged from 27 to 46 before 1952 but declined ranging from 8 to 22 – thus there is reduced biodiversity in CPRs. Third, the informal arrangements and social sanctions regarding use and protection/upgradation were abandoned, and the formal State interventions became slack and ineffective. As a result, many CPRs became 'open access' resources 'with everyone using them without any reciprocal obligations to maintain them' – about 90% of villages do not implement 'historical regulations, both formal and informal', hence pauperisation of the poor – 'a classic case of the vicious cycle of poverty and resource degradation reinforcing each other' (Jodha 1986.

Consequences of Public Policy Interventions: Main thrust of public policies involving CPRs was four: asset (land) redistribution, productivity increase (including forest areas), formal management systems, and biomass production projects (Jodha 1986):

(A) **Asset (Land) Redistribution** – Since most of the State Governments failed to redistribute ceiling surplus lands, these governments easily started re-distributing commons' infertile lands, which, in turn, led to the low crop yields (one-fourth

350   Commons for the Communities

to half of the yields from traditionally cropped lands) (Jodha 1992). Second, due to the individual privatisation of CPRs, a large proportion of common lands was cornered by the non-poor. Third, due to the unproductive quality of the redistributed land, and absence of support to develop such lands, 23–45% of the poor households were dispossessed of their new lands (Jodha 1986).

(B) **Increasing Forest Productivity** – Since CPRs were treated merely as physical resources, non-participation of the ordinary people, and public policy's focus on techniques rather than users' needs, limited species of plants and sometimes exotic species of plants (e.g. pilot projects) could not fulfil the mixed biomass needs of the local people. Further pilot projects demonstrating potential technologies, with the help of State grants, but without people's participation proved ineffective.

(C) **Formal Management Systems** – Gram panchayats were given responsibility to manage common property resources of villages, but they misused/diverted government grants. Further, 73rd Amendment in Indian Constitution ensured devolution of power to panchayats but, in practice, they failed to properly manage CPRs.

(D) **Biomass Production Projects** – These are usually sustained by the State subsidies, are technique-oriented, and lack people's participation. Further, such projects do not properly understand the nuances of CPRs.

Thus, the ultimate consequences of the colossal degradation of CPRs may be the 'elimination of vital biophysical processes of nature's regenerative activities (energy and material flows, etc.)' (Jodha 1991), hence the prevalence of the vicious cycle of CPRs' degradation and poverty. This has also been resulting into conflicts – individual, group, and caste conflicts (Box 9.1).

Box 9.1: Social Conflicts over Commons: A Case of Narmada River (MP and Gujarat), India

The most visible and widespread conflict occurred in the 1990s over the dam construction (especially the mega dam Sardar Sarovar Project (SSP)) on Narmada river (in Madhya Pradesh and Gujarat). In May 1990, 'the ecosystem people' – villagers (to be displaced) from Madhya Pradesh – protested and organised a *dharna* (sit-in) for many days at Gol Methi Chowk (New Delhi) – very close to the residence of the then Prime Minister of India, Shri V. P. Singh. He, ultimately, assured them to review the said Sardar Sarovar Dam. Subsequently 'the omnivores' (people from Gujarat to benefit most from the dam) organised counter *dharna* at the Boat Club (New Delhi). The Prime Minister immediately met their delegation and assured his government's commitment to complete the SSP.

After a few months, both groups engaged in a face-to-face conflict at the MP–Gujarat border. On 25 December 1990, several thousands of Narmada Bachao Andolan (Save Narmada Movement) activists organised a peaceful foot march (250 km) from Rajghat (Delhi) to Kevadia colony (Gujarat) – the site of SSP. Gujarat police stopped the marchers at the border village of Ferkuva. There some students, policemen in plain clothes, and others raised slogans against such protesting marchers and their leader Baba Amte (70 years plus), and in support of SSP. On 2 January 1991, 25 protesters (with their hands tied to symbolise peaceful protest) entered Gujarat but Gujarat police stopped them. On 3 January, two more groups of protesters joined them. On 5 January, Baba Amte and another group of 25 protesters also entered Gujarat State but were not allowed to proceed further. Then they sat on *dharna* at the Gordah River bridge. On 6 January 1991, NBA leader Medha Patkar and many activists went on hunger strike at the MP side of the border. It went on for many weeks, but Government of Gujarat did not pay heed to them. On 28 January 1991, Medha Patkar and other activists gave up their fast but the issue of displacement of people was duly highlighted at national and international levels.

Hence, Gadgil and Guha (1995) rightly point out the changing character of Indian State: during pre-1947 colonial period, Indian State was 'an instrument of subjugation and extortion of surplus on behalf of a colonial power'; during the 1950s and 1960s, Indian State was perceived as 'the authentic legatee of an all-class and genuinely mass-based national upsurge' – extending 'patronage to the entire population'; but during the 1980s and 1990s, Indian State lost its legitimacy and became 'an instrument of a narrow elite of omnivores'. But simultaneously, democracy has given voice and political clout to 'the ecosystem people' and 'the ecological refugees'. Hence, resource capture by the elite and the creation of a space for social protest have resulted into various conflicts between and among three ecological classes of modern India – *the imnivores, the ecosystem people, and the ecological refugees.* Such conflicts are for various common natural resources like (a) land for cultivation or grazing livestock; (b) water for domestic, agriculture, or industrial use; (c) mineral deposits; (d) fish stocks; (e) woody biomass; or (f) wildlife. Such conflicts are between the rich and the poor (e.g. trawler owners vs fisherfolk, tribals vs paper mills, peasants vs rich farmers/irrigation department, between Badaga-rich farmers in Nilgiri hills and poor scheduled caste labours (to maintain grazing grounds, or to occupy for cultivation); or between the poor and the poor (e.g. village common lands are ruined in a fight for fuelwood), or between the rich and the rich (e.g. in 1991 Chief Minister of Karnataka (Bangrappa) and Chief Minister of Tamil Nadu (Ms Jayalalitha) precipitated violence in a dispute over the use of Kaveri waters for the rich farmers and city dwellers of Karnataka and Tamil Nadu.

Source: Gadgil and Guha (1995).

Institutional Mechanisms for Governance of Commons

The institutional mechanisms for effective management/governance of commons depend on following 'contextual factors' (Subrata Singh 2004/2015):

(a) What is physically, legally, economically, and socially *feasible* in terms of improved communication and interaction with urban centres?
(b) What is economically, socially, and culturally *desirable*, in the changing individual preferences due to enhanced mobility of the villagers, high opportunity costs

352 Commons for the Communities

of social arrangements to manage local resources and gradual loss of common interests and group identity?

(c) What is institutionally *viable*, especially with the decrease in the perceived need to rely on local resources, and a revision of decision-making arrangements governing the resource?

Due to new processes of liberalisation, privatisation, and globalisation, there is a 'reconfiguration of governance' or common resources that may be held under one of three property regimes: communal property, private property, and State property. In India, there are primarily private or State property regimes – natural resources under State property regime are regulated by laws/guidelines (Singh 2004/2015). Third form, 'contractual property regime', has emerged with joint forest management wherein local communities (people's participation and forest officers form Van Suraksha Samitis (Forest Protection Committees) for managing the forest resource. Usually forests are managed by the forest department, and revenue lands are managed by the revenue department, and the resources are 'assumed commons', not commons in reality (Singh 2004/2015). In this regard, Subrata Singh (2004/2015, 8) perceptively remarks:

> In … India, the State has taken up large scale initiative in land reforms, often as a measure to provide the landless with some land for subsistence. The state has even gone in for the regularisation of encroachments of govt lands as welfare measures … (this) triggered more encroachments, thus reducing the availability of commons….powerful actors extend their control over commons.

However, the management mechanisms of common property resources or natural assets vary in different contexts, e.g. especially in remote villages, transitional villages, and urban areas (Table 9.2), as rightly points out Subrata Singh (2004/2015).

Myth of the Tragedy of Commons

It was thought by Garret Hardin (1968) that the inevitability of the 'tragedy of commons' can be avoided only either by privatisation or by State ownership. But, we argue that 'assurance problem' theory, and the theory of 'environmental entitlements' are gaining more acceptance over his view. They provide evidence that historically in many developing countries there have been successfully functioning common property regimes that are neither private nor State property regimes. This common property system has persisted due to comparative advantages. Here, at first, the hypothesis of 'tragedy of commons' is described; then the same will be critically analysed in the light of other theoretical perspectives; and, finally, some empirical evidences through case studies on this issue are provided, especially at local level, and thus its relevance is pointed out.

In 1968, Garrett Hardin wrote in *Science* an article titled 'The Tragedy of the Commons' with a sub-title: 'The population problem has no technical solution; it requires a fundamental extension in morality' (1968, 1243). His article begins with a quotation from J. B. Wiesner and H. P. York who concluded that the arms-race

Commons for the Communities   353

Table 9.2 A Comparison of Institutional Mechanisms for Managing CPRs in Different Areas

Attributes	Remote Villages	Transitional Villages	Urban Areas
(1) Orientation of use	Subsistence viewpoint, distribution orientation	Economic viewpoint, monetary orientation	Aesthetic viewpoint, conservation orientation
(2) Institution	Informal institutions	Formal/legitimate institutions	State/State-sponsored institutions
(3) Decision-making	General body	Executive committee	Elected representatives (1/10,000 or more)
(4) Conflict resolution	Within village	Resolved by support structures	State-sponsored support structure
(5) Protection	Rotational protection	Watchman/social watch	State
(6) Influence of market	Little/No influence	Moderate/increasing influence	Market-controlled
(7) Property rights (tenure)	Assumed commons	Legal tenure-uncontested regimes	State as tenurial right holder
(8) Governance	Self-governance	Governance through State-recognised legitimate institutions	Governance by State
(9) Dependence on biophysical resource	High dependence – economically and culturally across various strata	Dependence varies across various strata – poor are still dependent for subsistence and economic needs,	No direct dependence, dependence is for aesthetic reasons or for pleasure
(10) Reaction of stakeholders to damage (broken CPRs)	Livelihood crisis – revival	No impact, tend to seize rights on the resources in small and interior villages through force or influence	No impact
(11) Nestation	Self-governed, natural nesting – watershed, caste groups, marriage boundary	Apathy	State structures nested in larger governance structures

Source: Subrata Singh (2004/2015).

'dilemma has no technical solution'. Hardin, following that conclusion, argues that there is a class of human problems which are 'no technical solution problems' and he cites 'population problem' as one of these. Taking a pessimistic view, he imagines a 'finite' common pasture 'open to all' and assumes every herder to be a rational utility-maximiser who gets a positive utility by adding more animals even when the aggregate of all herders' activities far exceeds the sustainable yield of the said pasture. Then obviously occurs the 'tragedy':

> Each man is locked into a system that compels him to increase his head without limit - in a world that is limited. Ruin is the destination toward which all men

354 Commons for the Communities

rush, each pursuing his own best interest in a society that believes in freedom of the commons. Freedom in a commons brings ruin to all.

(Ibid., 1244)

His second argument is that, particularly in a welfare state, 'freedom to breed is intolerable' and an appeal to conscience to desist from exploiting a commons, or for responsible parenthood, has 'pathogenic effects' (ibid., 1246–1247). Therefore, he recommends 'mutual coercion, mutually agreed upon by the majority of the people affected' (ibid, 1247). So his alternative solution to the commons' problem is 'the institution of private property coupled with legal inheritance', though he admits that 'legal system of private property plus inheritance is unjust', yet he asserts: 'Injustice is preferable to total ruin' (ibid., 1247).

Assumptions of 'Tragedy of Commons' Hypothesis

Hardin had several assumptions which are noteworthy:

(a) Every individual is selfish and rationally utility-maximiser at the cost of commons since animals are held individually while the range is owned by 'everyone' or 'no one'.
(b) Common resources are finite and scarce, and 'carrying capacity' of any piece of range is fixed.
(c) The consumption pattern puts in far less than removes.
(d) Grazing commons is used as a metaphor for the general problem of over-population.
(e) There is mutual distrust that others will capture the benefits of the common resources.
(f) There is dominance of free-rider defecting strategy.
(g) All rural producers in a community practise the same livelihood, have the same interest in a resource, and can act fully independently of their fellow producers.
(h) There is a need for coercive enforcement, i.e. 'to legislate temperance'.
(i) In the ultimate analysis, solution lies in privatisation of common resources because only private property internalises the 'externalities' of non-exclusive resource exploitation.

The application of the hypothesis of 'tragedy of commons' is not limited to grazing lands or fisheries but extends to describe Sahelian famine of the 1970s, firewood crises in developing countries, urban crime, communal conflict in Cyprus, inability of US Congress of limiting its capacity to overspend, problem of acid rain, public–private sectors' relationships, and problems of international cooperation (Ostrom 1990).

A Critique of the 'Tragedy of Commons' Hypothesis

The 'tragedy of commons' hypothesis has been criticised by several social scientists and environmentalists on various theoretical and empirical counts. For instance, Hardin's diagnosis of over-population as the cause of degradation has been declared as unrealistic since population growth is more closely associated with poverty, illiteracy, malnutrition, lack of social security in old age, absence of health and sanitation, and cultural practices rather than with individual reproductive behaviour (World Bank 1984). Hence, Hardin's observation of no scientific/technical solution and seeking answers in moral reason is not sound.

Second, 'hard line' solutions to environmental problems (depletion taxes, residual taxes, etc.) are economically and socially regressive, and will only be effective if unaccompanied by genuine attempts to ensure fair and equitable distribution (O'Riordan 1976). But unfortunately, Hardin is against such equitable distribution of wealth and, in his view, in the absence of a world government controlling the reproduction, 'the ethics of a lifeboat' must govern since somebody has to be sacrificed in the interest of 'humanity'! For Hardin, thus 'humanity' means the rich and elite minority!

Third, the 'tragedy of commons' hypothesis is historically not supported. As Juergensmeyer and Wadley point out, care of common resources needs community efforts, and reciprocity of all members; the use of grazing and woodland of original English common lands predates feudalism to the period of Germanic laws when the concept of private property hardly existed – actually the communal rights and obligations governed the use of land (cited in O'Riordan 1976, 31). Hardin's hypothesis thus wrongly over-emphasises the absence of a sense of community and dominance of individual defective strategy. Actually, the divergence of individual concern and collective interest is more a result of inappropriate institutions to deal with an abundance of 'bads'. Therefore, Shelling correctly observes: 'Medicines are proprietary, but germs are free. I own the tobacco I plant in my field, but you may have the smoke free of charge' (1971, 78). Similarly Lane and Moorehead (1995) also criticise Hardin for undermining collective endeavour:

> However, by reiterating the primacy of the motivation for individual maximisation, he is adopting an economic model to explain behaviour and failing to acknowledge the existence of the benefits (mutual support, security) that come to individuals from collective behaviour in the public interest, as is displayed by traditional pastoral societies.
>
> (1995, 119)

Fourth, in response to massive criticisms, Hardin later conceded in 1988 that his theory only applies to 'open access' commons, and therefore, the tragedy is confined to the unmanaged commons. But Lane and Moorehead (1995) criticise him on this stand, too, that pastoral commons are not included in the three categories of commons Hardin describes – 'privatism', 'communism', and 'socialism'; further while talking of open access' systems, he (Hardin) does not refer to the property systems.

Finally, 'tragedy of the commons' hypothesis does not bother to understand the essence of property, and wrongly believes that it is a physical object like a forest, fishing pond, or a pasture. As Bromley (1991) correctly observes:

> By confusing the social dimension and the *concept* of property with a *physical object,* it is then easy for them to imagine that open access constitutes 'common property'…. But, of course, property *is not a physical object but is instead a social relation.*
>
> (1991, 30)

356   Commons for the Communities

Bromley there mentions four different types of property regimes to avoid such confusion:

(a) **State property** – Controlling agencies have right to determine use/access rules while individuals have *duty* to observe them.
(b) **Private property** – Individuals have *right* to undertake socially acceptable uses, and have *duty* to refrain from socially unacceptable uses; 'non-owners' have duty to refrain from preventing socially acceptable uses, and have a *right* to expect that only socially acceptable uses will occur.
(c) **Common property** – The management group ('owners') has *right* to exclude non-members and non-members have *duty* to abide by the exclusion; 'co-owners' have both *rights* and *duties* with respect to use rates and maintenance of the thing owned.
(d) **Non-property** – No defined groups of users or 'owners'; benefit available to anyone; individuals have both *privilege* and *no right* with respect to use rates and maintenance of asset that is an 'open access resource'.

However, in the context of pastoral systems in Africa, Scoones (1995) clarifies that because of the extent of spatial heterogeneity and temporal variability in resources, different resource tenures or property regimes co-exist and overlap:

> Different types of property regime may be more or less appropriate at different times and places. Empirical data from pastoral areas show no neat division between property regimes, but rather a complex set of overlapping rights that are continuously contested and renegotiated. These rights may shift over time and shift from place to place.
>
> (1995, 23)

There are instances of holistic transformation of large tracts of landscape made barren by top soil run-off due to heavy rainfall in Meghalaya (Cherrapunji receiving the highest amount of rainfall annually) through some catalyst agents (Box 9.2) and similarly in Karaikal district (Puducherry) efforts by district administration succeed in motivating the local people for restoring water bodies (Box 9.3).

Box 9.2: Holistic Transformation of Barren Land in Mawlyngkot, Meghalaya (India)

An Israeli philanthropist Aviram Rozin (and his family) founded Sadhna Forests (an NGO based in Auroville, Pondicherry) in order to improve food security in dryland areas through environmental transformation. In 2010, it started work in Haiti and in 2014 in Kenya, with the support of a grant from UNDP for wasteland reclamation. This turned arid land in Kenya into forests – a home of 166 species of trees/plants, and 75 species of birds now. Sadhna forest was invited to give shape to Meghalaya Community-Led Landscape Management Project (CLLMP), with a funding of Rs 660 crores (66 millions) by Japanese International Cooperation Agency (JICA) – its objective is

> to restore and conserve forest and natural resources of the village through sustainable forest management, livelihood improvement, and strengthening institutional development to enhance the capabilities

of the communities in contributing to conservation of environment, biodiversity, improvement of socio-economic conditions of people in Meghalaya.

Their team is training the community in Mawlyngkot village (East Khasi Hills) on conservation by using Swale technology. They are residing in the village (famous for Urlong tea) in tents and buses fitted with solar power, toilet, kitchen, etc. Rainwater from Meghalaya used to flow down to Bangladesh, the Swale concept manages water run-off from the top of slope, filters pollutants, and increases rainwater infiltration. Communities dug Swales, nitrogen-fixing plants are grown, and the soil is covered with mulch to prevent drying. As Aviram's wife Yurit says, 'we have to start with love and sympathy and no blame. Only then can we expect behavioural change in the community. A rich person is one who can give'. Villagers are inspired to learn better farming methods and forest regeneration there.

Source: *The Shillong Times* (7 January 2021).

Box 9.3: Restoration of 178 Water Bodies in Karaikal District, Puducherry (India)

In the Karaikal district of Puducherry Union Territory, the District Collector, Raja Vikranth restored 178 water bodies in 2019. He took cues from the ninth-century Chola dynasty method of making a network of channels and bunds to conserve rainwater for the use of people, animals, agricultural irrigation, etc. He ensured the participation of people, educational institutions, government officials, temples, etc. for various water reservoirs that are maintained by the stakeholders. Fifty ponds are used for aquaculture by the fisheries department, and many ponds are used for irrigation of crops. Many ponds are taken care of by public works department and local people. Due to water availability, cotton cultivation has increased across 100 acres. There is focus on holistic solutions to be learnt from history too. Earlier the peasants cultivated only one-fifth of their lands due to water scarcity; people had to walk miles to fetch water; and due to over-use water table had gone down 200–300 ft. District Collector Vikranth Raja launched 'Nam Neer' (our water) project under which out of 450 polluted, damaged, and dried up water bodies, 178 water bodies were desilted and revived in three months in 2019. Actually after digging of wells, ponds were neglected; after hand pumps were installed, wells were neglected; after piped water supply, hand pumps were neglected. For the ponds' revival, district collector introduced Employee Social Responsibility (ESR) engaging all government officials. At first, a pond attached to the temple Thirunallar was revived and it had a ripple effect. Then under MNREGA scheme, every village was to revive one pond – this restored 85 ponds! Temples revived 30 ponds; with corporate social responsibility (CSR) support 20 ponds and 81 km stretch of major canals were desilted – this facilitated the easy flow of water from Cauvery river to agricultural fields. In many cases, people and organisations offered free services

358   Commons for the Communities

or materials. In one village, Poovani cultivation was started after a 15 years' gap! As per CGWB, groundwater table increased by 10 ft during 2018–2019. People cleaned ponds to wish their favourite actors, family, friends, or teachers. There were also plantations around the ponds.

Source: the betterindia.com.

One recent example of restoring the commons (encroached upon) for collective purpose is from Shahpur Nanemau village in Sultanpur (UP, India) (see Box 9.4).

Box 9.4: Proper Use of Commons in Shahpur Nanemau Village in Sultanpur District, UP (India)

Till 2011, various common lands in Shahpur Nanemau village in Motigarpur Block, Distt. Sultanpur, UP (India) were illegally occupied by many families. In 2011, Government High School was sanctioned by the State government of UP with funding from Government of India under Sarva Shiksha Abhiyan, now renamed as Samagra Shiksha Abhiyan (SSA), after a lot of efforts by one senior government officer hailing from that village. In the beginning, there was a problem to locate the suitable land for the said school. One Thakur family from a neighbouring village Bir Shah Ka Pura had grabbed Gram Samaj land (commons) and was cultivating it for several decades. But taking a collective decision Gram Sabha approved that land for high school; it was okayed by the sub-divisional administration, too. Finally, High school building was built over there and school has been running there.

Second, the same senior government officer pursued for opening of a Primary Health Centre (PHC) in Shahpur Nanemau Gram Samaj land was partly encroached by a Brahmin family from the same village and another part was given on Patta (lease) to a dalit family. Hence, that dalit pattadhari (settlee) was given another piece of Gram Samaj land in exchange for his patta land, and the encroachment on other part was removed. The encroacher family intimidated some villagers, but PHC was sanctioned by government of UP in 2016–2017 and PHC building was constructed there. It has been operational and was quite useful during the pandemic COVID-19 in 2020–2021. Not only such villagers but also residents of a dozen gram sabhas are happy to get medical benefits from there. In both cases, people mobilised and, finally, the local administration helped them.

Source: Authors.

Alternative Theories

Regarding management of commons' resources, there are other theories too. Among them, theory of property rights, assurance problems theory, and theory of environmental entitlements are quite important. Since these theories' perspectives and assumptions differ from those of the 'tragedy of commons' hypothesis in fundamental ways, it becomes imperative to briefly mention about these.

[1] Theory of Property Rights

Demsetz (1967) and Behenke (1991) relate individual rights to land and mechanisms to use such rights to the levels of resource productivity, effects of population pressure, and application of rural technologies. Due to population pressure, land use is intensified – a shift from 'opportunistic grazing' (where pastures are exploited in periods of maximum production and then left to recover) to more continuous resource utilisation. According to Demsetz, common property regimes exist in those societies where natural resources have low value and the cost to control over their use is relatively high. As a resource acquires greater value or scarcity, the prevalence of individual maximising behaviour provides an incentive to overexploit it, and only then, the institutional innovations occur to conserve it – from non-exclusive to more exclusive forms of access. According to this theory, shifts to private property right occur only where the transaction costs or costs of 'policing', to use Demsetz's term, are exceeded by benefits accruing from the control of a resource. He cites an example of persistence of Native American communal hunting grounds on North American Great plains due to the fact that costs of containing roaming herds of bison were too high; the enclosure of the plains by cattlemen occurred when relatively low-cost barbed wire became freely available (Lane and Moorehead 1995).

According to this theory, common property regimes work only where resources are not scarce, and it does not matter that utility-maximising persons act as 'free-riders'. Thus, this theory partly accepts the privatisation notion of the 'tragedy of commons' hypothesis with a caveat if cost of 'policing' becomes less than benefits. But this theory does not explain as to why certain scarce and highly productive resources like Swiss Alpine meadows have been continuously existing as commons for thousands of years! (Lane and Moorehead 1995).

[2] Theory of Assurance Problems

Refuting the basis of the 'tragedy of commons' hypothesis, Runge (1984) argues that if expectations, assurance, and actions are coordinated to predict behaviour, there is less necessity for herd owners to pursue 'free-rider' strategies – i.e. the cooperative behaviour can be encouraged for utility maximisation. Runge (1984), Bromley and Cernea (1989) have pointed out following main tenets of this theory:

(a) Where communities have low and uncertain incomes, and are critically dependent on natural resources (especially rainfall), communal property systems are more efficient and cost-effective since they allow access to other areas, and thus function as a hedge against environmental risk.
(b) Institutions, customs, and conventions coordinate decisions in a community to promote voluntary support since production decisions by individuals are based on the expected decisions of others.
(c) Mobility is enhanced through reciprocity.
(d) Relative poverty puts a tough budget constraint on rural communities in terms of transaction costs (of policing, registering and adjudicating titles), thus the private property regime becomes, too, costly for a subsistence economy.

360 Commons for the Communities

(e) The more homogeneous a community, the more likely optimal outcomes are; the more heterogeneous a community, the more difficult cooperation becomes.
(f) Over-grazing does not necessarily arise from the strict dominance of a free-rider strategy, but from the inability of interdependent individuals to coordinate and enforce actions in situations of strategic interdependence.

Criticising Hardin's hypothesis, Bromley (1991, 157) again correctly remarks that private property regimes' proponents 'fail to understand that different ecological circumstances – and vastly different cultural contexts – must be dominant factors in the choice of institutional arrangements'. He cites an example of the nomadic pastoralists of Africa who are

> not mobile because they prefer wandering; rather they wander because their economic system demands it. To think that one 'solves' some problems by forcing them to be sedentary and by creating private property regimes over fixed Cartesian space is to confuse cause and effect.
>
> (1991, 157)

Thus this aspect has merit against Hardin's hypothesis.

[3] Theory of Environmental Entitlements

Leach, Mearns, and Scoones (1997) have developed this theory improving upon Amartya Sen's theory of entitlements. While Sen's theory is confined to command over resources through market channels, backed by formal-legal property rights, they extend entitlement framework to the whole range of socially sanctioned as well as formal-legal institutional mechanisms for gaining resource access and control. Secondly, instead of assuming that endowments are given, they focus on how both endowments and entitlements arise; they also include institutional skills that enable people to access official decision-making structures and other public goods.

They have cited three examples from India, Ghana, and South Africa. In southern Ghana, leaves of Marantaceae plants are commonly collected by women. The leaves become endowments depending on whether they lie inside the government forest or outside: (a) off-reserve, the leaves are common property of a village, with an actor's endowment depending on village membership; (b) if on the farmland, collection rights are acquired through membership of, or negotiation with the appropriate landholding family; (c) on reserve land, endowment mapping depends on the forest department's permit system. Women find leaf-collection in 'groups more effective, so collection depends on the membership of a regular group or on impromptu arrangements among kins and friends. There is frequently competition between groups for the best sites, as well as competition for leaves among members' (1997, 22). In disputes between individual women or between collection groups, or between them and forestry officials, 'a queen mother of leaf-gatherers', chosen by each village or neighbourhood's women gatherers, helps to mediate them' (1997, 22).

Similarly, they cite a case of seven hamlets in Rajasthan (India) which experienced soil degradation and deforestation due to new market opportunities, State policies of land reforms and population growth. People grazed their livestock there under an

unregulated open access system which, in turn, led to the over-grazing and encroachment of hills for cultivation. But, later Nayakheda Watershed Project, supported by
Udaipur based NGO, Seva Mandir, promoted soil and water conservation, and a system of enclosures to regulate the use of commons through a community-level committee. This resulted into not only common environmental good but also promoted
social cohesion and political action. Two points emerge from this study: First, people
are differentiated in terms of seven castes and tribes, their residence in three villages,
various occupations, and resource priorities but there is a dynamic interaction between
them; secondly, at first, Seva Mandir, worked on adult literacy, maternal and child
health care, leadership training, and 'institutional strengthening' for eight years, and
only then aimed at building up of capabilities-*social capital*.

Thus, this theory stands on sound footing against the 'tragedy of commons'
hypothesis.

Empirical Evidences: Seven Case Studies

In Africa, a dogma has prevailed among policy-makers that pastoralists would degrade
the commons if left to themselves since they own animals individually but use the
range as an 'open access'. Hence, there are three major processes under way in Africa
affecting the pastoral system: nationalisation of their resources, sedentarisation of
herders and titling, and privatisation. Now, we need to briefly discuss these processes
to see how they have managed commons:

[A] Nationalisation: Case of Mali

This is the first method of 'tragedy of commons' hypothesis. The classic 'tragedy of
commons' argument has legitimised the take-over of pastoral resources by the government in many countries of Africa. But the result is different. For instance, in Mali,
pastoral resources of the inland delta of river Niger were nationalised. Earlier under
customary practice, dry land flood pastures were divided into 30 territories allocated
to sub-clans of Fulani pastoralists, and outsiders could access only with the payment
of a fee. The resource managers for each territory set the dates on which crossings
into pastures took place. But nationalisation brought an inflexible, untracked policy
according to which animals moved on to flood pastures when they were dry – thus
preventing the regeneration of pasture resource itself. This case study conducted by
Moorehead in 1991 (cited in Lane and Moorehead 1995) shows that the 'tragedy of
commons' hypothesis is not tenable because after nationalisation the pastoralists could
not check outsiders' access to commons as they did in earlier systems of commonly
owned resources.

Similarly, in case of forests, which were managed by villagers for generations with
considerable restraint over rate and pattern of use of forest products, nationalisation
in many countries failed because of two reasons (Ostrom 1990): First, there were no
sufficient forest employees to enforce regulations; second, they were poorly paid, and
hence bribe became a means to supplement their income. Consequently, 'nationalisation created open access resources where limited-access common property resources
had previously existed' (Ostrom 1990, 23). The similar disastrous effects of nationalisation of former communal forests occurred in Thailand, Niger, Nepal, and India.

362 Commons for the Communities

[B] Sedentarisation and Land Titling: Case of Tanzania

Sedentarisation is the second method of the 'tragedy of commons' hypothesis of believing the incompetence of herders. Actually, many governments in Africa have often seen the herders as escaping their administration – going beyond national boundaries – or as potential threats to security or as of evaders of fiscal dues. Therefore, settlement of pastoralists is pursued as either an overt policy objective (e.g. villagisation in Tanzania) or as product of administrative action (e.g. famine response in Sahel) or as the inevitable consequence of land tenure reform and the push for privatisation sponsored by Western aid donors. In Tanzania, more than 1,00,000 acres of prime grazing lands were acquired by the government for a parastatal wheat scheme which undermined the Barabaig grazing system, adversely affected the environment and Barabaig welfare (Lane and Moorehead 1995).

Similarly, it was believed by the policy-makers in many African countries that only through more formal registered title, pastoralists' lands will be protected from grabbing, and/or rural land-users be encouraged to investment or it will induce lenders to finance. But a study by Kjarby (1979) of villages in Hanang district of Tanzania found that those Barabaigs who settled in villages in semi-arid plains were compelled to compromise by limiting their migration to the distance their animals could travel to and from their homestead in one day. But, unfortunately, the concentration of herds within the village had negative ecological effect, encouraged a tendency towards agropastoralism, and ultimately resulted in decrease in levels of production (Lane and Moorehead, 1995). Thus, both sedentarisation and land titling methods ultimately failed. As Place and Hazell (1993) also found in a household survey in Ghana, Kenya, and Rwanda, with a few exceptions, 'land rights are not found to be a significant factor in determining levels of investment in land improvements, use of input access, or the productivity of land' (1993, 10).

[C] Privatisation: Case of Botswana

The extreme policy under 'tragedy of commons' hypothesis is privatisation. In Botswana that the introduction of borehole drilling technology and the emergence of a rigid social order led to the monopoly of new water resources by the rich landowners who have got more share of national herd and controlled best grazing areas (White 1992, cited in Lane and Moorhead, 1995).

Similarly, in Pakistan-occupied Kashmir, privatisation of commons (Shamilat land) occurred in three stages – informal partitioning, incremental appropriation, and formal privatisation – facilitated both by regulation and backdoor influence or corruption. As a result, only selected influential individuals benefitted from this.

Thus, privatisation method of the 'tragedy of commons' thesis also did not give expected results.

[D] Case Study of Kottapalle, South India

In a case study of Kottapalle in South India, it was found that European by-laws commonly gave emphasis on regulation of the cropping, whereas South Indian villages emphasise regulation of livestock – the institution of village field guards. Secondly,

Commons for the Communities   363

there is a system of common property rights in canal water – practice of 'first come, first served' and thus: 'The villagers themselves have constituted an authority to impose rules of restrained access' (Wade, 1988, 18). The villagers have successfully created water rules, grazing rules, havesting rules, road maintenance, well repairs, etc. Wade (1988) also found that, unlike Hardin's assumption that individual herders have no information about the aggregate state of the commons and its nearness to the point of collapse, 'monitoring the condition of the commons, and of cheating, is relatively easy'. Thirdly, Hardin did not distinguish between commons where the resource is vital for an individual's survival, and those where it is not; but Hardin's logic is more likely to operate where the resource is not vital than where it is. To put in Wade's words: 'Where survival is at stake, the rational individual will exercise restraint at some point. In our villages, water and grazing are both vital' (1988, 205). Thus, the 'tragedy of commons' hypothesis does not work in such villages in South India.

[E]   Case Study of Somalia

In Somalia, pastoralists in management of the rangeland successfully cooperate, particularly where groups of users have no wide income disparities or where customary contract law – 'xeer' – is properly developed or where customary kin and clan structures facilitate cooperation on several issues. But, on the other hand, cooperation is undermined in three situations: (a) where government claims management sovereignty over the rangeland; (b) where individual rice herders or merchants introduce different management goals; and (c) where donor-aided projects undermine customary decision-making and enforcement processes (M. S. Said 1994, cited in Swift, 1995).

[F]   Case Study of Sukhomajri, India

In Sukhomajri, near Chandigarh, the hills, which were earlier vegetated, suffered from increase in the human and animal populations. Resulting into over-grazing and severe erosion – in the 1970s only 15% of the uplands had vegetation and erosion rate was 150–200 tons/ha – and 60% of lake was filled with sediment. Subsequently small earthen dams were built by the local people to irrigate fields, and farmers were given subsidised seeds and fertilisers; consequently, yields increased. Later water-users' society was also formed – with one representative from every family. Second, a coupon system was introduced and families with no land or little land could sell their water-rights or use the water to share-crop on others' land. Third, if someone's animal was found grazing on the hills, he lost his rights to water. Consequently, villagers replaced their goats with buffaloes that are stall-fed. The society maintained dams and catchments and was responsible for water distribution and record-keeping. Thus, on the one hand, villagers' livelihoods grew; on the other hand, hills became vegetated and rate of soil erosion had heavily fallen (Conway and Barbier, 1990). Though this project started as a conventional soil conservation project, it became a community project due to the intensive interaction between project staff and villagers through learning approach, a conscious decision to put 'people's priorities first', security of rights and gains for the poor, importance of self-help, and good project staff. This success story, too, disproves the 'tragedy of commons' hypothesis.

[G] Wildlife-Friendly People in Satyavedu Range, Chittoor, Andhra Pradesh (India)

For about a decade, the villagers of KVB Puram, B N Kandriga, Pichatur, Nagalapuram, and Satyavedu Mandals (Andhra Pradesh) came across deer sprinting near their habitations, and these were attacked by the stray dogs. But youth of these villages intervened collectively, and helped the deer from the stray dogs' attacks, and saved their life as they returned to their natural habitats in the sparse forests. Even at nights when the stray dogs start barking loudly, the young villagers wake up and shoo away the dogs. Due to such collective human efforts, in many villages, the stray dogs are no longer hostile to the deer. The local forest officers created wildlife awareness among the villagers. Consequently, the young villagers have also stopped hunting monitor lizards, pythons, and other reptiles. Further, the villagers stay away from the reserve forest zones, and voluntarily limit cattle grazing to the fringe areas. They have also provided a safer environment for forest animals like bears. Thus, man–animal conflict has been averted to a large extent. These collective efforts of the villagers have restored and enriched the forests and wildlife in that region in a genuine way (Umashankar 2019). This success story, too, disproves the 'tragedy of commons' hypothesis.

Conclusion

From the above discussion, the following conclusions may be drawn: There are two schools for avoiding the inevitability of the 'tragedy of commons'. First is the privatisation school that justifies establishment of full private property rights over commons. Second is the State ownership school. However, on the other hand, assurance problems theory and environmental entitlements theory are gaining more acceptance. These theories rightly challenge the 'tragedy of commons' hypothesis by giving evidence that, historically in many developing countries, there have been successfully functioning common property regimes that were neither private nor State property regimes. Actually, the common property system has persisted due to five comparative advantages: less transaction costs, better adapted to local conditions, adaptability to changing resource availability and social stability, hedge against individual failure, and high opportunity costs with changing established practices (Bromley 1992). However, in communal management, the smaller groups tend to have more cooperative behaviour than larger ones. Other factors like common understanding of interests, low discount rates, and low transaction costs may make it successful (Ostrom 1990). Hence, communal property use, ownership, maintenance, and upgradation are to be preferred to privatisation or statisation.

Secondly, States are already over-stretched in developing countries, hence they may not be able to provide necessary resources to make private property or State-controlled regimes work, as failure of nationalisation in Mali and of sedentarisation in Tanzania prove it. Further, in the present age of globalisation, State is usually to work more as a facilitator rather than a regulator. Hence, well-planned public action by the State and its agencies is called for 'to restrict the further curtailment' of CPRs through sensitisation, on the one hand, and the re-orientation of development policies for improving resource productivity and environmental stability with a common property (asset) perspective (Jodha 1986, 1992).

Thirdly, privatisation may not work in all societies and at all times. As the privatisation in Botswana led to monopoly of new water resources, or in Pakistan-occupied Kashmir, it benefitted only selected influential individuals, not the poor and needy. Obviously, privatisation (or even public–private partnership) seeks profit at the cost of mass welfare and upliftment – through short-term goals, ignoring long-term goals of intra-generational and inter-generational equity and justice.

Fourthly, the mining of CPRs should be strictly prohibited; instead their productivity should be enhanced through better and creative methods of conservation technology with the least costs and damages.

Fifthly, various user groups (e.g. water user groups, forest produce users) should be promoted as well as duly involved at all stages in order to break the vicious cycle of CPRs degradation and poverty, at the earliest.

Sixthly, Hardin's concept assumes that all commons are 'open access' systems but this is not true. Many commons, for instance, pastoral systems in Africa, or irrigation systems in South India, have evolved highly organised regulation mechanisms and rules about the use of common property resources, including sanctions from the community against individual over-exploitation.

Seventhly, not only competition and conflict but also cooperation through the use of indigenous knowledge regarding common resources does exist in villages of the developing countries like India.

Finally, the 'tragedy of commons' hypothesis has a cultural bias seen from the English commons' perspective, whereas in developing countries 'many common property resources are used without an accompanying tragedy. Users co-operate with one another rather than compete ... common property resource tragedies in the Hardin's sense seem not to be the rule but the exception' (Berkes 1989, 71). In fact, even the example given by Hardin (1968) reveals that in England common grazing lands were properly looked after for several centuries but subsequently declined for reasons other than any fault in the commons system as such.

Thus, the 'tragedy of commons' hypothesis is not valid empirically and theoretically in all societies at all periods of time. Local people have collective sensitivity to and wisdom for the common natural resources; however, they need support of the local administration/government too, in case of encroachment by the influential persons/families, because all commons ultimately belong to the local communities and they have full rights over these common natural resources.

Points for Discussion

(1) What are the benefits of commons to local people?
(2) Are there conflicts over use of commons? What are institutional mechanisms for their resolution?
(3) What are the assumptions of the 'Tragedy of Commons' hypothesis? How can it be criticised on empirical and theoretical grounds?
(4) Is there any alternative theory to the 'tragedy of commons' hypothesis? Illustrate with some empirical cases.

Bibliography

Aflatoon Ltd. Vs Governor of Delhi (1975), 4 SCC.

Agarwal, Anil and Sunita Narain (1982), *The State of India's Environment: First Citizen's Report*, New Delhi: Centre for Science and Environment.

Agarwal, Anil and Sunita Narain (1985), *The State of India's Environment: Second Citizen's Report*, New Delhi: Centre for Science and Environment.

Agarwal, Anil and Sunita Narain (1991), *Global Warming in an Unequal World*, New Delhi: Centre for Science and Environment.

Agarwal, Anil and Sunita Narain (1999), *State of India's Environment: Citizens' Fifth Report*, New Delhi: Centre for Science and Environment.

Agarwal, Beena (1992), The Gender and Environment Debate: Lessons from India, *Feminist Studies*, 18(1) (Spring), reprinted in Mahesh Rangrajan (ed.) (2007), *Environmental Issues in India*, New Delhi: Pearson India Education Services Pvt. Ltd., pp. 316–361.

Agarwal, Beena (2010), *Gender and Green Governance*, New Delhi: Oxford University Press.

Agarwal, Bina (1995), *Gender, Environment and Poverty Interlinks in Rural India: Regional Variations and Temporal Shifts 1971–1991*, UNRISD, Discussion Paper 62, Geneva: UNRISD.

Agarwal, Bina (2007), The Gender and Environment Debate: Lessons from India, in Mahesh Rangrajan (ed.), *Environmental Issues in India*, New Delhi: Pearson India Education Services Pvt. Ltd., pp. 316–361.

Agarwal, Priyangi (2021), Local Factors Play a Bigger Role in Fouling up Delhi Air, *The Times of India*, 17 February 2021.

Ahmad, Sohail (2018), Urbanisation and Health, in Sunita Narain et al. (eds.), *State of India's Environment 2018*, New Delhi: CSE, pp. 260–262.

Ahmed, A. (1997), Post-Colonial Theory and the 'Post' Condition, *The Socialist Register*, 1997.

Aiyar, Swaminathan and S. Anklesaria (2021), Yes, Glaciers are Melting But No Need for Panic, *The Times of India*, 11 April 2021.

Akula, V. K. (1995), Grassroots Environmental Resistance in India, in Taylor, B. R. (ed.), *Ecological Resistance Movements*, Albany, NY: State University of New York Press, pp. 127–145.

AlBdour, M. and M. M. Al Latayfeh (2004), Risk Factors for Pterygium in an Adult Jordanian Population, *Acta Ophthalomologica Scandinavica*, 82, pp. 64–67.

Amin, S. (1989), *Eurocentrism*, New York: Monthly Review Press.

Anderson, C. H. (1976), *The Sociology of Survival*, IL: Dorsey, Homewood.

Anderson, James R., Ernest E. Hardy, John T. Roach and Richard E. Witmer (1976), *A Land Useand Land Cover Classification System for Use with Remote Sensor Data*, Washington, DC: United States Government Printing Office.

Andharia, Janki and Chandan Sengupta (1998), The Environmental Movements: Global Issues and Indian Realities, in Murli Desai et al. (eds.), *Towards People-Centred Development, Part II*, Bombay: Tata Institute of Social Sciences, pp. 428–431.

Armer, Michael and Allan Schnaiberg (1972), Measuring Individual Modernity: A Near Myth, *American Sociological Review*, 37(June), pp. 301–316.

Arnold, David and R. Guha (1995), Introduction: Themes and Issues in the Environmental History of South India, in D. Arnold and R. Guha (eds.), *Nature, Culture, Imperialism*, New Delhi: Oxford University Press , pp. i–xi.

Arora, Saurabh, et al. (2019), Sustainable Development through Diversifying Pathways in India, *Economic and Political Weekly*, LIV(46), pp. 32–37.

Ashcroft, B., Griffiths, G. and Tiffin, H. (eds.) (1995), *The Post-colonial Studies Reader*, London: Routledge.

Ayres, Robert U. (1978), *Resources, Environment & Economics: Applications of the Materials/Energy Balance Principle*, New York: John Wiley & Sons.

Azad, Shivani (2021), Delhi-Doon e-way Extension Gets Nod, But 2,500 Old Trees to be Cut, *The Times of India*, 7 January 2021.

Bailes, Kendall E. (1985), *Environmental History: Critical Issues in Comparative Perspective*, Lanham, MD: University Press of America.

Bakker, Peter (2021), We Need to Move to Capitalism of True Value… You Will See All Capital Will Flow towards Sustainability, *The Times of India*, 9 April 2021.

Balakrishnan, M. (1984), Large Mammals and Their Endangered Habitats in Silent Forests in India, *Biological Conservation*, 29(3), pp. 277–286.

Bandyopadhayay, Jayanta (1992), From Environmental Conflicts to Sustainable Mountain Transformation, in D. P. Ghai and J. M. Vivian (eds.), *Grassroots Environmental Action: People's Participation in Sustainable Development*, London: Routledge.

Bandyopadhayay, Jayanta and Vandana Shiva (1987), Chipko: Rekindling India's Forest Culture, *The Ecologist*, 17(1), pp. 26–34.

Banerjee, Chandrajit (2019), Manage Stubble to Control Air Pollution: Recycling it Can Reduce Weedicide Use by 70% and Water Use by 40%, *The Times of India*, 7 November 2019.

Bavadam, Lyla (2018), A Plastic Ban, *Frontline*, 20 July, pp. 38–39.

Baviskar, Amita (1997), *In the Belly of the River*, Delhi: Oxford University Press (OUP).

Beauvoir, Simone de (1949), *The Second Sex*, London: Vintage Books.

Beck, Tony and Cathy Nesmith (2001), Building on Poor People's Capacities: The Case of Common Property Resources in India and West Africa, *World Development*, 29(1), pp. 119–133.

Beck, Ulrich (1992), *Risk Society*, London: Sage Publications Ltd.

Behenke, R. H. (1991), Economic Models of Pastoral Land Tenure in Cincotta, in R. P. Gay and Perrie, G. K. (eds.), *New Concepts in International Rangeland Development: Theories and Applications*, Logan: Utah State University.

Bell, Daniel (1973), *Coming of Post-Industrial Society: A Venture in Social Forecasting*, New York: Basic Books.

Bell, Derck (2011), Does Anthropogenic Climate Change Violate Human Rights? *Critical Review of International Social and Political Philosophy*, 14(2), pp. 99–124.

Bell, M. M. (2016), *An Invitation to Environmental Sociology*, Washington, DC: Sage Publications.

Bennett, John W. (1969), *Northern Plainsmen: Adaptive Strategy and Agrarian Life*, New York: Aldine.

Benton, T. (2001), *Environmental Sociology*, London: Sage Publications.

Berger, Peter L. and Thomas Luckmann (1966), *The Social Construction of Reality*, London: Penguin Books.

Berkes, F. (1989), *Common Property Resources*, London: Belhaven Press.

Bhabha, H. (1984), Of Mimicry and Man: The Ambivalence of Colonial Discourses, *October*, 28 (Spring), pp. 125–133.

Bhaskar, 11 August, 2019.

Bhaskar, 11 July, 2019.

Bhatt, C. P. (1983), Eco Development: People's Movement, in T. V. Singh and J. Kaur (eds.), *Himalayas Mountains and Men*, Lucknow: Print House.

Bhatt, Chandi Prasad (1990), The Chipko Andolan: Forest Conservation Based on People's Power, *Environment and Urbanisation*, 2(1), pp. 7–18.

Bhattacharya, Amit (2020), Pattern of Frequent Wet Spells this Season May Be Linked to Record Arctic Freeze, *The Times of India*, 6 March 2020.

Bhushan, Chandra (2020), To Reverse Mass Extinction of Species, *The Times of India*, 29 February 2020.

Bihar Land Acquisition Manual (2004), Patna: Malhorta Brothers.

Bilali, Hamid E. L. (2018), *Innovation-Sustainability Nexus in Agriculture Transition: Case of Agro-Ecology*, De Gruyter, https://doi.org/10.1515/opag-2009-0001.

Boggs, Danny J. (1985), When Governments Forecast, *Futures*, 17, pp. 435–439.

Boserup, Ester (1970), *Women's Role in Economic Development*, London: Earthscan.

Bromley, D. W. (1989), *Economic Interests and Institutions: The Conceptual Foundations of Public Policy*, Oxford: Basil Blackwell.

Bromley, D. W. (1991), *Environment and Economy: Property Rights and Public Policy*, Oxford: Blackwell Publishers.

Bromley, D. W. (1992), *Making the Commons Work*, San Francisco, CA: ICS Press.

Bromley, D. W. and M. M. Cernea (1989), *The Management of Common Property Natural Resource*, Washington, DC: World Bank.

Bryant, Raymond L. and Sinead Bailey (1997), *Third World Political Ecology*, New York: Routledge.

Bullard, Robert D. (1983), Solid Waste Sites and the Black Houston Community, *Sociological Inquiry*, 53, pp. 273–288.

Bullard, Robert D. (1990), *Dumping in Dixie: Race, Class, and Environmental Quality*, Boulder, CO: Westview Press.

Burch, W. R., Jr. (1971), *Daydreams and Nightmares: A Sociological Essay on the American Environment*, New York: Harper & Row.

Burke, T. (1995), View from the Inside: UK Environmental Policy Seen from a Practitioner's Perspective, in Tim S. Gray (ed.), *UK Environmental Policy in 1990's*, London: Macmillan.

Burningham, K. and G. Cooper (1999), Being Constructive: Social Constructionism and Environment, *Sociology*, 33, pp. 297–316.

Burton, Ian, Kates, Robert W. and White, Gilbert E. (1993), *The Environment as Hazard*, New York: Guilford Press.

Buttel, F. and P. Taylor (1994), Environmental Sociology and Global Environmental Change: A Critical Assessment, in M. Redclift and T. Benton (eds.), *Social Theory and Global Environment*, London: Routledge.

Buttel, F. H. (1987), New Directions in Environmentals Sociology, *Annual Review of Sociology*, 13, pp. 465–488.

Buttel, F. H. (1993), Environmental Sociology as Science and Social Movement, *Environment, Technology and Society* (Newsletter of the ASA Section of Environment and Technology), 73(Fall), pp. 101–111.

Buttel, Frederick H. (1976), Social Science and the Environment: Competing Theories, *Social Science Quarterly*, 57(2), pp. 307–323.

Buttel, Frederick H. (1986), Sociology and the Environment: The Winding Road toward Human Ecology, *International Social Science Journal*, 109, pp. 337–356.

370 Bibliography

Buttel, Frederick H. (2000), Ecological Modernization as Social Theory, *Geoforum*, 31(1), pp. 57–65.

Byravan, Sujatha (2019), Turning Down the Heat, *The Hindu*, 11 July 2019.

C.W.C. (1994), *Environmental Monitoring Committee: Annual Report (1992–93)*, New Delhi: Ministry of Water Resources, Govt. of India, P.II.

CAIT Climate Data Explorer, 2018.

Campbell, John (2010), Climate-induced Community Relocation in the Pacific: The Meaning and Importance of Land, in Jane Mc Adam (ed.), *Climate Change and Displacement*, Oxford: Hart Publishing Ltd., pp. 57–80.

Caney, Simon (2005), Cosmopolitan Justice, Responsibility and Global Climate Change, *Leiden Journal of International Law*, 18(4), pp. 747–775.

Caney, Simon (2010), Climate Change, Human Rights, and Moral Thresholds, in Stephen Gardiner et al. (eds.), *Climate Ethics*, Oxford: Oxford University Press.

Carolan, M. S. (2005), Society, Biology & Ecology, *Organisation and Environment*, 18, pp. 393–421.

Catton (Jr.) William R. and Dunlap, R. E. (1978), Environmental Sociology: A New Paradigm, *The American Sociologist*, 13(1), pp. 41–49.

Catton, William R. and Riley E. Dunlap (1978), Environmental Sociology: A New Paradigm, *The American Sociologist*, 13, pp. 41–49.

Catton, Willian R. and Riley R. Dunlap (1980), A New Ecological Paradigm for Post-Exuberant Sociology, *American Behavioural Scientist*, 24, pp. 15–47.

Cernea, M. M. (ed.), *Putting People First*, 2nd ed., New York: Oxford University Press.

Chakravorty, Shouvik (2019), A Reality Check on India's Renewable Energy Capacity, *The Hindu*, 15 August 2019.

Chambers, R. (1969), *Settlement Schemes in Tropical Africa: A Study of Organisation and Development*, New York: Routledge & Kegan Paul.

Chaturbhuj Pandey Vs Collector, Rajgarh, 1969 BLJR.

Chaturvedi, S., J. Bandyopadhyay and Shikui Dong (2017), Introduction, in Shikui Dong, J. Bandyopadhyay and S. Chaturvedi (eds.), *Environmental Sustainability from the Himalayas to the Oceans*, Cham: Springer International Publishing, pp. 1–16.

Chaulia, Sreeram (2021), New Year, New Vision, *The Times of India*, 1 January 2021.

Chiras, Daniel D. (2012), *Environmental Science*, 9th ed., Burlington, MA: Jones & Bartlett Learning.

Ciriacy-Wantrup, S. V. and Bishop, R. C. (1975), Common Property as a Concept in Natural Resources Theory, *Natural Resources Journal*, 15, pp. 713–727.

Clapp, J. and P. Danvergne (2011), *Paths to a Green World: The Political Economy of the Global Environment*, 2nd ed. Cambridge, MA: MIT Press.

Cohen, Robin and Paul Kennedy (2000), *Global Sociology*, London: Macmillan Press Ltd.

Collector Rajgarh Vs Harisingh Thakur, 1979 1, SCC.

Collins, H. M. and T. Pinch (1993), *The Golem: What Everyone Should Know about Science*, Cambridge: Cambridge University Press.

Conway, G. R. and Barbier, E. B. (1990), *After the Green Revolution*, London: Earthscan Publications Ltd.

Cronon, William (1983/2003), *Changes in the Land: Indians, Colonists and Ecology of New England*, New York: Hill and Wang.

Crush, J. (1995), *Power of Development*, London: Routledge.

Dainik Jagaran, 2 November 2017.

Dash, Dipak (2021), Bengaluru 'Most Liveable' of Mega Indian Cities, Delhi Ranks 13th, *The Times of India*, 4 March 2021.

Dayanand, Stalin (2020), The Myth of Sustainable Development in Mumbai's Infrastructure, *Economic and Political Weekly*, LV(3), pp. 16–20.

Delgado, F. G. and Vanessa Romo (2020), How the Women of Cuninico Are Leading Peru's Fight for Community Rights, *science.thewire.in*, 26 December 2020.

Demeritt, David (1998), Science, Social Constructivism and Nature, in Bruce Braun and Noel Castree (eds.), *Remaking Reality: Nature at the Millennium*, London: Routledge, pp. 172–192.

Demsetz, H. (1967), Towards a Theory of Property Rights, *American Economic Review*, 57, pp. 347–359.

Dickens, Peter (1996), *Reconstructing Nature: Alienation, Emancipation, and Division of Labour*, London and New York: Routledge.

Dogra, Bharat (1986), The Indian Experience with Large Dams, in E. Goldsmith and N. Hildyard (eds.), *The Social and Environmental Effects of Large Dams*, Vol. II, Cornwall: Wadebridge Ecological Centre, pp. 202–2012.

Dong, Shikui (2017), Himalayan Grasslands: Indigenous Knowledge and Institutions for Social Innovation, in Dong, Shikui, J. Bandyopadhyay and S. Chaturvedi (eds.), *Environmental Sustainability from the Himalayas to the Oceans*, Cham: Springer International Publishing, pp. 99–126.

Downs, A. (1972), Up and Down with Ecology: The Issue-Attention Cycle, *The Public Interest*, 28, pp. 38–50.

Dryzek, John S., et al. (2013), *Climate-Challenged Society*, Oxford: Oxford University Press.

DuBois, W. E. B. (1903), *The Souls of Black Folk*, Chicago: A.C. Mcdurg and Co.

Dunlap, R. E. (1997), Evolution of Environmental Sociology: A Brief History and Assessment of American Experience, in Michael Redclift and Graham Woodgate (eds.), *International Handbook of Environmental Sociology*, Cheltenham: Edward Elgar, pp. 21–39.

Dunlap, R. E. (2010), The Maturation and Diversification of Environmental Sociology, in Michael R. Redclift and G. Woodgate (eds.), *The International Handbook of Environmental Sociology*, Cheltenham (UK): Edward Elgar Publishing Ltd., pp. 15–32.

Dunlap, R. E. and W. R. Catton (1979), Environmental Sociology: A Framework for Analysis, in T. O'Riordan and R. C. D'Arge (eds.), *Progress in Resource Management and Environmental Planning*, Vol. I, Chichester: John Wiley & Sons, pp. 57–85.

Dunlap, R. E. and William Michelson (eds.) (1997), *Handbook of Environmental Sociology*, Westport, CT: Greenwood.

Durning, Alan (1992), *How Much is Enough? The Consumer Society and the Future of the Earth*, New York: W.W. Norton & Co.

Duru, M., et al. (2014), A Conceptual Framework for Thinking Now (and Organising Tomorrow): The Agro-ecological Transition at the Level of the Territory, *Cah. Agriculture*, 23, pp. 84–95, https://doi.org/10.1684/agr.2014.0691.

Eckersley, R. (1992), *Environmentalism and Political Theory*, London: UCL Press.

Eckholm, Erik (1982), *Down to Earth: Environment and Human Needs*, New York: WW. Norton and Co.

Eco-Equity (2008), *Greenhouse Development Rights*, www.gdrights.org.

Edou, Kim (2018), Minimum Waste, Maximum India, in Sunita Narain et al. (eds.), *State of India's Environment 2018*, New Delhi: CSE, pp. 234–235.

Ehrlich, Paul and A. W. Ehrlich (1968), *The Population Bomb*, New York: Sierraclub/Ballantine Books.

Elson, Diana (1995), *Male Bias in Development Process*, Manchester: Manchester University Press.

Escobar, A. (1997), Encountering Development: The Making and Unmaking of the Third World, in M. Rahnema and V. Bawtree (eds.), *The Post-Development Reader*, London: Zed Books.

Evans, Peter (1995), *Embedded Autonomy*, Princeton, NJ: Princeton University Press.

Evans, Peter (1996), Government Action, Social Capital and Development: Evidence on Synergy, *World Development*, 24(6), pp. 1119–1132.

372 Bibliography

Fanon, F. (1967), *The Wretched of the Earth*, New York: Grove Press.

Fanon, Franz (1963), *The Wretched of the Earth*, New York: Grove Weidenfeld.

FAO (2020), Three kgs of Food Crop are Used to Make Every kg of Meat You Buy, *The Times of India*, 5 March 2020.

Foster, John Bellamy (1999), The Canonization of Environmental Sociology, *Organization & Environment*, 12(4), pp. 461–467.

Foster, John Bellamy (2001), *Marx's Ecology*, Kharagpur: Cornerstone Publications.

Foster, John Bellamy (2012), The Planetary Rift and the New Human Exemptionalism: A Political – Economic Critique of Ecological Modernization Theory, *Organization & Environment*, September 23, 2012.

Foster, John Bellamy, et al. (2019), Imperialism in the Anthropocene, *Analytical Monthly Review*, July-August 2019, pp. 68–84.

Frank, A. G. (1975), *On Capitalist Underdevelopment*, New Delhi: Oxford University Press.

Freire, Paul (1996), *Pedagogy of the Oppressed*, London: Penguin.

Fusco, A. C. and J. A. Logan (2003), Analysis of 1970–1995 Trends in Tropospheric Ozone at Northern Hemisphere Midlatitudes with the Geos-Chem Model, *Journal of Geophysical Resources and Atmosphere*, 108, p. 4449, https://doi.org/10.1029/2002JD002742.

Gadgil, Madhav (1990), India's Deforestation: Patterns and Processes, *Society and Natural Resources*, 3, pp. 131–143, https://doi.org/10.1080/08941929009380713.

Gadgil, Madhav and Ramachandra Guha (1992), *This Fissured Land: An Ecological History of India*, Berkeley, CA: University of California Press.

Gadgil, Madhav and Ramachandra Guha (1994), Ecological Conflicts and Environmental Movements in India, in D. Ghai (ed.), *Development and Environment*, Oxford: Blackwell Publishers.

Gadgil, Madhav and Ramachandra Guha (1995), *Ecology and Equity: The Use and Abuse of Nature in Contemporary India*, London: Routledge.

Gadgil, Madhav and Ramchandra Guha (2007), Ecological Conflicts and the Environmental Movement in India, in Mahesh Rangrajan (ed.), *Environmental Issues in India*, New Delhi: Pearson India Educational Services Pvt. Ltd., pp. 385–428.

Gandhi, Indira (1982), *Man and this World, on Peoples and Problems*, 2nd ed., 1983, London: Hodder and Stoughton, pp. 60–67.

Gandhiok, Jasjeev (2021), 49% Urban Poor Used Unclean Fuel on Lockdown, *The Times of India*, 10 April 2021.

Gandhiok, Jasjeev (2021), This March was Delhi's Warmest in 11 Years, *The Times of India*, 1 April 2021.

Garner, R. (1996), *Environmental Politics*, London: Prentice-Hall.

Ghosh, Shinjini (2018), River Pollution, *The Hindu*, 29 September 2018.

Giddens, Anthony (1998), *The Third Way: The Renewal of Social Democracy*, Cambridge, Policy Press.

Giraldo, O. F. and P. M. Rosset (2018), Agro-ecology as a Territory in Dispute: Between Institutionality and Social Movements, *Journal of Peasant Studies*, https//doi.org/10.1080/03366150.2017.1353496.

Girija Dubey Vs State of Bihar, 1984, PLJR.

Gliessman, S. R. (2015), *Agro Ecology: The Ecology of Sustainable Food Systems*, Boca Raton, FL: CRC Press.

GOI (2000), *The Noise Pollution (Regulation and Control) Rules*, New Delhi: Ministry of Environment and Forests, Govt. of India.

GOI (2004), *Disaster Management in India*, New Delhi: Natural Disaster Management Division, Ministry of Home Affairs.

Goldsmith, E. and N. Hildyard (1984), *The Social and Environmental Effects of Large Dams*, Vol-I, Cornwall, UK: Wadebridge Ecological Centre, pp. 231–237.

Goodman, David and Michael Redclift (1991), *Refashioning Nature: Food, Ecology and Culture*, London: Routledge.

Gould, Kenneth Alan, Pellow, David N. and Schnaiberg, Allan (2003), Interrogating the Treadmill Production, in *Symposium on Environment and the Treadmill of Production*, University of Wisconsin, Madison, October 31 to November 1, 2003.

Gray, Tim S. (1997), Politics and the Environment in the UK and Beyond, in Michael Redclift and Graham Woodgate (eds.), *The International Handbook of Environmental Sociology*, Cheltenham: Edward Elgar, pp. 287–299.

Grove, Richard H. C. (1990), Colonial Conservation, Ecological Hegemony and Popular Resistance: Towards a Global Synthesis, in J. M. Mackenzie (ed.), *Imperialism and the Natural World*, Manchester: Manchester University Press.

Guha, Ramachandra (1988), Ideological Trends in Indian Environmentalism, *Economic and Political Weekly*, 23(49), December 03.

Guha, Ramachandra (1989), *The Unquiet Woods: Ecological Change and Peasant Resistance in the Himalaya*, Oxford: Oxford University Press.

Guha, Ramachandra (ed.) (1998), *Social Ecology*, New Delhi: Oxford University Press.

Guha, Ramchandra (2006), *How Much Should a Person Consume?* Ranikhet: Permanent Black.

Guoqing, Shi (2014), Management System of Involuntary Resettlement in Reservoir Projects in China, *ASCI Journal of Management*, 44(1), (Special), pp. 17–20.

Gupta, A. (1988), *Ecology and Development in the Third World*, London: Routledge.

Gupta, Dipankar (2000), *Mistaken Modernity: India between Worlds*, New Delhi: HarperCollins Publishers.

Haas, Peter M. (1990), Obtaining International Environmental Protection through Epistemic Consensus, *Millennium: Journal of International Studies*, 19(3), pp. 347–63.

Haigh, Martin J. (1988), Understanding 'Chipko': The Himalayan People's Movement for Forest Conservation, *International Journal of Environmental Studies*, 31, pp. 99–110.

Hajer, M. (1995), *The Politics of Environmental Discourse: Ecological Modernization and Policy Process*, New York: Oxford University Press.

Hall, A. (1996), Did Chico Mendes Die in Vain?, in H. Collinson (ed.), *Green Guerrillas*, London: Latin American Bureau.

Hall, Peter A. and David Soskice (2001), *Varieties of Capitalism: The Institutional Foundations and Comparative Advantage*, Oxford: Oxford University Press.

Hall, S. (1992), The West and the Rest, in Hall, S. and Gieben, S. (eds.), *Formations of Modernity*, Oxford: Polity Press and Open University Press.

Hall, S., D. Held and T. McGrew (1992), *Modernity and Its Futures*, Cambridge: Polity Press.

Haraway, Donna J. (1992), *Primate Vision: Gender, Race and Nature in the World of Modern Science*, London: Verso Books.

Harbans, N. Singh Vs State of Bihar, 1974, PLJR.

Hardin, Garret (1968), The Tragedy of Commons, *Science*, 162, pp. 1243–1248.

Hariss, John (2001), *Politicising Development*, New Delhi: Leftword.

Harper, Charles and Monica Snowden (2016), *Environment and Society*, 6th ed., New York: Routledge.

Harper, Charles L. and Snowden, Monica (2016), *Environment and Society: Human Perspectives on Environmental Issues*, New York: Routledge.

Harris, Marvin (1966), The Cultural Ecology of India's Sacred Cattle, *Current Anthropology*, 7(1), pp. 51–66.

Harvey, David W. (1989), *The Condition of Post Modernity*, London: Sage Publications Ltd.

Hawken, Paul (1993), *The Ecology of Commerce: A Declaration of Sustainability*, New York: Harper Business (HarperCollins Publishers).

Hawley, Amos H. (1950), *Human Ecology: A Theory of Community Structure*, New York: Ronald Press Co.

Bibliography

Helliwell, John F., et al. (eds.) (2020), *World Happiness Report 2020*, New York: Sustainable Development Solutions Network (for United Nations).

Hildingsson, Roger (2007), *Greening the (Welfare) State: Rethinking Reflexivity in Swedish Sustainability Governance*, Lund, Sweden: Lund University, Department of Political Science.

Hildingsson, Roger (2014), *Governing Decarburisation: The State and the New Politics of Climate Change*, Lund (Sweden): Lund University.

Hindustan Times, 22 March 2017.

Hindustan Times, 24 April 2019.

Hindustan Times, 26 March 2017.

Hindustan Times, 8 June 2017.

Hindustan Times, 9 August 2017.

Hobart, M. (ed.) (1993), *An Anthropological Critique of Development: The Growth of Ignorance*, London: Routledge.

Holland, Breena (2008), Justice and the Environment in Nussbaum's "Capabilities Approach" Why Sustainable Ecological Capacity is a Meta-capability, *Political Research Quarterly*, 61(2), pp. 319–332.

Hooks, G. and Smith, C. L. (2004), Treadmill of Destruction: Natural Sacrifice Areas and Native Americans, *American Sociological Review*, 69(4), pp. 558–575.

Hugo, Graeme (2010), Climate Change-Induced Mobility and the Existing Migration Regime in Asia and the Pacific, in Jane Mc Adam (ed.), *Climate Change and Displacement: Multidisciplinary Perspectives*, Oxford: Hart Publishing Ltd., pp. 9–35.

Huntington, S. P. (1996), The West: Unique, Not Universal, in *Foreign Affairs*, November/December 1996.

Hurst, P. (1990), *Rainforest Politics*, London: Zed Books.

Ibrar, Mohammad (2020), Sunder Nursery Sets Unique Mark with Awards, *The Times of India*, 19 December 2020.

IFOAM (2005), *Organic Agriculture Principles*, Bonn: International Federation of Organic Agriculture Movements.

Indian Express, 13 May 2016.

Indian Express, 24 October 2016.

Indian Express, 6 January 2016.

Inglehert, R. (1977), *The Silent Revolution*, Princeton, NJ: Princeton University Press.

IPCC (2007), *The Physical Science Basis*, Paris: World Meteorological Organization.

IPCC (R. K. Pachauri et al.) (2014), *Climate Change 2014: Synthesis Report (5AR)*, Geneva: IPCC.

Islam, S. and Hussain, Ismail (2015), *Social Justice in the Globalization of Production*, London: Palgrave Macmillan.

Jackson, Tim (2009), *Prosperity without Growth? The Transition to a Sustainable Economy*, London: Sustainable Development Commission.

Jain, Shobhita (1991), Standing Up for Trees: Women's Role in Chipko Movement, in S. Sontheimer (ed.), *Women and Environment*, London: Earthscan Publications.

James, C. L. R. (1984), *Beyond a Boundary*, New York: Pantheon.

Jamesons, Fredric (1984), *Post-Modernism or the Cultural Logic of Late Capitalism*, Durham, NC: Duke University Press.

Janicke, M. (1990), *State Failure*, Pennsylvania, PA: Pennsylvania University Press.

Jasanoff, Sheila and Marybeth L. Martello (eds.) (2004), *Earthly Politics: Local and Global in Environmental Governance*, Cambridge, MA: MIT Press.

Jha, Abhijay (2020a), Groundwater Level in Ghaziabad Has Gone Down by 12m in 4 Years, *The Times of India*, 27 December 2020.

Jha, Abhijay (2020b), City's Water Table Fell 17 Metres Since 2016, *The Times of India*, 28 December 2020.

Jodha, N. S. (1985), Pollution Growth and Decline of Common Property Resources in Rajasthan, India, *Population and Development Review*, 11, pp. 247–264.

Jodha, N. S. (1985a), Population Growth and the Decline of Common Property Resources in Rajasthan, India, *Population and Development Review*, 11(2), pp. 247–264.

Jodha, N. S. (1985b), Market Forces and Erosion of Common Property Resources in the Agricultural Markets in Semi-Arid Tropics, in *Proceedings of the International Workshops at ICRISAT Centre*, India, October 24–28.

Jodha, N. S. (1986), Common Property Resources and Rural Poor in Dry Regions of India, *Economic and Political Weekly*, 31(27), pp. 1169–1181.

Jodha, N. S. (1987), A Case Study of the Degradation of Common Property Resources in India, in P. Blaikie and H. Brookfield (eds.), *Degradation and Society*, London: Routledge, pp. 186–205.

Jodha, N. S. (1991), Sustainable Agriculture in Fragile Resource Zones: Technological Imperatives, *Economic and Political Weekly*, 26(13).

Jodha, N. S. (1992), *Common Property Resources and Dynamics of Rural Poverty in India's Dry Regions*, www.fao.org.

Jojola, T. S. (1984), The Conflicting Role of National Governments in the Tribal Development Process, *Antipode*, 16(2), pp. 19–26.

Jones, A. E. and E. W. Wolf (2003), An Analysis of Oxidation Potential of South Pole Boundary Layer and the Influence of Stratospheric Ozone Depletion, *Journal of Geophysical Resources*, 108, 4565, https://doi.org/10.1029/2003JD003379.

Kabeer, Naila (1995), *Reversed Realities: Gender Hierarchies in Development Thought*, London: Verso.

Kabeer, Naila (1999), Resources, Agency, Achievements: Reflection on the Measurement of Women's Empowerment, *Development and Change*, 30, pp. 435–464.

Kalin, Walter (2010), Conceptualising Climate-Induced Displacement, in Jane Mc Adam (ed.), *Climate Change and Displacement: Multi-disciplinary Perspectives*, Oxford: Hart Publishing Ltd., pp. 81–203.

Kalpavriksh (1985), Narmada Valley Project: Development or Destruction?, *The Ecologist*, pp. 1595–1596.

Kalpavriksh, et al. (1986), Narmada Valley Project: Development or Destruction?, in E. Goldsmith and N. Hildyard (eds.), *The Social and Environmental Effects of Large Dams*, Vol. II, Cornwall, UK: Wadebridge Environmental Centre.

Kandpal, P. C. (2018), *Environmental Governance in India*, New Delhi: Sage Publications.

Karindalam, Vivek (2020), Pettimudi Tragedy was a Disaster Waiting in the Wings to Happen', mirrornownews.com, *Times Nownews.com*, 11 August 2020.

Kawser, M. A. and Md. Abdus Samad (2016), Political History of Farakka Barrage and Its Effects on Environment in Bangladesh, *Bandung: Journal of the Global South*, 3, 4 January 2016.

Keck, M. E. and Kathryn Sikkink (1998), *Activists Beyond Borders: Advocacy Networks in International Politics*, Cornell: Cornell University Press.

Keshav Pal Vs State 1984, PL JR; AIR 1985, Patna.

Kingdon, John W. (1984), *Agendas, Alternatives and Public Policies*, Boston, MA: Little Brown & Co.

Kohler, J., et al. (2019), An Agenda for Sustainability Transitions Research: State of the Art and Future Directions, *Environmental Innovation & Societal Transitions*, 31, pp. 1–32.

Koll, Roxy Matthew (2019), Sea May Not Submerge all of Mumbai, but There Will be Long, Intense Flooding, *The Times of India*, 10 November 2019.

Kothari, A. (1989), Environmental Aspects of Narmada Valley Project, *Economic and Political Weekly*, XXXV(III).

Kothari, A. (1995), Development Aid: The Experience of Narmada Project, in A. Dasgupta and G. Lechner (eds.), *Development Aid Today*, New Delhi: Mosaic Books.

Kothari, A. and P. Parajuli (1993), No Nature without Social Justice: A Plea for Cultural and Ecological Pluralism, in W. Sachs (ed.), *Global Ecology*, London: Zed Books.

Kothari, A. and R. Bhartari (1984), Narmada Valley Project: Development or Destruction, *Economic and Political Weekly*, 19(22/23), pp. 907–909, 911–920.

376 Bibliography

Kothari, Rajni (1980), Environmental and Alternative Development, *Alternative Global Local Politics*, 5(4), pp. 427–475.

Kothari, Smitu and P. Parajuli (1993), No Nature without Social Justice: A Plea for Cultural and Ecological Pluralism, in W. Sachs (ed.), *Global Ecology*, London: Zed Books.

Kukreti, Ishan and Vibha Varshney (2018), Food Finders, in Sunita Narain et al. (eds.), *State of India's Environment 2018*, New Delhi: Centre for Science and Environment, pp. 172–176.

Kumar, Dinesh (2015), Environmental Movements in India, in Abhay Pratap Singh (ed.), *Development Process and Social Movements in Contemporary India*, New Delhi: Pinnacle Learning, pp. 312–337.

Kumar, Subodh (2017), Acid Rain-The Major Cause of Pollution: Its Causes and Effects, *International Journal of Applied Chemistry*, 13(1), pp. 53–58.

Kumari, Rima et al. (2019), *Deforestation in India: Consequences and Sustainable Solutions*, intechopen.com, https://doi.org/10.5772/intechopen.85804.

LaDuke, Winona (1999), *All Our Relations: Native Struggles for Land and Life*, Chicago, IL: Haymarket Books.

Lam, K. C. and P. K. Chan (2008), Socio-Economic Status and Inequalities in Exposure to Transportation Noise in Hong Kong, *Open Environmental Science Journal*, 2, pp. 107–113.

Lane, C. and Morehead, R. (1995), New Directions in Rangeland and Resource Tenure and Policy, in I. Scoones (ed.), *Living with Uncertainty*, London: Intermediate, Technology Publications Ltd.

Latour, B. (2004), Why has Critique Run Out of Steam? From Matters of Fact to Matters of Concern, *Critical Inquiry*, 30 (Winter), pp. 225–248.

Latour, Bruno (1993), *We Have Never Been Modern*, Cambridge, MA: Harvard University Press.

Leach, M., R. Mearns and I. Scoones. (1997), Environmental Entitlements: A Framework for Understanding the Institutional Dynamics of Environmental Change. *IDS Discussion Paper No. 359*, 1997 (March).

Lee, David (1980), On the Marxism View of the Relationship between Man and Nature, *Environmental Ethics*, 2, pp. 3–16.

Leiber, M. D. (ed.) (1977), *Exiles and Migrants in Oceania*, Honolulu, HI: University of Hawaii Press.

Leopold, Aldo (1949/1986), *A Sand County Almanac: Essays on Conservation from Round River*, New York: Sierra Club and Ballantine Books.

Litfin, Karen (ed.) (1999), *The Greening of Sovereignty in World Politics*, Cambridge, MA: MIT Press.

Llabres, M. and S. Augusti (2006), Picophytoplankton Cell Death Induced by UV Radiation; Evidence for Oceanic Atlantic Communities, *Limnological Oceanography*, 51, pp. 353–359.

Lyotar, Jean-Francois (1979), *The Post-Modern Condition: A Report on Knowledge*, Manchester: University of Manchester.

Madeleine (2019), Gaia Greek Goddess: 5 Facts, *theoi.com*, 13 September 2019.

Marglin, Stephen A. (1991), Understanding Capitalism: Control Vs Efficiency, in B. Gustaffson (ed.), *Power and Economic Institutions*, Aldershot: Edward Elgar Co.

Mahesh.com (1998), Silent Valley National Park, *Internet*, 14 July 1998.

Marsden, Terry, Philip Lowe and Sarah Whatmore (eds.) *Labour and Locality: Uneven Development and the Rural Labour Process*, London: Routledge.

Martinez-Alier, Joan (2002), *The Environmentalism of the Poor*, Geneva: United Nations Research Institute for Social Development.

Massey, D. (1991), A Global Sense of Place, *Marxism Today*, June 1991.

Mc Cune, N., et al. (2017), Mediated Territoriality: Rural Workers & the Efforts to Scale Out Agro-ecology in Nicaragua, *Journal of Peasant Studies*, https://doi.org/10.1080/0306150.2016.1233868.

McNeill, J. R. (2000), *Something New under the Sun: An Environmental History of the Twentieth Century*, London: Penguin/Allen Lane.

McNeill, J. R. (2007), The Green Revolution, in Mahesh Rangrajan (ed.), *Environmental Issues in India*, New Delhi: Pearson India Education Services Pvt. Ltd., pp. 184–194.

Meadows, D. H., Meadows, D. L., Randes, R. and Behrens, W. W. (1972), *Limits to Growth: A Report for the Club of Rome's Project on Predicament of Mankind*, New York: Universe.

Meadows, D. H., et al. (1972), *The Limits to Growth*, Washington, DC: Universe Books.

Mehmet, O. (1995), *Westernising the Third World*, London: Routledge.

Merchant, Carolyn (1990), *The Death of Nature: Women, Ecology and the Scientific Revolution*, New York: HarperCollins Publishers.

Merzian, Richie (2020), Climate Change has Lengthened Australian Summers by a Month, *The Times of India*, 2 March 2020.

Mies, Maria and Shiva, Vandana (1993/2014), *Ecofeminism*, London: Zed Books.

Mills, C. Wright (1959), *The Sociological Imagination*, New York: Oxford University Press.

Milton, Kay (1996), *Environmentalism and Cultural Theory*, London: Routledge.

Modi, Renu (2013), Displaced from Private Property: Resettlement and Rehabilitation Experiences from Mumbai, *Economic and Political Weekly*, June 8, XLVIII(23), pp. 71–74.

Mohanty, Chandra T. (1991), Under Western Eyes: Feminist Scholarship and Colonial Discourses, in Mohanty, C., Russo, A. and Torres, L. (eds.), *Third World Women and the Politics of Feminism*, Bloomington, IN: Indiana University Press.

Mol, Arthur P. J. (1995), *The Refinement of Production: Ecological Modernization Theory and the Dutch Chemical Industry*, Utrecht: International Books.

Mol, Arthur P. J. and Spaargaren, Gert (2000), Ecological Modernization Theory in Debate: A Review, *Environmental Politics*, 9(1), pp. 17–49.

Morrison, Denton E. (1976), Growth, Environment, Equity and Scarcity, *Social Science Quarterly*, 57, pp. 292–306.

Muggah, Robert (2008), *Relocation Failures in Sri Lanka*, London: Zed Books.

Mukerjee, Radhakamal (1945), *Social Ecology*, London: MacMillan & Co.

Mukerjee, Radhakamal (1952), *The Dynamics of Morals*, London: Macmillan & Company.

Mumbai Mirror (2020), A Judge Pours His Heart Over Trouble Caused by a Leaning Tree, *Mumbai Mirror*, 17 January 2020.

Munshi Singh Vs Union of India, 1973, 2, PCC.

Murphy, Raymond (1997), *Sociology and Nature: Social Action in Context*, New York: Routledge.

Myrdal, Gunnar (1968), *Asian Drama: An Inquiry into the Poverty of Nations* (2 vols), New York: Pantheon.

Nair, Reshmy (2014), RFCTLARR Act 2013: A Critical Review, *ASCI Journal of Management*, 44(1) (Special), pp. 82–100.

Narain, Sunita (2018), Hold a Straw Poll, in Sunita Narain, et al. (eds.), *State of India's Environment*, New Delhi: Centre for Science and Environment, pp. 8–12.

NBS (National Bureau of Standards) (1971), *The Social Impact of Noise*, Washington, DC: US Environmental Protection Agency.

Neale, R. E., et al. (2003), Sun Exposure as a Risk Factor for Nuclear Cataract, *Epidemiology*, 14, pp. 707–712.

Nepal, Padam (2009), *Environmental Movements in India: Politics of Dynamism and Transformations*, Delhi: Authors Press.

Neumann, I. B. (1998), European Identity, EU Expansion, and the Integration Exclusion Nexus, *Alternatives*, 23(3), pp. 397–416.

Nicholson, M (1987), *The New Environmental Age*, Cambridge: Cambridge University Press.

Nussbaum, Martha C. (2006), *Frontiers of Justice: Disability, Nationality, Species Membership*, Cambridge, MA: Harvard University Press.

O'Connor, James (1996), The Second Contradiction of Capitalism, in T. Burton (ed.), *The Greening of Marxism*, New York: Guilford Press, pp. 197–221.

378 Bibliography

O'Connor, M. and C. Spash (eds.) (1999), *Valuation and the Environment: Theory, Methods and Practice*. Cheltenham: Edward Elgar.

Omvedt, Gail (1984), Ecology and Social Movements, *Economic and Political Weekly*, XIX(44).

Oommen, T. K. (2015), Radhakamal Mukerjee on Social Ecology: Filling up Some Blanks, *Sociological Bulletin*, 64(1), pp. 15–35.

OPHI (2020), *Global MPI 2020*, Oxford: Oxford Poverty and Human Development Initiative, and UNDP. ophi.org.uk/ophi_stories/20.

O'Riordan, T. (1976), *Environmentalism*, London: Pion Ltd.

Ortner, Sherry B. (1974), Is Female to Male as Nature is to Culture? In M. Z. Rosaldo and L. Lamphere (eds.), *Women, Culture and Society*, Stanford, CA: Stanford University Press, pp. 68–87.

Osborn, F. (1948), *Our Plundered Planet*, Boston: Little, Brown.

Ossewaarde, Martin J. (2018), *Introduction to Sustainable Development*, New Delhi: Sage Publications.

Ostrom, Elinor (1990), *Governing the Commons: The Evolution of Institutions for Collective Action*. Cambridge: Cambridge University Press.

Ostrom, Elinor (1992), *Crafting Institutions for Self-governing Irrigation Systems*, San Francisco, CA: Institute for Contemporary Studies.

Oza, G. M. (1981), Save Silent Valley as a World Heritage Site, *Environmental Conservation*, 8(1), p. 52.

Padel, Felix and Samarendra Das (2010), Cultural Genocide and the Rhetoric of Sustainable Mining in East India, *Contemporary South Asia*, 18(3), pp. 333–341.

Pandey, Kundan (2018), In a Stunted State, in Sunita Narain et al. (eds.), *State of India's Environment 2018*, New Delhi: CSE, pp. 168–171.

Parris, Kristen and Robert McCauley (2016), *Noise Pollution and the Environment*, Australian Academy of Science, science.org.au.

Parsons, Talcott (1951), *The Social System*, New York: Free Press.

Patker, Medha (2019), No Rift between Development and Natural Continuity, *Vagarth*, August, 2019, pp. 65–69.

Pavel, S., et al. (2004), Disturbed Melanin Synthesis and Chronic Oxidative Stress in Dysplastic Naevi, *European Journal of Cancer*, 40, pp. 1423–1430.

Pellow, David N. (2002), *Garbage Wars: The Struggle for Environmental Justice in Chicago*, Cambridge, MA: The MIT Press.

Perminski, B. (2013), *Mining-induced Displacement and Resettlement*, New York: Columbia University Press.

Place, F. and Hazell, P. (1993), Productivity Effects of Indigenous Land Tenure Systems in Sub-Saharan Africa, *American Journal of Agricultural Economics*, February, pp. 10–19.

Portney, K. E. (1992), *Controversial Issues in Environmental Policy*, London: Sage.

Prabhu, Swati (2014), *50FAQs on Climate Change*, New Delhi: The Energy and Resources Institute.

Preston, P. W. (1997), Global Changes and New Political-Cultural Identities, in Preston, P. W. (ed.), *Political/Cultural Identity: Citizens and Nations in a Global Era*, London: Sage Publications Ltd.

Pt. Lila Ram Vs Union of India, 1975, 2, SCC.

Radcliffe, S. (1994), (Representing) Post-Colonial Women: Authority, Difference and Feminisms, *Area*, 26(1), pp. 25–32.

Rajagopalan, R. (2009), *Environment and Ecology*, Delhi: Oxford University Press.

Rakotondrazafy, Vatosoa (2020), Madagascar's Small Fisherfolk Respect the Sea — They Strive to Protect Ocean Life, *The Times of India*, 26 December 2020.

Ram Chandrariah Vs Land Acquisition Officer, Sagar, 1973, BBCJ, IV-105.

Rangan, H. (1996), From Chipko to Uttaranchal, in R. Peet and M. Watts (eds.), *Liberation Ecologies: Environment, Development, Social Movements*, London: Routledge.

Rattansi, A. and Westwood, S. (eds.) (1994), *Racism, Modernity and Identity on the Western Front*, London: Polity Press.

Rex, John (1976), *Key Problems of Sociological Theory*, London: Routledge & Kegan Paul.

Rist, Gilbert (1997), *The History of Development: From Western Origins to Global Faith*, London: Zed Books.

Robbins, Paul (2012), *Political Ecology: A Critical Introduction*, 2nd ed., West Sussex: Wiley Blackwell.

Robbins, Paul, P. R. Blaikie and H. Brookfield (1987), *Land Degradation and Society*, London: Methuen.

Rockstrom, Johan, et al. (2009), A Safe Operating Space for Humanity, *Nature*, 461, pp. 472–475.

Rostow, Walt Whitman (1960), *Stages of Economic Growth: A Non-communist Manifesto*, New York: Cambridge University Press.

Roszak, Theodore (1979), *Person/Planet: The Creative Disintegration of Industrial Society*, London: Wiley-Blackwell.

Roy, D. and G. Sen (1992), The Strength of a People's Movement, in G. Sen (ed.), *Indigenous Vision*, New Delhi: Sage Publications.

Runge, C. F. (1984), Institutions and Free Rider: The Assurance Problem in Collective Action, *Journal of Politics*, 46, pp. 154–181.

Sabatier, Paul A. (1999), *Theories of Policy Process*, Boulder, CO: West View Press.

Sahoo, Tanmayee, Usha Prakash and Mrunmayee M. Sahoo (2014), Sardar Sarovar Dam Controversy – A Case Study, *Global Journal of Finance and Management*, 6(9), pp. 887–892.

Said, E. (1978), *Orientalism*, New York: Vintage Books.

Said, E. (1986), Knowing the Oriental, in Donald James and S. Hall (eds.), *Politics and Ideology*, Maidenhead and Berkshire, UK: Open University Press.

Sainath, P. (1996), *Everyone Loves a Good Drought*, New Delhi: Penguin Books.

Sanghavi, S. (1995), Reopening Sardar Sarovar Issues: Significant Gains of Narmada Struggle, *Economic and Political Weekly*, XXX(11), March 18.

Sanghvi, S. (1994), 'Nation', 'Nationalism' and Mega Projects, *Economic and Political Weekly*, XXIX(10), March 5.

Schelling, T. E. (1971), On the Ecology Mircromotives, *The Public Interest*, 25, pp. 61–98.

Schnaiberg, Allan (1980), *The Environment: From Surplus to Scarcity*, New York: Oxford University Press.

Schnaiberg, Allan and Gould, Kenneth Alan (1975), *Environment and Society: The Enduring Conflict*, New Jersey: Blackburn Press.

Scoones, I. (1995), New Directions in Pastoral Development in Atka, in I. Scoones (ed.), *Living with Uncertainty*, London: I.T.P. Ltd.

Scott, James C. (1976), *The Moral Economy of the Peasant: Rebellion and Subsistence in Southeast Asia*, New Haven, CT: Yale University Press.

Scott, James C. (1985), *Weapons of the Weak: Everyday Forms of Peasant Resistance*, New Haven, CT: Yale University Press.

Searles, P. S., et al. (2001), A Meta-Analysis of Plant Field Studies Simulating Stratospheric Ozone Depletion, *Oecologia*, 127, pp. 1–10.

Sen, Amartya (1999), *Development as Freedom*, New York: Anchor.

Sen, Gita and Caren Grown (1987), *Development Crises and Alternative Visions: Third World Women's Perspectives*, London: Routledge.

Sengupta, Sushmita (2017), Sukhomajiri Falls Apart, *Down to Earth*, 16–18 February, pp. 24–25.

Sethi, H. (1993), Survival and Democracy: Ecological Struggles in India, in P. Wignaraja (ed.), *New Social Movements in the South* Chapter 6, New Delhi: Sage Publications.

Seudder, T. (1973), The Human Ecology of Big Projects: River Basin Development and Resettlement, in B. Siegel (ed.), *Annual Review of Anthropology*, Palo Alto, CA: Annual Reviews, pp. 45–55.

380 Bibliography

Shah, Sonia (2020), When We Cut Down Forests that Bats Live in … They Come to Our Backyards, Says US Journalist Sonia Shah, *The Times of India*, 8 March 2020.

Shambhu Nath Vs State of Bihar (1989), PLJR.

Shankar, Kunal (2021), How a Retired Professor Took on a Mining Giant-and Won, *www.aljazeera.com*, 7 January 2021.

Shapiro-Garza, Elizabeth (2020), An Alternative Theorisation of Payments for Ecosystem Services from Mexico: Origins and Influence, *Development and Change*, 51(1), pp. 196–223.

Sharma, Devinder (2018), Kheti Par Machine ka Bojh, *Dainik Bhaskar*, 3 October 2018.

Sharma, Subhash (2015), *Development and Its Discontents*, Jaipur and New Delhi: Rawat Publications.

Sharma, Subhash (2016), *Why People Protest: An Analysis of Ecological Movements*, 2nd ed., New Delhi: Publications Division (GOI).

Sharma, Subhash (2017), Environmental Pollution, *Employment News*, 16–22 December.

Shiva, Vandana (1987), Ecology Movements in India, *Alternatives*, 11, pp. 255–73.

Shiva, Vandana (1988), *Staying Alive: Women, Ecology and Development*, London: Zed Books.

Shiva, Vandana (1991a), *Ecology and the Politics of Survival: Conflicts over Natural Resources in India*, London: Sage Publications.

Shiva, Vandana (1991b), *The Violence of Green Revolution: Third World Agriculture, Ecology and Politics*, London: Zed Books.

Shiva, Vandana and Jayanta Bandyopadhyay (1988), The Chipko Movement, in J. Ives and D. C. Pitt (eds.), *Deforestation: Social Dynamics in Watersheds and Mountain Ecosystems*, London: Routledge.

Shrivastava, Aseem and Ashish Kothari (2012), *Churning the Earth: The Making of Global India*, New Delhi: Penguin Books.

Shue, Henry (1993), Subsistence Emission and Luxury Emission, *Law and Policy*, 15, pp. 39–59.

Shuurman, F. J. (1993), Introduction: Development Theory in 1990's, in Schuurman, F. J. (ed.), *Beyond the Impasse*, London: Zed Books.

Shyamnanda Prasad Vs State of Bihar, 1993, 4, PCC.

Simon, Julian L. (1981), *The Ultimate Resource*, Champaign, IL: University of Illinois Press.

Simon, Julian L. and Herman Kahn (eds.) (1984), *The Resourceful Earth*, New York: Basil Blackwell.

Singh, J. S., S. P. Singh, A. K. Saxena and Y. S. Rawat (1984), India's Silent Valley and its Threatened Rainforest Ecosystems, *Environmental Conservation*, 11(3), pp. 223–233.

Singh, Paras (2021), Why Water Disruption is Now 24 x 7 Headache, *The Times of India*, 16 January 2021.

Singh, Shekhar and Pranab Banerji (eds.) (2002), *Large Dams in India*, New Delhi: IIPA.

Singh, Subrata (2004/2015), Common Property Resource Management in Transitional Villages, Paper presented at *Tenth Biennial Conference of the International Association for the Study of Common Property*, Oaxaca, Mexico, 9–13 August 2004 (revised in 2015), www.researchgate.net/publication/42762697.

Singh, Tavleen (1998), Luddite Sisters, *India Today*, XXIII(25), June 16–22.

Singh, Y. (1973), *Modernisation of India Tradition*, New Delhi: Thomson Press (India).

Sinha, D. and Y. Mittal (2017), Skeletal Existence, *Down to Earth*, pp. 56–57, 16–18 February.

Sismondo, Sergio (1993), Some Social Contributions, *Social Studies of Science*, 23(3), August, pp. 515–553.

Slater, D. (1995), Challenging Western Visions of the Global, *The European Journal of Development Research*, 7(2), pp. 366–388.

Smith, M. R. and Leo Marx (1994), *Does Technology Drive History? The Dilemma of Technological Determinism*, Cambridge, MA: MIT Press.

Snidal, D. (1995), The Politics of Scope: Endogenous Actors, Heterogeneity and Institutions, in R. O. Keohane and E. Ostrom (eds.), *Local Commons and Global Interdependence*, London: Sage Publications.

Sobrino, C., et al. (2004), UV-B Radiation Increases Cell Permeability and Damage Nitrogen Incorporation Mechanisms in Nannochloropsis Gaditana, *Aquatic Science*, 66, pp. 421–429.

Solesbury, W. (1976), The Environmental Agenda, *Public Administration*, 54, pp. 379–397.

Solomon, Keith R. (2008), Effects of Ozone Depletion and UV-B Radiation on Humans and the Environment, *Atmosphere Ocean*, 46(1), pp. 185–202, https://doi.org/10.3137/ao.460109.

Spaargaren, Gert, Arthur P. J. Mol and F. H. Buttel (eds.) (1996), *Environment and Global Modernity*, London: Sage Publications.

Speth, James and Peter M. Haas (2006), *Global Environmental Governance*, Washington, DC: Island Press.

Spivak, G. C. (1987), *In Other Worlds*, London: Routledge.

Spivak, G. C. (1991), Neo-colonialism and the Secret Agent of Knowledge, *Oxford Literary Review*, 13(1–2).

Spivak, G. C. (1999), *A Critique of Post-Colonial Reason*, Cambridge, MA: Harvard University Press.

State of Bihar Vs Parasuram Prasad (1976), BLJR.

State of Bihar Vs S.K. Thakur (1981), AIR, Patna.

State of Bihar Vs Thakur K.P.Singh (1968), BLJR.

STEPs Centre (2010), *Innovation, Sustainability, Development: A New Manifesto*, Brighton, UK, https://steps-centre.org/wp-content/uploads/steps-manifesto-small-file.pdf.

Steward, J. H. (1972), *Theory of Culture Change: Methodology of Multilinear Evolution*, Urbana, IL: University of Illinois Press.

Stojanov, R. (2008), Environmental Factors of Migration, in R. Stojanov and J. Novasak (eds.), *Development, Environment and Migration: Analysis of Linkages and Consequences*, Loamouc, Palacky University.

Subbiah, A. R. (2004), *State of the Indian Farmer*, New Delhi: Academic Foundation.

Suich, Helen, et al. (2016), An Introduction to People in Nature, in Davidson-Hunt, I. J., et al. (eds.), *People in Nature: Valuing the Diversity of Interrelationships between People and Nature*, Gland (Switzerland): IUCN.

Sushila Devi Vs State of Bihar (1963), BLJR.

Swaminathan, M. S. (1983), The Silent Valley: Development with Eco-conservation, in T. V. Singh and J. Kaur (eds.), *Himalayas Mountains and Men – Studies in Ecodevelopment*, Lucknow: Print House.

Swaminathan, M. S. (2007), Agriculture on Spaceship Earth, in Mahesh Rangrajan (ed.), *Environmental Issues in India*, New Delhi: Pearson India Education Services Pvt. Ltd., pp. 161–183.

Swift, J. (1995), Dynamic Ecological Systems and the Administration of Pastoral Development, in I. Scoones (ed.), *Living with Uncertainty*, London: I.T.P. Ltd.

Taylor, P. J. and F. H. Buttel (1992), How Do We Know We Have Global Environmental Problems? Science and the Globalisation of Environmental Discourse, *Geoforum*, 23, pp. 405–416.

Temper, Leah and Joan Martinez-Alier (2013), The God of Mountain and Godavarman: Net Present Value, Indigenous Territorial Rights and Sacredness in a Bauxite Mining Conflict in India, *Ecological Economics*, 96, pp. 79–87.

Terminski, B (2013), *Development-induced Displacement and Resettlement: Theoretical Frameworks and Current Challenges*, Geneva: Onlineresearch.

382 Bibliography

Thapar, Romila (2007), Forests and Settlements, in Mahesh Rangrajan (ed.), *Environmental Issues in India*, Noida: Pearson India Education Services Pvt. Ltd., pp. 33–41.

The Hindu, 17 November 2016.

The Hindu, 26 November 2015.

The Hindu, 28 September 2016.

The Hindu, 3 October 2018.

The Hindu, 30 December 2015.

The Morning India, 5 October 2019.

The Morning India, 6 November 2019.

The Morning India, 8 November 2019.

The Shillong Times (2021), Meghalaya's Barren Landscapes Poised for a Big-time Change, *The Shillong Times*, 7 January 2021.

The Times of India, 1 April 2017.

The Times of India, 1 November 2017.

The Times of India, 10 August 2017.

The Times of India, 12 March 2020.

The Times of India, 12 November 2019.

The Times of India, 14 November 2019.

The Times of India, 15 December 2015.

The Times of India, 15 March 2020.

The Times of India, 16 October 2017.

The Times of India, 2 October 2017.

The Times of India, 2 October 2019.

The Times of India, 4 April 2017.

The Times of India, 4 March 2017.

Thiong, N. (1986), *Decolonising the Mind*, London: James Currey.

Thompson, E. P. (1978), *The Poverty of Theory and Other Essays*, New York: Monthly Review Press.

Thurow, Lester (1980), *The Zero-Sum Society: Distribution and the Possibilities for Change*, New York: Basic Books.

Times Evoke (2021), Every Bite We Waste, *The Times of India*, 3 April 2021.

Tiwary, R. (2006), Conflicts over International Waters, *Economic and Political Weekly*, 41(17), pp. 1684–1692.

TOI (2021), Mercury Touches New 15-Yr February High, *The Times of India*, 26 February 2021.

TOI (2021), Pollution in Delhi Cost 54,000 Lives, Rs. 58,000 Cr in Monetary Losses, *The Times of India*, 19 February 2021.

Trimbur, T. J. and M. Watts (1976), Are Cultural Ecologists Well Adapted? *Proceedings of the Association of American Geographers*, 79(1), pp. 88–100.

Ulrich, R. S. (1984), View Through a Window may Influence Recovery from Surgery, *Science*, 224(4641), pp. 420–421.

Ulrich, R. S., et al. (1991), Stress Recovery during Exposure to Natural and Urban Environments, *Journal of Environmental Psychology*, 11(3), pp. 201–230.

Umashankar, K. (2019), A. P. Villagers Turn Wildlife-Friendly, *The Hindu* (Patna. ed.), 19 August 2019.

UN (1997), *Report of the Inter-governmental Working Group of Experts on IDPS*, Geneva: UNHCR.

UN (2020), *World Happiness Report 2020*, New York: Sustainable Development Solution Network.

UNDP (2003), *Human Development Report 2003*, Oxford: Oxford University Press.

UNDP (2019), *Human Development Report 2019*, New Delhi: Oxford University Press.

UNDP (2020), *Human Development Report 2020*, New Delhi: Oxford University Press.

UNEP (2006), Environmental Effects of Ozone Depletion and its Interaction with Climate Change: Progress Report 2005, *Photochemical Protobiological Science*, 5, pp. 13–24.

Vanderheiden, Steve (2008), *Atmospheric Justice*, New York: Oxford University Press.

Vazquez-Brust, D., J. Sarkis and A. M. Smith (2014), Managing the Transition to *Critical* Green Growth: The 'Green Growth State', *Futures*, https://doi.org/10.1016/j.futures.2014.10.005.

Venkatesh, S. (2018), Upward Curve of Extreme Climate, in Sunita Narain et al. (eds.), *State of India's Environment' 2018*, New Delhi: Centre for Science and Environment, pp. 116–121.

Vishwamohan (2020a), Five Urban Himalayan Towns in High Water Availability Region are Running Dry, *The Times of India*, 2 March 2020.

Vishwamohan (2020b), Seven States Saw Drop in Rainfall in Last 30 Years, *The Times of India*, 9 March 2020.

Vishwamohan (2020c), Warming Could Force 45 M Indians to Migrate from Homes in 30 Yrs, *The Times of India*, 19 December 2020.

Vishwamohan (2021a), 25% of Govt. Waterbodies in 5 States Dried Up, *The Times of India*, 28 March 2021.

Vishwamohan (2021b), Extreme Weather Events Claimed 1,400 Lives in 2020, *The Times of India*, 5 January 2021.

Vishwamohan (2022), Sadhguru: Time for Actual Work to Boost Soil Health, *The Times of India*, 16 June 2022.

Von Wehrden, Henrik, et al. (2011), The Ecological Consequence of Nuclear Accidents, *Conservation Letters*, researchgate.net.

Vousdoukas, Michalis, et al. (2020), Beach Bummer: Half of World's Sandy Beaches Could Vanish by 2100, *The Times of India*, 4 March 2020.

Wade, R. (1988), *Village Republics*, Cambridge: Cambridge University Press.

Wade, Robert (1987), The Management of Common Property Resources: Collective Action as an Alternative to Privatization or State Regulation, *Cambridge Journal of Economics*, 11, pp. 95–106.

Wainwright, Joel (2008), *Decolonizing Development: Colonial Power and the Maya*, Oxford: Wiley – Blackwell.

Ward, Barbara and Rene Dubos (1972), *Only One Earth: The Care and Maintenance of a Small Planet*, London: Penguin Books.

Warrington, Siobhan (2014), Research on Development-Induced-Displacement: Oral History from the Displaced, *ASCI Journal of Management*, 44(1) (Special), pp. 20–23.

WCED (1987), *Our Common Future*, New York: Oxford University Press.

Wezel, A., et al. (2011), Agro-ecology as a Science, a Movement and a Practice, *Sustainable Agriculture*, 2, pp. 27–43, https://doi.org/10.1007/978.94-007-0394-03.

Woodgate, Graham (1997), Introduction, in Michael Redclift and Graham Woodgate (eds.), *The International Handbook of Environmental Sociology*, Cheltenham: Edward Elgar, pp. 1–17.

Woodgate, Graham (2010), Introduction, in Michael R. Redclift and Graham Woodgate (eds.), *The International Handbook of Environmental Sociology*, Cheltenham: Edward Elgar, pp. 1–8.

Woolgar, Steve (1988), *Science, the Very Idea*, Chichester: Tavistock.

World Bank (1984), *World Development Report 1984*, Washington, DC: World Bank.

Yogendra (2022), Ganga Mukti Andolan: Aam Janon Ke Sangharsh Ki Apporv Gatha, *Bhaskar (daily)*, 26 July 2022.

Young, R. (1990), *White Mythologies: Writing History and the West*, London: Routledge.

Zetter, R. (2007), More Labels, Fewer Refugees: Remaking the Refugee Label in an Era of Globalisation, *Journal of Refugee Studies*, 20(2), pp. 173–192.

Index

A
Aarey Milk Colony Forest 111
Acid Rain 5–6, 24, 36, 75, 139, 231–232, 259, 271, 306–307, 321, 327–330, 345–346, 354
Acid Deposition 327
Agarwal, Anil 86, 114, 260–261
Agarwal, Bina 31–34, 258, 347
Agro-Climatic Zones 13, 38, 132
Ahmed, A. 299
Akula, V.K. 190, 191, 194, 206
Al Gore 12, 273, 300
Alternative Justice Theory 128, 130
Amin, Samir 141, 142
Amte, Baba 202, 218, 221, 249, 251, 350, 351
Andharia, Janki 184, 186
Appropriate Technology/Technologists 25, 186, 194, 195, 206, 252, 253
AQI (Air Quality Index) 14, 308, 320, 345
Aranya (Forest) 91–94, 132
Aranyani (Goddess of Forest) 90, 93
Arcadian Tradition 21
Aristotle 3
Arnold, D. 204
Arora, Saurabh 113, 114, 119, 121
Assurance Problems Theory 358, 364

B
Bahuguna, Sundarlal 70, 194, 202, 206, 208, 210–212, 216, 218, 221, 249, 251– 253
Bandyopadhyay, Jayanta 189, 190, 192–194, 204, 206
Banerji, Pranab 136, 162, 163, 174, 178
Bavadam, Lyla 343
Baviskar, Amita 209
BCE-Before Common Era 16, 22
Beauvoir, Simon De 63
Beck, Tony 348
Beck, U. 40, 54, 56, 233
Bell, Daniel 45

Bell, M.M. 16, 37, 269
Benton, T. 26, 85
Berger, Peter L. 2, 4, 42
Berger, Peter L. and Thomas Luckmann 2, 4, 42
Berkes, F. 365
Bhabha, H. 136, 137
Bhatt, Chandi Prasad 193, 194, 205, 206, 208, 210–212, 214, 216, 218, 221, 222, 249, 250, 252, 253, 257,
Bhopal Gas Tragedy 96, 229
Bhushan, Chandra 12
Bio-Diversity 297
Bio-environmentalists 103, 104, 132
Bio-Regionalism 41
Bisnois 247
Boserup, E.136
Bromley, D.W. 355, 356, 359, 360, 364
Brundtland Commission 95, 96, 122, 132, 228
Buddha, Gautam 25, 90
Bullard, Robert D. 5, 79
Bullet Train 111–113
Burch, W.R. 46
Burningham, K. 26
Buttel, F.H. 4, 7, 29, 42, 46, 55–57, 59, 230

C
Campbell, John 155–157
Caney, Simon 294
Carolan, M.S. 26, 27
Carson, Rachel 17, 227, 244, 245
Catton, William R. 4, 7, 26, 40, 45, 46, 48
CDR (Common but Differentiated Responsibility) 100, 240, 242, 291, 293, 302, 305
CE-Common Era 13, 16, 22, 23, 91
Cernea, M.M. 359
CFCs 232, 330, 333
Chambers, R. 156
Chaturvedi, S. et al. 18

Chernobyl 5, 18, 19, 64, 96, 186, 229, 292, 326

Chipko Andolan 188, 244, 248,

Chiras, Daniel 117

Climate Change 6, 10, 12, 15, 18, 26, 29, 36, 39, 44, 80, 81, 91, 92, 99, 108, 110, 127, 128, 139, 152, 153, 156, 231, 239, 240, 242, 246, 249, 254, 261–276, 278, 280–294, 298–307, 330, 334, 337, 338, 340, 344,

Climate Justice 293–295, 302

CMS COP 12

Common Property Resources-CPRs 24, 34, 77, 139, 189, 347, 350, 352, 361, 365

Commoner, Barry 84, 244, 245

Commons 11, 26, 34, 44, 77, 78, 82, 96, 104, 111, 121, 166, 189, 204, 346–355, 358, 359, 361–365

Compensation for Ecosystem Services (CES) 118, 120

Comte, Auguste 1

Constructivism 28–30

Convention on Conservation of Migratory Species 12

Conway, G.R. 363

Cooley, Charles 2

Copper, G. 71, 86, 87, 329

Cormack, Carol P. Mac 31

Critical Life Issues 35, 183, 203, 208, 221, 243, 260

Crusading Gandhians 70, 195, 212, 245, 252, 253, 255, 256

Crush, J. 139

Cultural Ecology 74–76, 88

D

Daly, Herman 18, 103

Das, S. 114, 115

Dayanand, Stalin 111–113

Deep Ecology 21, 69, 88, 123, 195, 231, 254–257, 261

Deforestation 12, 47, 80, 101, 110, 155, 181, 190, 193, 195, 205, 208, 218, 220, 242–245, 251, 257, 262, 270, 273, 293, 300, 306, 307, 334–338, 345, 346, 360

Delgado, F.G. 319

Demeritt, David 29, 30

Dependency Theory 62, 73, 84, 8, 142

Development 4, 10, 12, 13, 16, 17, 22, 24, 25, 26, 32–36, 41, 46, 47, 49, 54, 56, 57, 61, 65, 67–70, 73, 75–77, 79–82, 84, 88–115, 117–123, 125, 126, 128, 132, 134–155, 157–162, 164–166, 169, 171, 172, 175–177, 180–182, 184–186, 190–192, 195, 198–203, 206, 209–212, 215–223, 228, 230, 231, 235, 238, 240–246, 249, 251, 253, 258, 260, 264, 274, 280, 289, 290, 292–295, 297, 304, 312, 317, 322, 333–335, 338, 340, 344, 345, 348, 356, 364

Development Rights 294, 295

Dharati Mata (Mother Earth) 91

Dickens, Peter 7

Displacement 23, 35, 36, 47, 49, 83, 114, 115, 132, 134, 139, 150–156, 159–161, 169, 170, 172–176, 179–182, 184–186, 198, 201, 202, 206, 207, 218, 248, 263, 273, 292, 351

Dogra, Bharat 161

Dong, Shikui 17, 18

Dryzek, John S. 267, 293, 295, 296, 302

DuBois, W.E.B. 136, 137

Dunlap, Riley E. 4–7, 25–27, 40, 45, 46, 48

Durkheim, Emile 1, 2, 4, 5, 40, 45, 46

Durning, Alan 232, 234

E

Earth Summit–Rio 6, 99, 101, 230, 239, 240, 293

Eco-Feminism 31

Eco-goals 20

Ecological Refugees 183, 246, 351

Ecological Marxists 82, 251–253

Ecological Modernisation-Eco-Modernisation 41, 44, 54–59, 62, 88, 121, 128, 129

Ecological Synthesis 48, 49

Ecologically Sensitive Areas (ESA) 193, 249, 250

Ecologist 9, 21, 27, 28, 39, 41, 42, 72, 73, 75, 76, 78, 104, 105, 113, 118, 127, 206, 228, 243, 292

Ecology 4, 8–11, 14, 16, 20–22, 25, 28, 29, 33, 35, 37, 38, 40, 41, 50, 55, 58, 62, 69, 70–81, 83–85, 88, 90, 94, 96, 106, 111–113, 122, 123, 125, 132, 139, 150, 158, 159, 190, 195, 196, 231, 243, 248, 249, 254–257, 261, 264, 296–298, 306

Eco-Marxism 82, 83, 85

Eco-Scarcity 70–72

Eco-Strategies 20

Eco-System People 183, 246, 350, 351

Eco-System Services 12, 17, 18, 25–27, 118, 120, 252, 297, 303, 326, 327, 340

Eco-Zone 91, 92, 132

Ehrlich, Paul 46, 71, 103, 105, 244–246

Elson, D. 136

Environmental Agonostics 27

Environmental Discourse Theory 259, 261

Environmental Entitlements 352, 358, 360, 364

Environmental Impact Assessment 100, 113, 162, 200, 201, 230

386 Index

Environmental Performance Index (EPI) 115–117, 126, 127
Environmental Pragmatism 27
Environmental Rights, Justice & Equity 260, 261
Escobar, A. 138
Evans, Peter 57

F

Forced Migration 151, 248, 265
Fossil Fuels 16, 23, 25, 254, 267, 282, 284, 292, 304, 327,
Foster, John Bellamy 46, 85, 257
Frank, A.G. 25, 141–143
Franz, Fanon 123
Fukishima Daiichi 19

G

Gadgil, Madhav 63, 90, 91, 123, 183, 184, 189, 190, 194, 195, 197–199, 201, 204–206, 209, 211, 215, 217, 242, 243, 245, 246, 249–251, 253, 260, 261, 335, 351
Gaia 63, 90, 91, 123
Gandhi, Mahatma 16, 25, 88, 188, 194, 246, 252, 255, 291, 301, 345
Ganga 17, 91, 158–159, 174, 175, 180, 185, 187–188, 255, 275, 316, 321–322, 343
Garner, R. 229
Geddes, Patrick 7
Gender and Development 135, 136
Gender and Nature 31, 33
Ghaziabad 15, 16, 281, 309, 312, 315, 321, 341
Giddens, Anthony 7, 55–56
Global Climate Risk Index 266
Global Environmental Politics 222–223
Goa 21, 22, 87, 154, 185–186
Goldsmith, E. 136, 162, 228
Goodman, David 224
Gorbachev, M. 19
Gorz, Andre 82
Gould, Kenneth Allan et al. 46–49, 51, 52
Grasslands 17–18, 44, 78, 113
Gray, Tim S. 59, 230
Greater Noida 15, 309
Green Capitalism 257
Green Revolution 13–14, 23–24, 38, 61, 72, 75–76, 119, 121, 260, 273, 296, 298
Greenpeace International 19, 118, 225–226, 239, 244, 255, 307
Groundwater 13–15, 24, 34, 36, 47, 72, 87, 159, 161, 184, 302, 305, 312–315, 317, 322, 342–344, 358
Grown, C. 136
Guha, Ramchandra 231, 244, 246, 257, 261

Guoqing, Shi 175
Gupta, Dipankar 60

H

Haigh, Martin J. 54–55, 57, 194–195, 267
Hildingsson, R. 58–59, 128–129
Hall, Peter 124
Hardin, Garret 77, 244–246, 352
Harper, Charles and Monica Snowden 42, 44
Harris, John 143
Harris, Marvin 76, 142
Harvey, D.W. 137
Hazare, Anna 53, 186, 250
High Yielding Varieties-HYV 13–14, 23, 119
Hildyard, N. 136, 162
Holistic Sustainability Framework (HSF) 104, 106, 132, 150
Hooks, G and C.L. Smith 53–54
Hug the Tree Movement 33, 188, 192, 248
Hugo, G. 151–152
Human Development Report (HDR) 97, 145–146
Human Ecology 9, 16, 41, 88
Human Exeptionalism Paradigm-HEP 7, 45–46, 89
Huntington, S.P. 242

I

Imperialist Tradition 21
Institutionalism-Institutionalists 103–105, 132
Integrated Life System 44, 88
Inter-Generational Equity 38, 96, 98, 106, 122, 132, 222, 228, 365
International Forum on Globalisation (IFG) 104–105
IPCC-Intergovernmental Panel on Climate Change 12, 127, 156, 240, 265, 267, 273, 275, 282
Islam, S and Ismail Hussain 50

J

Jackson, Tim 122–124
Jaitapur 19
James, C.L.R. 137
Jameson, F. 136
Jan Vikas Andolan 251
Janicke, M. 54–55, 57
Japan 19, 24, 61, 78, 116, 124, 132, 146, 159, 177, 229–230, 240, 266, 273, 280–281, 283, 285, 287, 292, 326
Jasanoff, Sheila 258, 260–261
Jha, Abhijay 15
Jodha, N.S. 78, 347–350, 364

K

Kabeer, Naila 136
Kalin, W. 153
Kalpavriksh 192, 199
Kandpal, P.C. 22
Kawser, M.A. 159
Khejri 189, 247–248, 305
King, Ynestra 31, 63, 69
Kohler, J. 128, 130
Koll, R.M. 265
Kothari, A. 199–201, 204, 207, 209, 292
Kothari, Rajni 17
Kothari, Smitu 162
Kundankulam 19, 186, 304
Kyoto Protocol 124, 230, 239–242, 289–291, 293, 304, 306

L

Land Acquisition Act 165–168, 204,
Land Titling 362
Lane, C. 355, 359, 361–362
Large Dams 155, 160–166, 173, 176–178, 180–182
Latour, B. 2, 26, 29, 30
Leach, M. 360
Lee, David 82–83
Lieber, M.D. 155
Leopold, Aldo 123
Loss of Biodiversity 6, 50, 55, 74, 150, 231, 243, 245, 251, 262, 306–307, 339–341, 34–346
Lyotard, Jean-Francois 136–137

M

Market Liberalism 103, 105, 124, 132
Marshal, Alfred 142
Martinez-Alier, Joan 115, 243, 252, 254
Marx, Karl 2, 25, 45, 76, 82–83, 86, 104, 141
Maya Culture 81
Mc-Neill, John 16, 23–24
Mead, George Herbert 2
Meadows, D.H. 4, 71, 228, 244,–245
Merchant, Carolyn 31–32, 223
Merton R.K. 30, 230
Meta Fix 102
Mexico 23, 76, 81, 92, 96, 118–120, 132, 155, 237, 266, 269, 272, 279, 324, 341
Mies, Maria 63–67, 69, 256, 268,
Millennium Development Goals 106–107, 149, 294
Mills, C. Wright 2
Milton, Kay 227
Modernisation 21, 40, 41, 44, 54–60, 62, 65, 67–68, 70–72, 88, 121, 128–129, 137, 140–141, 144, 161, 190, 204, 223, 233, 256, 322, 348

Modi, Renu 136, 177
Mohanty, Chandra T. 136
Mol, Arthur P.J. 46, 53–55, 59
Montreal Protocol 230, 231, 240, 259, 333, 334
Moorehead, R. 355, 359, 361, 362
MPI (Multi-Dimensional Poverty Index) 147
Muggah, Robert 152–154, 156
Mukerjee, Radhakamal 8–9
Multiple Knowledge Systems 10–11
Murphy, Raymond 40, 55
Myrdal, Gunnar 335

N

Naess, Arne 21, 254, 256
Namami Gange 316, 322
Narain, Sunita 178, 180, 198, 206, 211, 254, 260–261, 263, 275, 282
Narmada Bachao Andolan 151, 184–186, 188, 198, 200–201, 209, 221, 244–245, 249, 255, 260, 350
Narmada River 159, 181, 198, 350
NDC (Nationally Determined Contributions) 242, 289, 290
Neumann, I.B. 29
New Ecological Paradigm, NEP 4, 7, 45–46
New International Division of Labour (NIDL) 50
NGOs-Environmental 57, 127, 151, 199, 208, 213, 225, 230, 244, 261,338, 346
NIMBY 6
Niyamgiri 114–115, 175
NOIDA 15, 309, 312, 341

O

O'Connor, James 83, 85
O'Connor, M. 254
ODA (Official Development Assistance) 101, 110
ODF (Open Defecation Free) 291, 320
Ogoni Ethnic Group (Nigeria) 36
Omvedt, Gail 194, 206, 208
Oommen, T.K. 9–10
O'Riordan 355
Ortner, Sherry 2, 31
Orwell, George 16
Osborn, Fairfield 94
Ossewaarde, Martin J. 122, 266–267
Ostrom, Elinor 78, 354, 361, 364
Oxford Poverty & Human Development Initiative (OPHDI) 147
Ozone Depletion 5–6, 75, 139, 231, 261, 303, 306–307, 330–333, 345

P

Padel, F 114–115
Panch Mahabhutas (Five Basic Elements) 134

388 Index

PAP (Project-Affected People) 163, 201
Parajuli, P. 162, 204
Parali (Stem of Paddy) 14, 308–310, 320, 344–345
Paris Climate Agreement 307
Parris, K. 325
Parsons, Howard 82
Parsons, Talcott 46, 59, 60, 140
Patker, Medha 164
Payments for Ecosystems (PES) 118, 120
Pellow, David N. 51–52
Per Capita Equity 295,
Perminski, B. 136, 155
Political Ecology 28, 41, 70–74, 77–81, 85, 88
Pollution 4, 6–7, 10, 12, 14, 17, 20, 24, 27, 36–37, 44, 48–50, 52, 55–56, 59, 65, 83, 86–87, 95, 99–100, 103–104, 108, 110, 112–114, 121, 126–128, 139, 151–152, 183–185, 228, 231–231, 234, 243, 245–246, 252, 254, 261, 263, 266–267, 282, 284, 286, 291, 299–300, 303, 306–312, 314–325, 327–330, 332, 337, 340–346
Population 5, 12–13, 16–18, 24, 34, 36–37, 43–45, 70–71, 75, 78, 88, 94, 96, 98–100, 103–105, 110, 116, 125, 127, 142, 147–149, 151, 154–156, 159–160, 162, 175–178, 183, 187, 191, 199, 206, 210, 228, 232, 239, 242, 246, 255, 256, 264–266, 270, 280–282, 292, 314, 331–332, 335, 340, 342, 349, 351–354, 359–360, 363
Positivist Theory 259, 261
Prakash, Anil 187
Privatisation 24, 34, 38, 50, 64–66, 68, 89, 165, 224, 233, 246, 335, 348, 350, 352, 354, 359, 361–362, 364–365
Property Rights 47, 81, 118, 347, 353, 358–360, 363–364
Putnam, Robert 143

R
Radiation 12, 19, 24, 186, 229, 268, 307, 315, 325–326, 330–332, 345
Rajagopalan, R. 248
Rajaji Tiger Reserve 150, 339
Real Ecological Sustainability (RES) 115, 121, 132
Realism 30
Realist-Constructionist Debate 1, 20
Redclift, Michael 224
Rehabilitation 134, 155, 159–161, 164, 168–174, 176, 179–180, 182, 185, 201–202, 211, 217–218, 220–221, 235, 250
Renaud, F. 151–152
Rex, John 1
RFCTLARR 168, 173, 181
Rio Declaration 99–100

Rishis/Munis (Sages) 94
Risk Society 40, 56–57, 233
Rist, Gilbert 378
Robbins, Paul 27
Rostow, W.W. 59, 140
Roszak, Theodore 228
Runge, C.F. 359
Russia 19, 73, 146, 186, 229, 238, 254, 269, 279–281, 283–287, 291, 312, 323, 341

S
Said, Edward 137
Sainath, P. 77, 288–289
Sanghavi, S. 162, 207, 210, 215
Sarna (Sacred Groves of Tribals) 94, 102, 174, 180
Sauer, Carl Ortwin 75
Save Narmada Movement 164, 188, 198, 217–218, 221–222, 350
Seudder, T. 136, 156
Schnailberg, Allan 47
Schnailberg, Allan and Kenneth 47
Schumacher, E.F. 25
Scoons, I 356, 360
Scott, James C. 78–79
Sen, Amartya 135–144, 360
Sen, Gita & C. Grown 136
Sengupta, Chandan 184, 186,368
Sethi, Harsh 197–198, 205, 208, 213, 217
Shallow Ecology 254–257 261
Shapira–Garza, Elizabeth 118, 120
Sharma, Devinder 345
Sharma, Subhash 21, 35, 41, 43, 60, 70, 72, 83, 85, 161, 183, 203, 207, 212, 214–215, 220, 228, 243, 254, 256–257, 260–261, 310, 320
Shell Oil Company 36–37
Sheshadri Naidu, D. 21
Shiva, Vandana 32, 33, 35–36, 46, 63, 65, 67, 69–70, 184, 194, 256
Shodhganga 136, 175
Shuurman, F.J. 135
Silent Spring 17, 227
Silent Valley Movement 48, 184–185, 188, 196, 205, 208, 216, 221, 249, 252, 257, 260
Simon, Julian L. 5, 46, 71
Singh, Shekhar & P. Banerji 136, 162–163, 174, 178
Singh, Subrata 351–353
Singh, Yogendra 188
Slater, D. 135–136, 138
Social Development 4, 46, 95, 136, 149–150, 182, 195
SDR-Social Development Report 149
Social Greens 103–105, 132

Social Impact Assessment (SIA) 5, 39, 45, 113, 166, 169–172, 200
Soskice David 124
Sovacool, B 128, 130
Spivak, G.C. 81, 137
SSP (Sardar Sarovar Project) 159, 162, 181, 199–202, 215, 350
State of India's Bird Report-SIBR 13
Steward, Julian 75–76
Steady State Economy 18, 51, 103
Stojanov, R. 152
Suich, H. et al. 11
Sukhomajri 317, 363
Sustainable Development 4, 9, 56–57, 79, 90, 94–98, 100–103, 110–111, 113–115, 117, 121–123, 125, 128, 132, 138, 150, 164, 177, 186, 228, 240–242, 289, 294, 312, 340
Sustainable Development Goals-SDGs 24, 102, 107–109, 126, 231, 235, 241, 280, 334
Sustainability Transition 127, 128, 130–132, 296
Swaminathan, M.S. 24, 38, 196–197, 208–209, 211, 213–214, 296

T
Temper, L. 115
Thapar, Romila 91–93
The New Normal 263–264
Thiong' N. 137
Third World Network (TWN) 104
Three Mile Island 5, 19, 228
Thurow, Lester 231, 242
Tinkathia 84
Tiwary, R. 159
Tolman, Charles 82
Total Fertility Rate–TFR 13, 149
Tragedy of Commons 77–78, 347, 352, 354–355, 358–359, 361–365
Trans National Corporations (TNCs) 50, 104
Transnational Alliance Theory 259
Treadmill of Production 41, 47–48, 50–54, 88

U
Ukraine 18–19, 96, 229, 292, 326
Ulrich, R.S. 126
UNDP (United Nations Development Programme) 97, 132, 144–145, 147, 333, 348, 356

UNEP (UN Environment Programme) 37, 94, 104–105, 118, 124, 127, 240–241, 333
UNESCO 92, 124–125
United Nations Conference on Human Environment 16, 94, 228, 240
US 5, 19, 47, 49, 52–54, 81, 103, 132, 187, 229, 234–235, 238–239, 241–242, 244, 246, 256, 268, 273, 282, 291–292, 299, 321, 324–325, 327, 329, 331, 341, 354
USSR 18, 19, 61, 96, 186, 226, 229, 292, Uttarakhand Sangharsh Vahini 70, 206, 211, 245, 252, 253

V
Vastu/Vaastu 91
Vasudhaiv Kutumbkam 91, 134
Vedanta 86–87, 114–115
Vedic 20, 90, 92–94
Vietnam 23, 52, 155, 236
Vishwamohan 266, 273, 279, 284, 338

W
Wade, Robert 77, 363,
Wainwright, Joel 81–82
Ward, Barbara 228
Warrington, Siobhan 176
WCED (World Commission on Environment and Development) 94–98, 105, 132, 240
Weber, Max 2, 39, 45–46
West Bengal 19, 36, 147, 154, 158, 162, 179–180, 187–188, 270, 277–279, 313, 316, 322, 337–338
White, Gilbert 75
Woodgate, Graham 12
Woolger, Steve 28
World Bank 97, 103, 105, 114, 118, 144, 155, 159, 161, 163–164, 181, 199, 201, 213, 217, 235, 263, 294, 333, 355
World Economic Forum 103, 105
World Happiness Report 126, 133, 301
World Social Forum 104

Y
Yamuna 91, 189, 313–314, 316–317, 321, 343

Z
Zero Growth Economy 18
Zetter, R. 152

Printed in the United States
by Baker & Taylor Publisher Services